Natural Products for Chronic Diseases: A Ray of Hope

Natural Products for Chronic Diseases: A Ray of Hope

Editors

Syed Shams ul Hassan
Mohamed M. Abdel-Daim
Tapan Behl
Simona Gabriela Bungau

MDPI • Basel • Beijing • Wuhan • Barcelona • Belgrade • Manchester • Tokyo • Cluj • Tianjin

Editors

Syed Shams ul Hassan
School of Pharmacy
Shanghai Jiao Tong
University
Shanghai
China

Mohamed M. Abdel-Daim
Department of
Pharmaceutical Sciences
Batterjee Medical College
Jeddah
Saudi Arabia

Tapan Behl
School of Health Science and
Technology
University of Petroleum and
Energy Studies
Bidholi
India

Simona Gabriela Bungau
Pharmacy
University of Oradea
Oradea
Romania

Editorial Office
MDPI
St. Alban-Anlage 66
4052 Basel, Switzerland

This is a reprint of articles from the Special Issue published online in the open access journal *Molecules* (ISSN 1420-3049) (available at: www.mdpi.com/journal/molecules/special_issues/marine_NP_diseases).

For citation purposes, cite each article independently as indicated on the article page online and as indicated below:

LastName, A.A.; LastName, B.B.; LastName, C.C. Article Title. *Journal Name* **Year**, *Volume Number*, Page Range.

ISBN 978-3-0365-6955-0 (Hbk)
ISBN 978-3-0365-6954-3 (PDF)

Contents

About the Editors

Syed Shams ul Hassan

Syed Shams ul Hassan has completed his Undergraduate from Department of Pharmacy Gomal University D.I.Khan, Pakistan. He has achieved his Master's degree from Zhejiang University and Ph.D. from Shanghai Jiao Tong University, (both universities under top 50 world ranked) in the field of Medicinal chemistry through fully funded scholarship by Chinese Government Scholarship. The main focus of his Ph.D. was on the isolation and purification of natural products and evaluating their pharmacological potential under the supervision of Prof. Dr. Hui-Zi Jin. He has published more than 70 WOS indexed articles. He also serves as leading guest editor and associate editor for high index journals such as Molecules, *Biomolecules, Frontiers in Pharmacology, Frontiers in neuroscience,* and *Biomedical Research International.*

Mohamed M. Abdel-Daim

Dr. Mohamed M. Abdel-Daim is an associate professor of Pharmacology, Faculty of Veterinary Medicine, Suez Canal University, Ismailia, Egypt. Additionally, he serves as an associate professor at the Department of Zoology, Science College, King Saud University, Saudi Arabia. He obtained his Ph.D. from Kobe University Graduate School of Medicine, Japan. He published over than 500 original studies and review articles at many ISI scientific journals, related to Pharmacology, Toxicology, Ecotoxicology, Drug safety, Nutrition and Food Supplements, Alternative Medicine, Toxicology, Neuroscience, Biochemistry, and Molecular Biology. His current Scopus h-index is 60. He is a member of the Research Ethics Council at the Academy of Scientific Research & Technology, Egypt. Owing to the recognition of his research works, he has been awarded various research prizes, including the State Encouragement Award from the Academy of Scientific Research & Technology, Cairo, Egypt in 2014 and the First Class Medal of Excellence from the Presidency of the Arab Republic of Egypt on the national science day in August 2017. He served as an associate editor for many peer-review journal including "Environmental Science and Pollution Research Ecotoxicology and Environmental Safety" and had been a guest editor for many peer-review journal Special Issues. He also served as a reviewer for over 200 journals, as well as numerous national and international research grants. He has supervised over 30 Master and Ph.D. theses. He has published more than 500 WOS indexed articles. He also serves as senior editor for many high indexed journals including "Environmental Science and Pollution Research".

Tapan Behl

Dr. Tapan Behl is currently involved in guiding bachelors, masters and Ph.D. students. His research group is working mainly on diabetes and associated complications, rheumatoid arthritis, cancer, obesity, Parkinson's Disease, and other neurological complications. He has published more than 200 WOS indexed articles. He also serves as guest editor for many Special Issues in different high-index journals.

Simona Gabriela Bungau

Simona Gabriela Bungau became a full professor in 2005, having a continuous teaching and research activity, started in 1992 and going through all the stages of a university teaching career. After two Bachelor's degrees (in Chemistry-physics and Pharmacy), she obtained her Doctorate (Ph.D.) in Chemistry at "Babes-Bolyai" University, Cluj-Napoca, Romania, in 2003. The particularly rich research activity was presented in over 400 Web of Science indexed publications, and over 10 books published in Romanian publishing houses and 2 in international publishing houses. The initial fields of interest were Chemistry, respectively, Analytical Chemistry, then expanding to Pharmacology, Botany, Public health, Ecology and Sustainability/Environmental protection. Obtaining the qualification of Ph.D. supervisor in the field of Pharmacy came naturally, in 2015, by presenting the Habilitation Thesis at "Carol Davila" University of Medicine and Pharmacy, Bucharest, Romania. Participating in over 20 research projects, numerous Erasmus teaching programs, member of societies in her native country and abroad, member of Editorial board of well known journals, Mrs. Bungau continues her activity for over 30 years at the same university, namely the University of Oradea, Romania.

Preface to "Natural Products for Chronic Diseases: A Ray of Hope"

Developed over centuries to encode biological processes, nature's products are the predecessors of medicine. Out of the desire to cure diseases, conventional remedies gave a substantial territory to modern chemical techniques much more competent and capable in the development of new drugs. The designated chemotypes of natural components (β-lactam antibiotics obtained from Penicillium sp. and quinoline/iso-quinoline alkaloids obtained from tree bark are prototypical examples of therapeutically appreciated active substances) have proven their ability to act on the broad spectrum of pathogens that continue to significantly impact the health of the human population, the latter being a huge economic burden worldwide. Products of natural origin, considered to be the core of pharmaceutical armaments, are ready for access compared to synthetic products; moreover, low concern in maintaining an archetype of discovery, including fermentation, isolation, structural determination, and biological testing of new pharmacologically active natural compounds, must be taken into account. Most studies in the field have been performed to determine the pharmacokinetics and pharmacognostic potential of natural products, and fortunately, researchers have successfully isolated many new compounds. Such substances have been shown to have antibacterial, antiviral, antifungal, antimalarial, antitumor, anti-inflammatory, antioxidant, immunosuppressive, cardiovascular activity, etc. However, the mode of action of many compounds, through which they interfere with human pathogenesis, has not been clarified so far, and this knowledge is essential in knowing and establishing the possibility of transforming chemical molecules into drugs.

Due to these circumstances, the following principles have been pursued in preparing the manuscripts for this Special Issue:

1. Capacity of medicinal substances, which have been examined in vitro and in vivo, against pathogenic microbes;
2. Mode of action of these medicinal compounds and the basic mechanisms by which natural products function or act;
3. Natural products derived with efficient bioactivities;
4. Biomolecules derived from natural products used in experimental or clinical studies;
5. The new role of natural products in drug targeting.

We are most grateful to the numerous contributors (List of contributors) who undertook the painstaking job of a literature survey, experimental work, writing, and condensing of material. We are also thankful to the all reviewers and editors who have taken part in the anonymous peer review process which is one the most important part of the scientific publications. We also pay gratitude to the assistant editor Lisa Xu for her step by step guidance and through out support for making this Special Issue successful.

Syed Shams ul Hassan, Mohamed M. Abdel-Daim, Tapan Behl, and Simona Gabriela Bungau
Editors

molecules

MDPI

Editorial

Natural Products for Chronic Diseases: A Ray of Hope

Syed Shams ul Hassan [1,2,*], Mohamed M. Abdel-Daim [3,4], Tapan Behl [5] and Simona Bungau [6,7]

1 Shanghai Key Laboratory for Molecular Engineering of Chiral Drugs, School of Pharmacy, Shanghai Jiao Tong University, Shanghai 200240, China
2 Department of Natural Product Chemistry, School of Pharmacy, Shanghai Jiao Tong University, Shanghai 200240, China
3 Department of Pharmaceutical Sciences, Pharmacy Program, Batterjee Medical College, P.O. Box 6231, Jeddah 21442, Saudi Arabia
4 Pharmacology Department, Faculty of Veterinary Medicine, Suez Canal University, Ismailia 41522, Egypt
5 School of Health Sciences and Technology, University of Petroleum and Energy Studies, Dehradun 248007, Uttarakhand, India
6 Department of Pharmacy, Faculty of Medicine and Pharmacy, University of Oradea, 410028 Oradea, Romania
7 Doctoral School of Biomedical Sciences, University of Oradea, 410087 Oradea, Romania
* Correspondence: shams1327@yahoo.com

Citation: Hassan, S.S.u.; Abdel-Daim, M.M.; Behl, T.; Bungau, S. Natural Products for Chronic Diseases: A Ray of Hope. *Molecules* **2022**, *27*, 5573. https://doi.org/10.3390/molecules27175573

Received: 23 August 2022
Accepted: 25 August 2022
Published: 30 August 2022

This Special Issue includes many high advanced quality papers that focus on natural products with their potent pharmacological potential targeting various areas of diseases. The papers in this Special Issue present new insights into natural products with potent anticancer, anti-inflammatory, antioxidant, anti-bacterial, analgesic, anti-diabetic, and enzyme inhibitory activities.

Secondary metabolites from nature, predominantly plants, are still a research hotspot because of their promising novel scaffolds against chronic diseases. Plant-derived bioactive compounds were proved to have promising anticancer activities. Recently, many researchers have driven their research interest toward plants to evaluate the use of plant-derived bioactive compounds against different kinds of cancer. Hassan and his team [1] evaluated one guaiane-type sesquiterpene dimer vieloplain F from Xylopia Vielana species against melanoma by targeting B-Raf kinase. The results indicated that vieloplain F has good anticancer activity against melanoma by displaying a higher binding energy of −11.8 kcal/mol against B-Raf protein compared to the FDA-approved drug vemurafenib. Further MD simulations and MM-GBSA showed that vieloplain F had the most remarkable binding propensity to active site residues. In addition, the absorption, distribution, metabolism, excretion, and toxicity (ADMET) profile of the FDA-approved medicine vemurafenib was hepatotoxic, cytochrome-inhibiting, and non-cardiotoxic compared to vieloplain F, which at this moment has been selected for further investigation due to its potential effects against melanoma. Majid et al. [2] isolated a new triterpenoid nummularic acid (NA) from Ipomoea batatas and evaluated its anticancer activity against prostate cancer (PCa). The results showed that significant ($p < 0.05$ and $p < 0.01$) time and dose-dependent reductions in the proliferation of PCa cells, reduced migration, invasion, and an increased apoptotic cell population were recorded after NA treatment (3–50 μM). Further profound mechanistic studies revealed that NA treatment considerably increased the cleavage of caspases and downstream PARP, upregulated BAX and P53, and downregulated BCL-2 and NF-κB, inducing apoptosis in PCa cells. Khan et al. [3] evaluated the effects of DL-propargylglycine (PAG, inhibitor of CSE), aminooxy acetic acid (AOAA, inhibitor of CBS), and L-aspartic acid (L-Asp, inhibitor of 3-MPST) against breast cancer (BC) by determining the role of endogenous H_2S in the growth of BC by performing in vitro and in vivo experiments. The results showed that the combined dose (PAG + AOAA + L-Asp) group showed exclusive inhibitory effects against BC cells' viability, proliferation, migration, and invasion compared to the control group. Further, treated cells exhibited increased apoptosis and a reduced level of phospho (p)-extracellular signal-regulated protein kinases such as p-AKT, p-PI3K, and p-mTOR. Moreover, the combined group exhibited potent inhibitory effects on the growth of BC xenograft tumors in nude mice without apparent toxicity.

Natural products have a broad pharmacological spectrum because of their complex scaffolds. Huneif et al. [4] isolated two compounds from wild strawberries and evaluated their anti-diabetic and antioxidant activity. The results showed that both compounds have good anti-diabetic activity against α-glucosidase, α-amylase, and antioxidant activity against DPPH free radicals. Al-Joufi et al. [5] evaluated the anti-diabetic, antioxidant, and anti-microbial potential of the Anabasis articulata plant. The results showed that different extracts (methanolic and n-hexane) displayed remarkable anti-diabetic activity against α-glucosidase, α-amylase, antioxidant activity against DPPH free radicals and anti-microbial activity against *Shigella dysentery* (*S. dynasties*), *Escherichia coli* (*E. coli*), and *Salmonella typhi* (*S. typhi*). Ahmed et al. [6] evaluated the vegetable plant Pleurospermum candollei by investigating its phytochemical profile and biological activities such as antioxidant, anti-bacterial, thrombolytic and enzyme inhibitory characteristics (tyrosinase, α-amylase, and α-glucosidase). The results displayed that methanolic and n-hexane extracts showed remarkable pharmacological activities in terms of antioxidant, anti-bacterial, thrombolytic and enzyme inhibitory characteristics. In addition, pure compounds also displayed good docking results against targeted proteins. Mahmood et al. [7] evaluated the anti-inflammatory, analgesic and antioxidant capacity of New (2*S*,3*S*)-2-(4-isopropylbenzyl)-2-methyl-4-nitro-3-phenylbutanals. The results revealed that two compounds have potent anti-inflammatory activity in vitro against COX $\frac{1}{2}$ and 5-LOX, antioxidant activity against DPPH free radicals and in vivo analgesic activity. Faheem et al. [8] investigated the effects of natural compounds, berbamine, bergapten, and carveol on paclitaxel-associated neuroinflammatory pain. The results revealed that all the compounds attenuated thermal hypersensitivity and increased the threshold for pain sensation. The compounds also increased the protective glutathione (GSH) and glutathione S-transferase (GST) levels in the sciatic nerve and spinal cord while lowering inducible nitric oxide synthase (iNOS) and lipid peroxidase (LPO). Furthermore, the compounds also inhibited cyclooxygenase-2 (COX-2), tumor necrosis factor-alpha (TNF-α), and nuclear factor kappa B (NF-κb) overexpression.

Glensk et al. [9] isolated bioactive compounds echimidine and its C-7 isomers from Echium plantagineum L. and evaluated their hepatotoxic effect on rat hepatocytes. The results revealed that the compounds at 3 to 300 μg/mL caused the concentration-dependent inhibition of hepatocyte viability, with mean IC_{50} values ranging from 9.26 to 14.14 μg/mL. This study revealed that under standard HPLC acidic conditions, echimidine co-elutes with its isomers, echihumiline and to a lesser degree with hydroxy myoscorpine, obscuring the actual alkaloidal composition, which may have implications for human toxicity. Khan et al. [10] evaluated the effects of shrimp peptide hydrolysate on intestinal microbiota restoration and immune modulation in cyclophosphamide-treated mice. The results showed that shrimp peptide hydrolysate significantly restored goblet cells and intestinal mucosa integrity, modulated the immune system, and increased the relative expression of mRNA and the tight-junction associated proteins occludin, Zo-1, claudin-1, and mucin-2).

Marine drugs possess an undoubtedly diverse range of sources as they are distributed over 70% of the earth's surface, possess a wide range of variations in structure and present a promising target in the discovery of newer and better treatment approaches. In the past seven decades, many structurally diverse drug products and their secondary metabolites have been isolated from marine sources which have successfully presented an exceptional potential in the treatment of various diseases ranging from acute to chronic conditions. Hence, Bhatia et al. [11] highlighted the significant role of marine-derived drugs in the management of chronic diseases such as diabetes, cancer, cardiovascular and neurodegenerative disorders.

Principally, the Special Issue "Natural Products for Chronic Diseases: A Ray of Hope" provides a current perspective of the natural products from the marine and terrestrial area and the rapidly developing research area, as evident from the resistance to the available drugs and wide variety of chronic diseases. Considering the challenges in this exciting field of natural products drug discovery, this issue not only complements our knowledge on bioactive compounds but also may uncover some novel ideas and motivation for the further investigation of various prospective biologically active compounds impacting medical practice.

Funding: This research received no external funding.

Acknowledgments: We express our sincere thanks to the contributing authors, the reviewers who reviewed the submitted manuscripts and contributed to the quality of the manuscripts, and the editorial staff of *Molecules* for their support throughout the process.

Conflicts of Interest: The authors declare no conflict of interest.

References

1. Shams ul Hassan, S.; Qamar Abbas, S.; Ali, F.; Ishaq, M.; Bano, I.; Hassan, M.; Jin, H.-Z.; Bungau, S.G. A Comprehensive In Silico Exploration of Pharmacological Properties, Bioactivities, Molecular Docking, and Anticancer Potential of Vieloplain F from Xylopia vielana Targeting B-Raf Kinase. *Molecules* **2022**, *27*, 917. [CrossRef] [PubMed]
2. Majid, M.; Farhan, A.; Asad, M.I.; Khan, M.R.; Hassan, S.S.U.; Haq, I.-U.; Bungau, S. An Extensive Pharmacological Evaluation of New Anti-Cancer Triterpenoid (Nummularic Acid) from Ipomoea batatas through In Vitro, In Silico, and In Vivo Studies. *Molecules* **2022**, *27*, 2474. [CrossRef]
3. Khan, N.H.; Wang, D.; Wang, W.; Shahid, M.; Khattak, S.; Ngowi, E.E.; Sarfraz, M.; Ji, X.-Y.; Zhang, C.-Y.; Wu, D.-D. Pharmacological Inhibition of Endogenous Hydrogen Sulfide Attenuates Breast Cancer Progression. *Molecules* **2022**, *27*, 4049. [CrossRef]
4. Huneif, M.A.; Alqahtani, S.M.; Abdulwahab, A.; Almedhesh, S.A.; Mahnashi, M.H.; Riaz, M.; Ur-Rahman, N.; Jan, M.S.; Ullah, F.; Aasim, M.; et al. α-Glucosidase, α-Amylase and Antioxidant Evaluations of Isolated Bioactives from Wild Strawberry. *Molecules* **2022**, *27*, 3444. [CrossRef] [PubMed]
5. Al-Joufi, F.A.; Jan, M.; Zahoor, M.; Nazir, N.; Naz, S.; Talha, M.; Sadiq, A.; Nawaz, A.; Khan, F.A. Anabasis articulata (Forssk.) Moq: A Good Source of Phytochemicals with Antibacterial, Antioxidant, and Antidiabetic Potential. *Molecules* **2022**, *27*, 3526. [CrossRef] [PubMed]
6. Ahmed, M.; Khan, K.-R.; Ahmad, S.; Aati, H.Y.; Ovatlarnporn, C.; Rehman, M.S.; Javed, T.; Khursheed, A.; Ghalloo, B.A.; Dilshad, R.; et al. Comprehensive Phytochemical Profiling, Biological Activities, and Molecular Docking Studies of Pleurospermum candollei: An Insight into Potential for Natural Products Development. *Molecules* **2022**, *27*, 4113. [CrossRef]
7. Mahmood, F.; Khan, J.A.; Mahnashi, M.H.; Jan, M.S.; Javed, M.A.; Rashid, U.; Sadiq, A.; Hassan, S.S.U.; Bungau, S. Anti-Inflammatory, Analgesic and Antioxidant Potential of New (2S,3S)-2-(4-isopropylbenzyl)-2-methyl-4-nitro-3-phenylbutanals and Their Corresponding Carboxylic Acids through In Vitro, In Silico and In Vivo Studies. *Molecules* **2022**, *27*, 4068. [CrossRef] [PubMed]
8. Faheem, M.; Khan, A.; Saleem, M.W.; Shah, F.A.; Ali, F.; Khan, A.W.; Li, S. Neuroprotective Effect of Natural Compounds in Paclitaxel-Induced Chronic Inflammatory Pain. *Molecules* **2022**, *27*, 4926. [CrossRef] [PubMed]
9. Gleńsk, M.; Dudek, M.K.; Kinkade, P.; Santos, E.C.S.; Glinski, V.B.; Ferreira, D.; Seweryn, E.; Kaźmierski, S.; Calixto, J.B.; Glinski, J.A. Isolation of Echimidine and Its C-7 Isomers from Echium plantagineum L. and Their Hepatotoxic Effect on Rat Hepatocytes. *Molecules* **2022**, *27*, 2869. [CrossRef] [PubMed]
10. Khan, A.I.; Rehman, A.U.; Farooqui, N.A.; Siddiqui, N.Z.; Ayub, Q.; Ramzan, M.N.; Wang, L.; Xin, Y. Effects of Shrimp Peptide Hydrolysate on Intestinal Microbiota Restoration and Immune Modulation in Cyclophosphamide-Treated Mice. *Molecules* **2022**, *27*, 1720. [CrossRef] [PubMed]
11. Bhatia, S.; Makkar, R.; Behl, T.; Sehgal, A.; Singh, S.; Rachamalla, M.; Mani, V.; Iqbal, M.S.; Bungau, S.G. Biotechnological Innovations from Ocean: Transpiring Role of Marine Drugs in Management of Chronic Disorders. *Molecules* **2022**, *27*, 1539. [CrossRef] [PubMed]

molecules

MDPI

Article

A Comprehensive In Silico Exploration of Pharmacological Properties, Bioactivities, Molecular Docking, and Anticancer Potential of Vieloplain F from *Xylopia vielana* Targeting B-Raf Kinase

Syed Shams ul Hassan [1,2], Syed Qamar Abbas [3], Fawad Ali [4], Muhammad Ishaq [1,2], Iqra Bano [5], Mubashir Hassan [6,7], Hui-Zi Jin [1,2,*] and Simona G. Bungau [8]

1 Shanghai Key Laboratory for Molecular Engineering of Chiral Drugs, School of Pharmacy, Shanghai Jiao Tong University, Shanghai 200240, China; shams1327@yahoo.com (S.S.u.H.); mishaqjnj@yahoo.com (M.I.)
2 Department of Natural Product Chemistry, School of Pharmacy, Shanghai Jiao Tong University, Shanghai 200240, China
3 Department of Pharmacy, Sarhad University of Science and Technology, Peshawar 25000, Pakistan; qamar0613@yahoo.com
4 Department of Pharmacy, Kohat University of Science and Technology, Kohat 26000, Pakistan; fawadalee@gmail.com
5 Faculty of Bio-Sciences, SBBUVAS, Sakrand 67210, Pakistan; iqrashafi05@yahoo.com
6 Institute of Molecular Biology and Biotechnology, The University of Lahore, Nisbet Road, Lahore 54000, Pakistan; mubashirhassan_gcul@yahoo.com
7 Battelle Center for Mathematical Medicine, The Research Institute at Nationwide Children's Hospital, Columbus, OH 43205, USA
8 Department of Pharmacy, Faculty of Medicine and Pharmacy, University of Oradea, 410028 Oradea, Romania; simonabungau@gmail.com
* Correspondence: kimhz@sjtu.edu.cn; Tel.: +86-130-65711080

Citation: Hassan, S.S.u.; Abbas, S.Q.; Ali, F.; Ishaq, M.; Bano, I.; Hassan, M.; Jin, H.-Z.; Bungau, S.G. A Comprehensive In Silico Exploration of Pharmacological Properties, Bioactivities, Molecular Docking, and Anticancer Potential of Vieloplain F from *Xylopia vielana* Targeting B-Raf Kinase. *Molecules* **2022**, *27*, 917. https://doi.org/10.3390/molecules27030917

Academic Editor: Silvia Schenone

Received: 27 December 2021
Accepted: 25 January 2022
Published: 28 January 2022

Abstract: Compounds derived from plants have several anticancer properties. In the current study, one guaiane-type sesquiterpene dimer, vieloplain F, isolated from *Xylopia vielana* species, was tested against B-Raf kinase protein (PDB: 3OG7), a potent target for melanoma. A comprehensive in silico analysis was conducted in this research to understand the pharmacological properties of a compound encompassing absorption, distribution, metabolism, excretion, and toxicity (ADMET), bioactivity score predictions, and molecular docking. During ADMET estimations, the FDA-approved medicine vemurafenib was hepatotoxic, cytochrome-inhibiting, and non-cardiotoxic compared to the vieloplain F. The bioactivity scores of vieloplain F were active for nuclear receptor ligand and enzyme inhibitor. During molecular docking experiments, the compound vieloplain F has displayed a higher binding potential with −11.8 kcal/mol energy than control vemurafenib −10.2 kcal/mol. It was shown that intermolecular interaction with the B-Raf complex and the enzyme's active gorge through hydrogen bonding and hydrophobic contacts was very accurate for the compound vieloplain F, which was then examined for MD simulations. In addition, simulations using MM-GBSA showed that vieloplain F had the greatest propensity to bind to active site residues. The vieloplain F has predominantly represented a more robust profile compared to control vemurafenib, and these results opened the road for vieloplain F for its utilization as a plausible anti-melanoma agent and anticancer drug in the next era.

Keywords: guaiane dimer; melanoma; molecular docking; ADMET; MM-GBSA

1. Introduction

Melanoma is the most aggressive and deadly form of skin cancer [1–4]. It is the seventh leading cancer in women and fifth in men in the USA [5,6]. It is predicted that the five-year survival rate for patients with stage IV or advanced melanomas is less than

15%. This cancer can grow and spread for an indefinite period due to mutations in the cells [7]. The primary treatments are BRAF, C-Kit, and NRAS inhibitors. BRAF inhibitors, such as vemurafenib, are the most effective FDA-approved treatments for BRAF positive melanoma. However, the negative side of the drug vemurafenib is that patients started getting resistance after six months of therapy, making therapy no longer effective [8]. More effective treatments are urgently needed as the prevalence of melanoma rises in the United States and other developed nations.

There has been an increase in scientific interest in medicinal plants [9], In the last several decades, plants with their intriguing secondary metabolites have been explored for their strong anticancer properties [10,11]. The family Annonaceae includes the genus Xylopis. Recently, it has been described for its extraordinary broad range of pharmacological spectrum encompassing rheumatism, analgesia, bactericidal, fungicidal, antioxidant antitumor and anti-inflammatory properties. *Xylopia vielana* is the only species identified in China [12–16]. Natural products have been shown to influence BRAF kinases; however, it is unclear whether or not they communicate with the gene product or even the transposable elements that regulate the gene. [5].

The manufacturing of a new medicine is a lengthy and complex procedure. Selecting a suitable lead molecule is one of the crucial steps during drug development [17]. However, unexpected toxicity and adverse drug reactions, on the other hand, caused around 40% of the drug candidates to fail [18]. Computer-assisted in silico approaches have become increasingly important in the initial phases of drug development as they are more cost-effectively [17,19]. This method decreases the number of animals killed and reduce failures in the ultimate stage [19]. Often, unexpected toxicity is detected late in drug development. An in silico technique for predicting toxicological parameters is an alternative to animal testing [20]. The guaiane dimers from *Xylopia vielana* due to their complex unique structures are already proved to provide great effects during in silico studies against CoX-2 [12]. During the in silico studies such as ADMET, the initial profile of the compounds has been created. If any of the compounds provide any severe toxicity or create drug-drug interaction with some metabolizing cytochromes, the time should not be wasted on that kind of compounds.

New dimers were discovered in our recent study on the *Xylopia vielana* to aid in the quest for powerful anticancer medicines [13,15]. The in vitro anticancer potential of these guaiane dimers has been reported, which shows that some of these guaiane dimers, especially vieloplain F have displayed potent anticancer activity with IC_{50} values of 9.5 μM. However, its profound mechanistic studies and pharmacological profile were incomplete. Vieloplain F's pharmacological characteristics were studied in this study, with an emphasis on drug-likeness, bioactivities, administration, distribution, metabolic activity, excretion, and toxicity. The guaiane dimer blockage of BRAF kinases was further studied using molecular docking, MD modeling, and MM-GBSA calculations to identify its carcinogenic mechanism.

2. Results

2.1. Chemical Structures of Vieloplain F and Vemurafenib

The chemical structure of a guaiane dimer vieloplain F was drawn on ChemBioDraw (v13.0) and the control drug vemurafenib was downloaded from PubChem and was redrawn on ChemBioDraw (Figure 1).

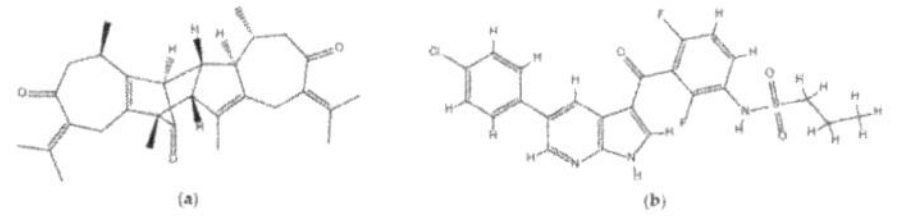

Figure 1. Chemical structures of (**a**) vieloplain F; (**b**) vemurafenib.

2.2. Estimation of Activity Spectra for Substances (PASS)

Over 4000 biological activities are predicted by PASS Online (Way2Drug, Moscow, Russia), including pharmacological effects, modes of action, toxic and unfavorable consequences, linkages with metabolic enzymes and transporter, and impact on expression of genes. Table 1 shows the best outcomes from all of the predicted activities for vieloplain F. The biological activity as antineoplastic had the most significant predicted activity (Pa) for vieloplain F, with Pa > 0.8. The probable activity (Pa) values were higher than Pa > 0.5, and the probable inactivity (Pi) scores were extremely near to 0, demonstrating that the compound is highly expected to demonstrate these activities.

Table 1. Best predicted bioactivities of compounds by PASS online.

Biological Activities	Pa	Pi
Anti-neoplastic	0.862	0.006
Anti-leukemic	0.592	0.009
Testosterone 17beta-dehydrogenase (NADP+) inhibitor	0.589	0.094

2.3. Toxicological and Pharmacokinetic Assets

Absorption, distribution, metabolism, excretion, and toxicity (ADMET) estimations were primarily directed at all the guaiane dimers (vieloplains A–G). Many novel compounds have been exemplified during initially ADMET screening due to violations, including inhibiting different cytochromes, skin permeation, and P-gp inhibition. Only vieloplain F was selected for further studies because of no violations.

2.3.1. Pharmacokinetic Characteristics

The physico-chemical characteristics of both compounds are discussed in Table 2. According to Table 2, the lipophilicity, insolubility, size, insaturation, polarity, and flexibility of vemurafenib and vieloplain F were studied and classified into six sections with appropriate ranges for oral bioavailability (Figure 2a). The oral bioavailability graph of the vieloplain F is shown in Figure 2a and vemurafenib in Figure S1 (Supplementary Materials), which is based on the six sections stated in the physicochemical characteristics section. The results of the compound vieloplain F was within these limits, demonstrating that vieloplain F has a favorable physiochemical profile, which is one of the factors that must be monitored in pharmaceuticals and clinical studies.

Table 2. Predicted physicochemical parameters and lipophilicity properties of vieloplain F and vemurafenib.

Properties	Parameters	Vieloplain F	Vemurafenib
Physicochemical Properties	MW [a] (g/mol)	446.62	489.92
	Rotatable bonds	0	7
	HBA [b]	3	6
	HBD [c]	0	2
	Fraction Csp3	0.63	0.13
	TPSA [d]	51.21	100.30
Lipophilicity Log $P_{o/w}$	iLOGP	3.93	3.04
	XLOGP3	4.02	4.97
	MLOGP	6.35	3.41
	Consensus	5.07	4.89

[a] Molecular weight, [b] H-bond acceptor, [c] H-bond donor, [d] Topological polar surface area.

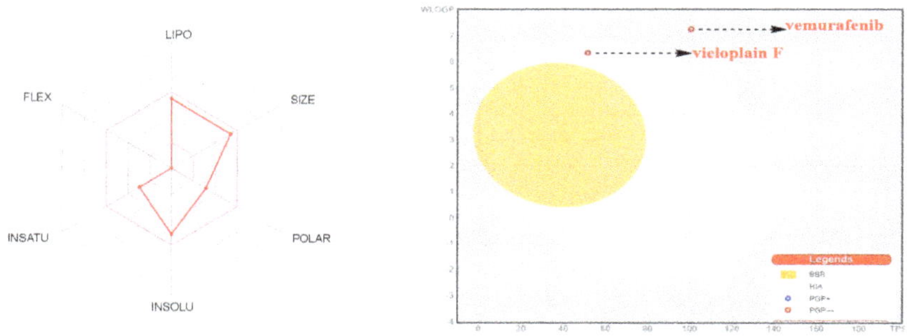

(a) (b)

Figure 2. (**a**) Bioavailability radar chart for vieloplain F. The pink zone represents the physicochemical space for oral bioavailability, and the red line represents the oral bioavailability properties. (**b**) Predicted BOILED-Egg plot from *swiss ADME* online web tool for all the compounds.

HIA and CNS absorption are important parameters checked for every biomolecule before its entry for drug formulation in the pharmaceutical or clinical trials field [21]. The blood–brain barrier penetration is essential as if the compounds that act on the central nervous system (CNS) must cross through the blood–brain barrier, and the inactive compounds on the CNS should not intersect to avoid adverse effects on the CNS [22]. As mentioned in Table 3, the compound vieloplain F displayed a high gastrointestinal absorption (HIA) with no BBB permeability, indicating that vieloplain F shows low occurrence for adverse CNS effects. The compound vieloplain F HIA absorption ratio was elevated than vemurafenib.

Table 3. Predicted pharmacokinetics parameters of vieloplain F and vemurafenib.

Properties	Parameters	Vieloplain F	Vemurafenib
Absorption	Water solubility	−5.971	−4.656
	GI [a]	100	98.45
	Log K_p (skin permeation) cm/s	−6.17	−5.76
Distribution	BBB [b]	−0.37	−1.686
	CNS permeation (Log PS)	−1.027	−2.976
	V_D [c] (human)	−0.013	−0.461
Metabolism CYP2D6	CYP1A2 inhibitor	No	No
	CYP2C9 inhibitor	No	Yes
	CYP2C19 inhibitor	No	Yes
	CYP3A4 inhibitor	No	Yes
	CYP2D6 inhibitor	No	No
Excretion	Total Clearance (log mL/min/kg)	0.053	0.136
	Renal OCT2 substrate	No	No

[a] Gastrointestinal, [b] Blood-brain barrier, [c] Volume of distribution.

Figure 2b shows the BOILED-EGG curve [23]. The BBB penetration and GI absorption (HIA) of the substances may be predicted by this method. There are two areas: one for the GI absorption zone (HIA) and the other for BBB penetration (yolk). Neither GI absorption nor BBB penetration is indicated if any component is found in the gray zone. Because neither vieloplain F nor the control medication vemurafenib showed that they are P-gp substrates, they are not sensitive to the efflux mechanism of P-gp, which is used by many cancers' cell lines to develop resistance to drugs. Vemurafenib, the reference medicine, was shown in gray, whereas vieloplain F was shown in white, as can be seen in Figure 2b.

The skin permeation Log K_p of vieloplain F, compared to vemurafenib, was lower (Table 3), as mentioned by [24]. The greater the negative value of K_p, the less permeant the molecule is to the skin. It also forecasts the five main cytochromes (CYP) isoform, which is an additional benefit. These enzymatic isoforms play a crucial role in the excretion of pharmaceuticals, and they handle the metabolism of about 75% of the medicines available on the market. There are major drug-drug interactions caused by inhibiting any of these isoforms [21,24]. Compared to the control vemurafenib, the vieloplain F did not block any cytochrome isoform, as shown in Table 3, and was rapidly metabolized. Drug-medication interactions may occur when three cytochrome isoforms are inhibited by the control drug vemurafenib: CYP2C19, CYP2C9, and CYP3A4. Dosing rates for achieving steady-state concentrations depend on drug clearance, which is determined by adding up the excretion rates from the liver and kidney. Moreover, vieloplain F's clearance value was insufficient. Organic cation transporter 2 (OCT2) intermediates may have an influence on the unfavorable interactions that occur when OCT2 inhibitors and substrates are used together. It has been hypothesized that the compound vieloplain F would act as a non-substrate for OCT2.

2.3.2. Toxicity Assessment

It is critical to evaluate the toxicological profile of a medicine before it reaches the clinical trials stage or the production phase of the pharmaceutical business before it is approved [25]. A variety of toxicities were assessed for each molecule, including those affecting human health and those affecting the environment. (Table 4). Using the Ames test, a compound's mutagenic potential may be assessed. Both substances were classified as non-Ames dangerous, meaning that they are uncertain to be carcinogenic, according to the results. In humans, the maximum tolerated dose (MTD) serves as an indicator of a chemical's toxicity level. In comparison to vieloplain F, the MTD for vemurafenib was significantly greater. It is possible that inhibition of the potassium channels encoded by the hERG would cause a catastrophic ventricular arrhythmia. Several studies have shown that both vieloplain F and vemurafenib can inhibit hERG II, but not hERG I. However, vemurafenib was expected to be hepatotoxic, which would likely result in drug-induced liver damage. Vieloplain F was also predicted to be non-hepatotoxic. Skin hypersensitivity is a possible adverse effect of dermally given products, and none of the chemicals tested has been shown to cause skin sensitization in humans.

Table 4. Predicted toxicity profile of vieloplain F and vemurafenib.

Parameters	Vieloplain F	Vemurafenib
Ames Toxicity	No	No
Max. Tolerated Dose (human) (log mg/kg/day)	0.013	0.601
hERG I Inhibitor	No	No
hERG II Inhibitor	Yes	Yes
Oral Toxicity (LD50) (mg/kg)	1640	2316
Oral Toxicity classification *	IV	V
Hepatotoxicity	No	Yes
Skin Sensitization	No	No
Bioaccumulation Factor Log_{10} (BCF)	2.489	0.674
Daphnia magna $LC_{50} - Log_{10}$ (mol/L)	7.127	6.969
Fathead Minnow LC_{50}·Log10 (mmol/L)	−4.972	−4.154
Tetrahymena pyriformis $IGC_{50} - Log_{10}$ (mol/L)	1.998	2.203

* Class I: fatal if swallowed ($LD_{50} \leq 5$); class II: fatal if swallowed ($5 < LD_{50} \leq 50$); class III: toxic if swallowed ($50 < LD_{50} \leq 300$); class IV: harmful if swallowed ($300 < LD_{50} \leq 2000$); class V: may be harmful if swallowed ($2000 < LD_{50} \leq 5000$); and class VI: non-toxic ($LD_{50} > 5000$).

In addition, for the prediction of lethal dose (LD_{50}), the compound vieloplain F received a score greater than 300 mg/kg and was classified as class 4; therefore, it is considered "harmful if swallowed" ($300 < LD_{50} \leq 2000$), while the control vemurafenib has a score of over 2000, and classified as class 5; therefore, it is considered "may be harmful if swallowed" ($2000 < LD_{50} \leq 5000$), see Table 4. Vieloplain F and vemurafenib, according to their toxicological qualities, are not considered being at risk for protein toxicity, and the substances are classified as classes 4 and 5.

GUSAR, an online web server (Way2Drug, Moscow, Russia), and the environmental toxicity of each molecule were all considered. GUSAR predicted the environmental toxicity, where 96-h fathead minnow 50% lethal concentration, 48-h *Daphnia magna* 50% lethal concentration, and *Tetrahymena pyriformis* 50% growth inhibition concentration and bioconcentration factors were evaluated. The results are depicted in Table 4. The chemicals vieloplain F and vemurafenib fit into the application area of models in all circumstances when it comes to environmental toxicity prediction using GUSAR. Because of its superior safety profile when compared to vemurafenib, guaiane dimer vieloplain F was shown to have a lower risk of liver damage.

2.4. Drug-Likeness Prediction

The drug-likeness explains the compound's potential as a drug molecule candidate. As shown in Table 2 and Table S1, the compound vieloplain F met the requirements of drug-likeness and passed their filters, such as the Veber filter (rotatable bonds ≤ 10 with TPSA ≤ 140) [26]. Furthermore, the compound vieloplain F was also checked for Lipinski's rule of five (MW ≤ 500, MLOGP ≤ 4.15, N or O ≤ 10, NH or OH ≤ 5 and Log $P_{o/w} \leq 5$), except for one violation of Lipinski's rule of five with Log $P_{o/w}$ greater than 5 (Table 2) [27]. In addition, both vieloplain F and vemurafenib displayed a good bioavailability score of 0.55, within the range of F > 10% in rats which further proves the Veber and Lipinski's rule of five predictions [28]. Importantly, neither vieloplain F nor vemurafenib elicited an indication for the pan assay interference substances (PAINS), demonstrating that neither medication contains any fragments that might cause false positive biological tests. The molecules that must be synthesized in the laboratory are critical if they are to be produced in large quantities. The structure's intensity is graded into three categories: moderate (scores 1–4), medium (scores 4–7), and difficult (scores 8–10) [24]. The combination vieloplain F has received a score of 6.46 (Table S1, Supplementary Materials), which shows that vieloplain F, according to its complicated structure, is difficult to synthesize in the laboratory, but vemurafenib is easier to produce. In conclusion, these in silico studies for modeling the physicochemical and pharmacokinetic properties of vieloplain F revealed that the pattern of vieloplain F was better than that of vemurafenib in terms of CYP inhibition, hepatotoxicity, and pharmacokinetic properties. Vieloplain F was also found to have superior pharmacokinetic properties.

2.5. Prediction of the Bioactivity Score

Using the Molinspiration Chemoinformatics tools, the projected bioactivity scores of the substances vieloplain F and vemurafenib were calculated. These results are shown in Table 5. According to the bioactivity score, vieloplain F is moderately active in the presence of a G-protein-coupled receptor (GPCR) ligand, an intracellular signaling regulator, a kinase inhibitor, as well as protease inhibitors, while vemurafenib is moderately active. According to studies by [29] the nuclear receptor has a dual function in inflammation and immunity. Vieloplain F is a more dynamic and high-scoring molecule than the control vemurafenib, according to the research. Enzyme inhibition levels show both substances were active. Vieloplain F's activity score profile shows it is physiologically active and has a physiologic impact. There has been no inactivity, as predicted by the bioactivity score.

Table 5. Bioactivity score of the compounds vieloplain F and vemurafenib according to the Molinspiration software.

Compounds	GPCR Ligand	Ion Channel Modulator	Kinase Inhibitor	Nuclear Receptor Ligand	Protease Inhibitor	Enzyme Inhibitor
Vemurafenib	0.45	0.25	0.64	0.02	0.12	0.34
Vieloplain F	−0.12	−0.18	−0.54	0.17	−0.04	0.06

2.6. Prediction of Cardiac Toxicity

The FDA requires that every biomolecule be tested for hERG safety before it may be used as a therapeutic candidate. The hERG blockage has been connected to deadly cardiac arrhythmias. Using pred-hERG results to predict cardiac toxicity, the likelihood map for vieloplain F and vemurafenib as a control is shown here (Figure 3). Attributions to hERG blockage, both positive and negative, are shown in the figure. Increasing the number of contour lines and the intensity of the green color shows that an atom or fragment has made a more positive contribution to the hERG blockage. With a 50% confidence level, the pred-hERG projected that Vieloplain F would be non-cardiotoxic, whereas the control vemurafenib was projected as having a 60% confidence level that it may be cardiotoxic. The findings have revealed that our isolated molecule, vieloplain F, is less hazardous to the heart than the control drug, vemurafenib, for cardiovascular toxicity.

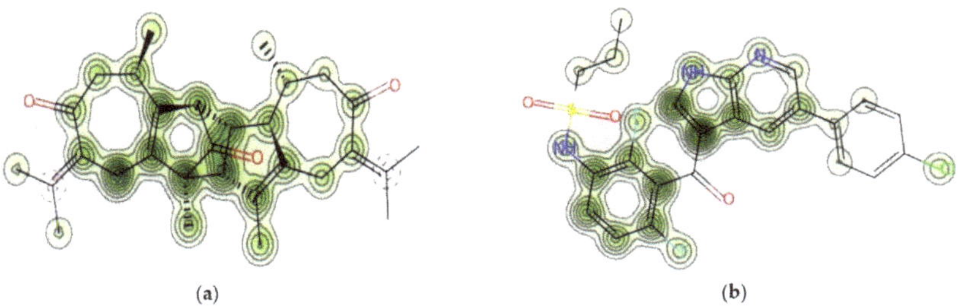

Figure 3. Cardiac toxicity of drugs derived from pred-hERG in a map format: (**a**) vieloplain F; (**b**) vemurafenib.

2.7. Biomolecular Macromolecules: Epoxidation and Reactivity Prediction

The prediction of epoxidation and reactivity to biological macromolecules of vieloplain F and control vemurafenib is depicted in Figure 4. The possible sites of epoxidation of the compound vieloplain F have been shown in Figure 4b. Here, the double bond between atoms 18–19 and 31–7 was more prone to epoxidation, where the probability score was 0.19. In Figure 4c,e, cyanide and GSH's reactivity showed no probability scores for any atoms. In Figure 4d, reactivity to DNA is shown for atoms 11, 26, and 28 with the probability scores of 0.050, 0.029, and 0.030. In Figure 4f, reactivity to protein was only shown for one atom 29 with a probability score of 0.047. The possible sites of epoxidation of the control vemurafenib are shown in Figure 4h. Here, the epoxidation was predicted on one double bond between atoms 19–20 with a probability score of 0.031. The epoxidation was also predicted on atoms between 20–22 and 24–26 with the probability score of 0.031 and 0.035. In Figure 4i, cyanide reactivity showed no probability scores for any atoms. In Figure 4j, reactivity to DNA is shown for nine atoms, among which the highest probability score was predicted for atoms 6 and 32 with the score of 0.21. The reactivity to DNA for the rest atoms 1–3, 9, 15, 20, and 24 shown probability scores with 0.11, 0.066, 0.081, 0.038, 0.029, 0.075, and 0.028. In Figure 4k, reactivity to GSH was shown for atoms 3, 6, 9, 20, and 32 with the probability score of 0.031, 0.056, 0.053, 0.050, and 0.045. In Figure 4l, reactivity to protein was shown for the atoms 1, 3, 6, 9, 15, and 20 with probability scores of 0.14,

0.057, 0.050, 0.17, 0.078, and 0.034, in which the highest score was predicted for atom 9. The results declared that the guaiane dimer vieloplain F has better reactivity results than control vemurafenib, which also has better reactivity results, and prove that natural compounds have better epoxidation and reactivity profiles.

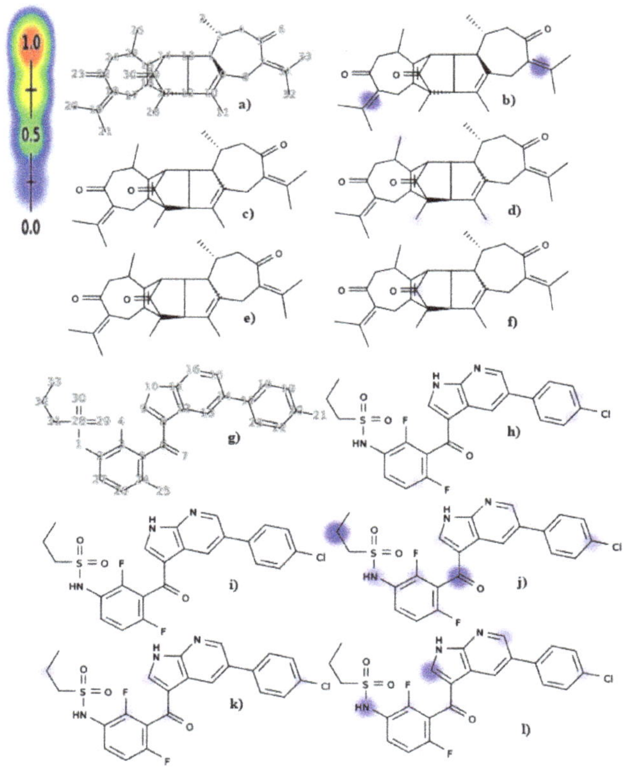

Figure 4. Prediction of epoxidation and reactivity to biological macromolecules by XenoSite. (**a**) The input structure of vieloplain F; (**b**) prediction of epoxidation; prediction of reactivity to: (**c**) cyanide; (**d**) DNA; (**e**) glutathione (GSH); (**f**) protein; (**g**) the input structure of vemurafenib; (**h**) prediction of epoxidation; prediction of reactivity to: (**i**) cyanide; (**j**) DNA; (**k**) glutathione (GSH); (**l**) protein.

2.8. Prediction of Endocrine Disruption Potential

The multi-color-coded table obtained from the online web tool of Endocrine Disruptome is given in Figure 5. There were fourteen nuclear receptors with eighteen targets. For the compound vieloplain F, according to the results (Figure 5a), fifteen targets showed low probability as they were coded in green (sensitivity > 0.75). Two targets, estrogen receptor alpha antagonist and progesterone receptor, were coded in yellow, indicating that vieloplain F had a medium probability of binding (0.50 < sensitivity < 0.75), and one target, glucocorticoid receptor antagonist, was coded in red, indicating a high probability of binding (sensitivity < 0.25). The control vemurafenib showed the most negligible results compared to vieloplain F Figure 5b. Initially, only 11 targets were coded green (sensitivity > 0.75) and showed a low probability. Three targets were encompassed in the yellow zone, such as estrogen receptor alpha antagonist, estrogen receptor beta antagonist, and peroxisome proliferator-activated receptor gamma, indicating that vemurafenib had a medium probability of binding (0.50 < sensitivity < 0.75). Four targets were encompassed in orange color (0.25 < sensitivity < 0.50)—glucocorticoid receptor, glucocorticoid receptor antagonist, liver X receptor alpha, and retinoid X receptor alpha—indicating a medium probability of binding. The control did not show any high probability of binding to the nuclear receptor. The results have revealed that vieloplain F has a strong profile as it has only one target encoded in red color, but still, the docking score of that target is −10.0, which is a high score for binding. According to docking rules, more negative scores mean more affinity towards binding. As mentioned in Figure 5b, the control vemurafenib has very high orange zone scores up to −12.0, indicating a higher possibility for bindings. From all the results, it has been clear that the control vemurafenib has shown more human nuclear receptors binding affinity than vieloplain F.

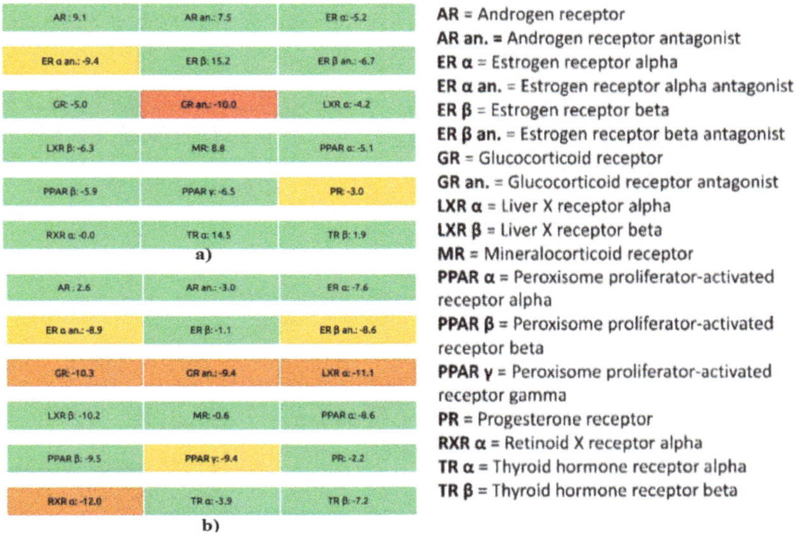

Figure 5. Endocrine disruption potential of compounds as obtained from Endocrine Disruptome. (**a**) Vieloplain F, (**b**) vemurafenib. Red color describes the high probability of binding, Orange and Yellow describes the medium probability of binding while the Green color describes the low probability binding.

2.9. Prediction of Cell Line Cytotoxicity

In silico prediction of cell line cytotoxicity for cancer cells is shown in Table 6. Both the compounds vieloplain F and control vemurafenib showed the highest scores for the melanoma cell line (Sk-Mel-28). In this set of predictions, probable activity (Pa) was higher than probable inactivity (Pi), and we only selected Pa > 0.5 because the differences were significant; these results can be recorded as probable cytotoxic activities for both the compounds. Vieloplain F was recorded with the highest score of (Pa) 0.722 against melanoma.

Table 6. In silico prediction of cell line cytotoxicity by CLC-pred.

Compound	Cell Line	Cell Line Full Name	Tissue	Tumor Type	Pa	Pi
Vemurafenib	SK-MEL-28	Melanoma	Skin	Melanoma	0.618	0.010
	A2058	Melanoma	Skin	Melanoma	0.581	0.004
	M14	Melanoma	Skin	Melanoma	0.546	0.012
	PA-1	Ovarian carcinoma	Ovarium	Carcinoma	0.520	0.005
Vieloplain F	SK-MEL-2	Melanoma	Skin	Melanoma	0.722	0.006
	K562	Erythroleukemia	Hematopoietic and lymphoid tissue	Leukemia	0.598	0.015

2.10. Docking of the Compounds with B-Raf Kinase Structure (PDB: 3OG7)

2.10.1. Structural Analysis of B-Raf Kinase

B-Raf kinase is a class of transferase that consists of two domains having 289 amino acids. The overall statistical VADAR analysis showed protein architecture, containing 39% helices, 22% β sheets, and 38% coils. Moreover, Ramachandran plots indicated that 99.6% of residues were present in the allowed region, which shows the precision of phi (φ) and psi (ψ) angles among the coordinates of B-Raf kinase (Figure S2, Supplementary Materials).

2.10.2. Binding Energy Evaluation of the Compounds

To predict the best conformational position within the active region of the target protein, vieloplain F and vemurafenib were docked and analyzed based on docking energy value (kcal/mol). Moreover, generated docked complexes were examined based on the hydrogen/hydrophobic interaction pattern. Docking results justified that the compound vieloplain F showed good energy values (−11.8 kcal/mol) as compared to standard vemurafenib (−10.2 kcal/mol) against target protein (Table 7). The comparative analysis showed that guaiane dimer vieloplain F may have good therapeutic potential against B-Raf kinase protein and can be considered an emerging candidate against melanoma.

Table 7. Chemical structures, amino acid residues, and docked results of the vieloplain F and vemurafenib with B-Raf kinase (PDB: 3OG7).

Compound Name	Chemical Structure	Binding Energy (kcal/mol)	Amino Acid Residues Involved in the Bonding
Vieloplain F	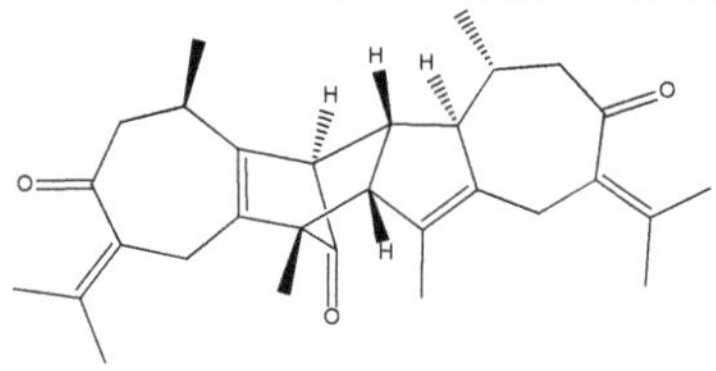	−11.8	Ser164, Asp165, Asp125, Lys127, Trp150, Ser147, Ile148, Met151, Val155.
Vemurafenib	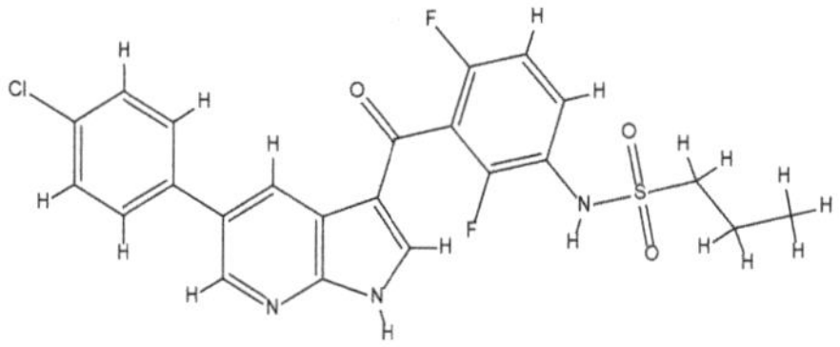	−10.2	Asp143, Lys127, Asn129, Ser88, Phe132, Ser87, Cys84, Trp83, Ala33, Ile15, Ile79, Lys35, Phe20.

2.10.3. Protein–Ligand Complex Analysis

The compounds vieloplain F and vemurafenib were bound against the target protein in different conformations. In vieloplain F docking results, a single hydrogen bond was observed at SER164. The benzene oxygen atom forms a hydrogen bond, with SER164 having a bond length of 2.29 Å. The comparative results showed that common residues were observed in docking studies, strengthening our docking results (Figure 6a). In vemurafenib docking results, hydrogen bonds were observed at CYS84, CYS532, LYS483, GLY596, ASP594, and PHE595. The residual comparison showed that our compound binds within the target protein's active site as with standard drug vemurafenib with different conformations (Figure 6b).

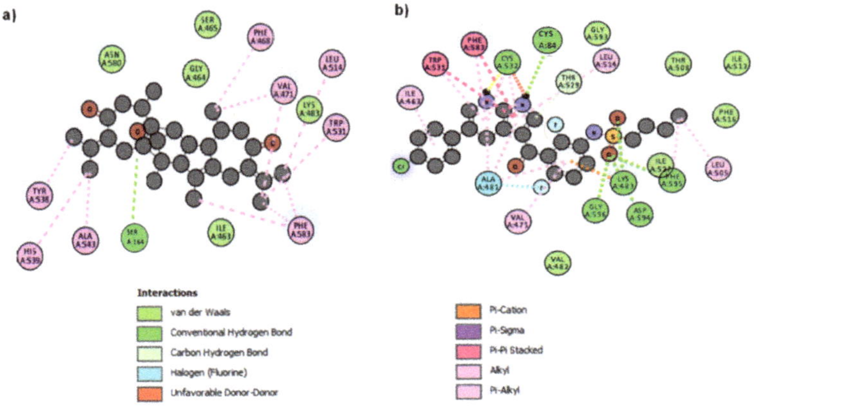

Figure 6. (**a**) 2D interactions of the compound vieloplain F with B-Raf (PDB: 3OG7); (**b**) 2D interactions of the control vemurafenib with B-Raf (PDB: 3OG7).

2.11. Molecular Dynamics and Simulation

MDS was performed for the vemurafenib-B-Raf kinase structure complex and vieloplain-F- B-Raf kinase structure complex up to 100 ns. The parameters explored for analyses include RMSD, RMSF, protein–ligand contact, RoG, and binding free energy.

2.11.1. RMSD Analysis

The RMSD study was done to find the simulation results stabilities. The RMSD graph of protein (left Y-axis) can give the understanding of its structural conformation during the simulation while ligand RMSD (right Y-axis) denotes the stability of ligand toward the specific protein and its binding site pocket. In the case of vemurafenib-B-Raf kinase complex, initially, RMSD showed robust stabilization and remained stable throughout the simulation period up to 100 ns. The vieloplain-F–B-Raf kinase complex showed that both vieloplain-F and B-Raf Kinase attained stability at 20 ns and remained constant throughout the simulation (Figure 7A). At first, the vieloplain-F–B-Raf kinase complex RMSD showed fluctuations of 0.2 nm up to 20 ns and then remained stable through the simulation period up to 100 ns (Figure 7B). Both complexes revealed a good level of interaction throughout the simulation, with less deviation in structure.

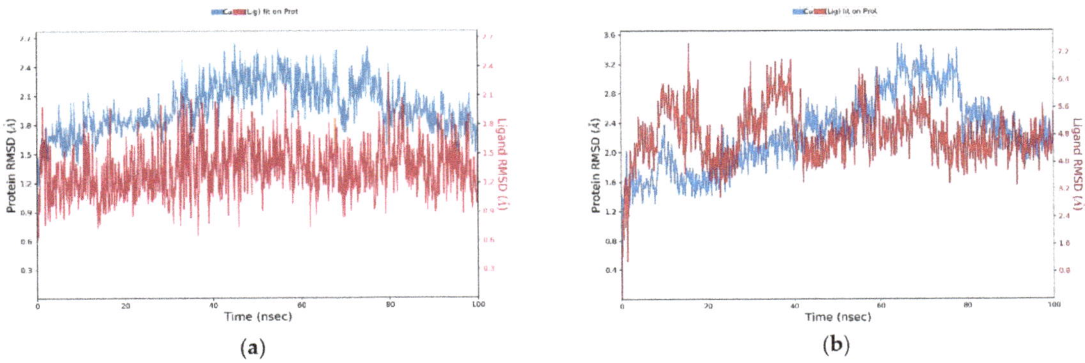

Figure 7. Root mean square deviation (RMSD) of the (**a**) vemurafenib–B-Raf kinase; (**b**) vieloplain-F–B-Raf kinase.

2.11.2. RMSF Assay

The RMSF defines the deviation of the particle in the macromolecule. It specifies the protein structure flexibility and rigidity. The residues with higher peaks belong to loop areas or N- and C-terminal zones, as typically N and C fluctuate the most, recorded by MD trajectories (Figure 8). The stability of ligand binding to the protein is shown by low RMSF values of binding site residues. The percentages of Helix and Strand in vemurafenib–B-Raf kinase were determined to be 30.30% and 15.44%, respectively, and the overall secondary structure elements (SSE) were found to be 45.76%. In the case of vieloplain-F–B-Raf kinase, the percentages of Helix and Strand were 30.19% and 15.24%, respectively, and the total SSEs were 45.44 %. Protein SSEs are more rigid than the unstructured part of the protein, showing slight fluctuations in Figure 9.

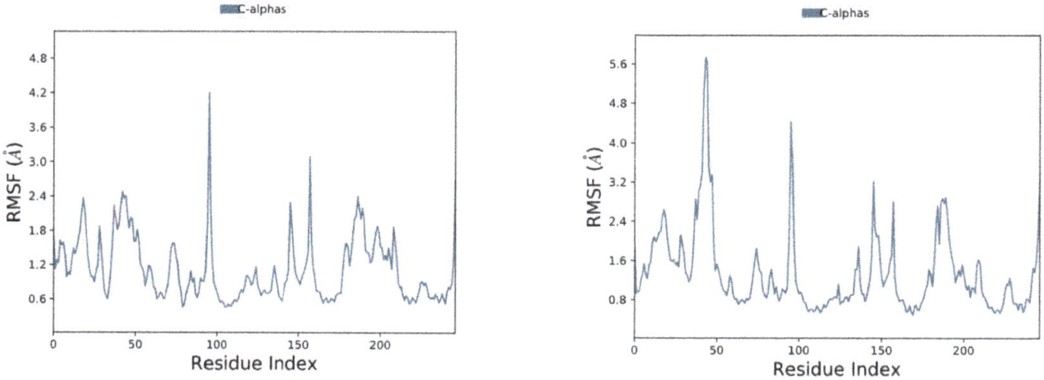

(**a**) (**b**)

Figure 8. Root mean square fluctuation (RMSF) of protein complexes (**a**) vemurafenib–B-Raf kinase, (**b**) vieloplain-F–B-Raf kinase.

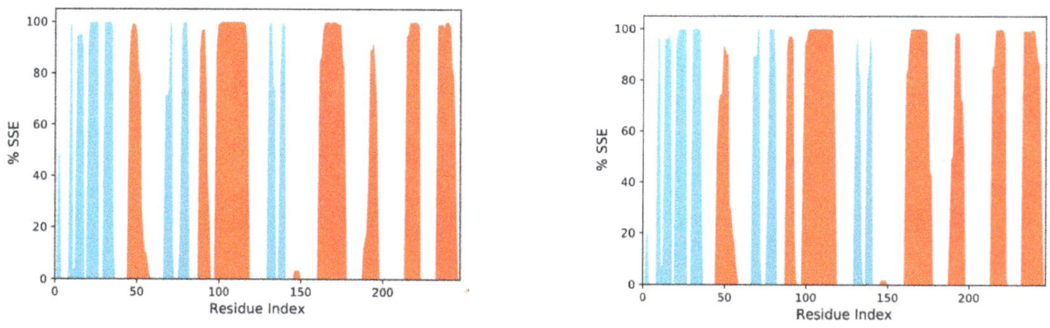

a) **b**)

Figure 9. Protein secondary structure element distribution by residue index throughout the protein structures complexed with ligand (**a**) vemurafenib–B-Raf kinase, (**b**) vieloplain-F–B-Raf kinase. The red color represents α-helices, and the blue represents β-strands.

2.11.3. Protein–Ligand Interaction

The interaction of the target protein with the ligand was monitored during the simulation. These interactions were categorized into four types: (i) hydrogen bonds, (ii) hydrophobic, (iii) ionic, and (iv) water bridges.

This study found that the most significant ligand–protein interactions were hydrogen bonds, water bridges, and hydrophobic interactions. The vemurafenib–B-Raf Kinase complex showed the most important hydrophobic interactions with ALA_481, LEU_514, TRP_531, and PHE_583, whereas GLN_530, CYS_532, ASP_594, PHE_595, and GLY_596 were chief in terms of H-bonds (Figure 10A). In vieloplain-F–B-Raf kinase, the hydrophobic contacts PHE_583, TYR_538, TRP_531, LEU_514, ARG_462, and PHE_468 were the most vital, while SER_465 and LYS_483 were dynamic interactions for H-bonds (Figure 10b).

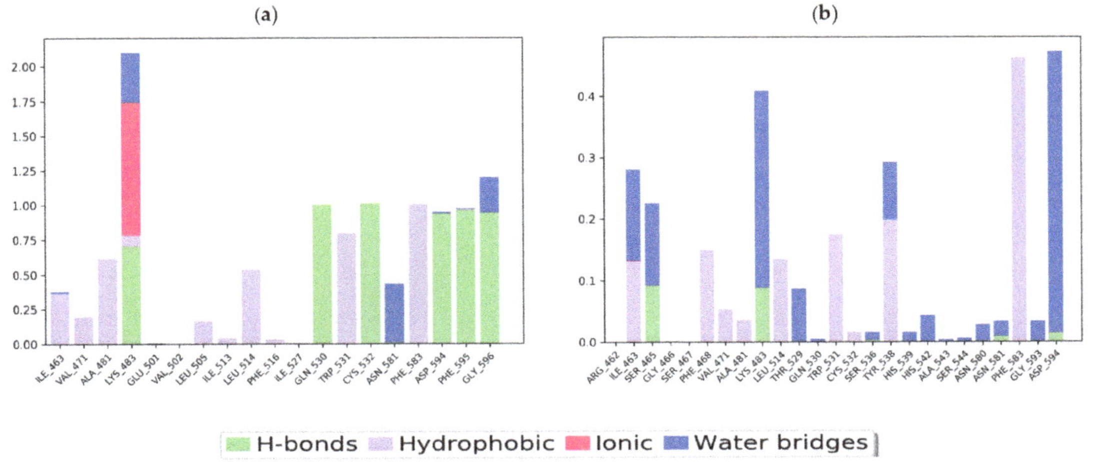

Figure 10. Protein–ligand contact histogram: (**a**) vemurafenib–B-Raf kinase, (**b**) vieloplain-F–B-Raf kinase.

2.11.4. RoG Analysis

The folding and compactness of the protein are often arbitrated with the assistance of RoG. It is a crucial method of showing the influence of ligand on the three-dimensional conformational structural changes after the interaction with ligand. RoG with high valve depicts the molecule loose packing and folding nature of the protein after the interaction with the ligands. Both the complexes were in their native structures as there was not much variation observed throughout the 100 ns in the RoG graph (Figure 11). However, there were a few minor variations in RoG due to conformational changes in the secondary structure of protein during the MDS process. The RoG graph of the complexes shows that the ligand remains tightly bound to the protein's active site.

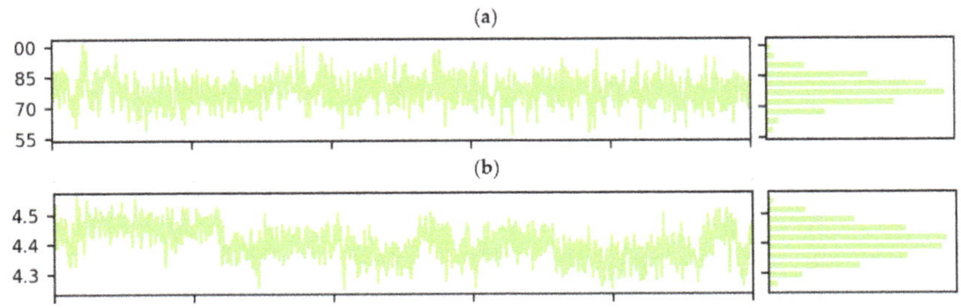

Figure 11. The radius of gyration was calculated for (**a**) vemurafenib–B-Raf kinase, (**b**) vieloplain-F–B-Raf kinase.

2.12. MM-GBSA Calculations

The average binding free energy (ΔG) of the vemurafenib- B-Raf kinase and vieloplain-F-B-Raf Kinase was calculated at 0 and 100 ns by the MM-GBSA approach (Figure 12). The average values of ΔG for ligands vemurafenib were −138.8836 kcal/mol (0 ns) and

−115.1949 kcal/mol (100 ns). The average dG for vieloplain-F were −144.2660 kcal/mol (0 ns) and −158.7822 kcal/mol. It was found that both compounds showed minimum VDW and hydrogen bonding energies that meant all these drugs are maximum potential to bind with active site residues.

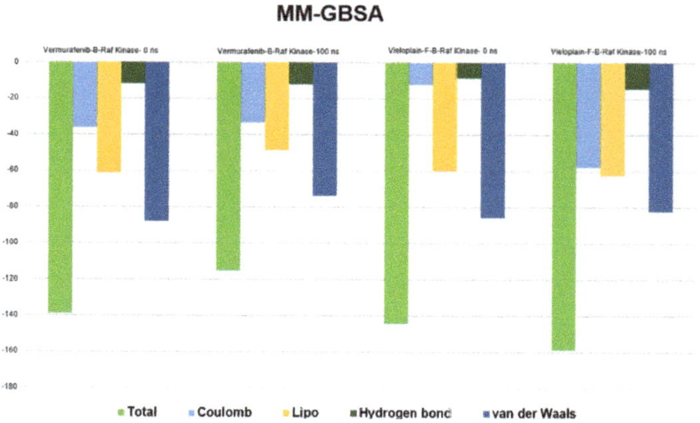

Figure 12. MM-GBSA calculated before and after the simulation.

3. Discussion

This work examines the utility of selecting vieloplain F as a possible biomolecule therapeutically active against melanoma by using publicly available techniques to investigate its bioactivity, ADMET, drug-likeness, molecular docking, and simulations. In silico interpretations of pharmacological spectra illustrated a new road to the most promising directions while also assisting in the initial phases of research by filtering out the biomolecules with a potentially low pharmacological profile [30]. According to the PASS prediction, vieloplain F could be considered an anti-neoplastic and anti-leukemic agent as sesquiterpene-type guaiane dimers are already reported for their attractive potencies against cancer [12,15] and reversal of resistance towards cancer [16]. Furthermore, the compound vieloplain F was also predicted as a testosterone 17-beta-dehydrogenase (NADP+) inhibitor, which means that vieloplain f could also be investigated for its participation against androgen and estrogen metabolism. The compounds also predicted high scores for NADP+ and evidenced by our studies that this compound has good effects against melanoma and it can be considered for its studies against different kind of skin diseases.

ADMET is crucial for every biomolecule before its biotransformation into a drug [21]. According to the ADMET profile of vieloplain F, the absorption and distribution of the compound were moderate. The compound is highly soluble in GIT and has no solubility for BBB, showing that this compound cannot create any adverse effects related to CNS. The inactive compounds on the CNS should not intersect to avoid adverse effects on the CNS [22]. The compound vieloplain F has a high ratio of GIT absorption as compared to control vemurafenib.

Furthermore, vieloplain F revealed that it is not a P-gp (P-glycoprotein) substrate; therefore, vieloplain F is not susceptible to the efflux mechanism of P-gp, which many cancer cell lines utilize as a drug resistance mechanism. CYP enzymes play a crucial role in drug excretion, and these isoforms are metabolizing almost 75% of market available drugs. Inhibition of any of these isoforms results in causing some significant pharmacokinetics-based drug-drug interactions [12,24]. Vieloplain F did not inhibit any of the CYP enzymes, but the control vemurafenib inhibited 3 CYP enzymes which meant that it can create drug-drug interactions for those CYP enzyme-targeted drugs. One of the significant drawbacks of vemurafenib was its causing hepatotoxicity [31] and cardiotoxicity [32]. The compound vieloplain F did not show any hepatotoxicity and is revealed to be cardioprotective. Blocking the hERG K^+ channel can cause QT prolongation and potentially fatal arrhythmia [33].

As a result, vieloplain F was projected to be a non-inhibitor of hERG with no cardiac adverse effects. Finally, the toxicity profile acquired from pkCSM was adequate.

Acute toxicity was defined as the harmful effects of a single exposure to a drug over a short period [34]. In general, mice and rats were used to measure acute toxicity. The compound vieloplain F was predicted as non-toxic and categorized in class 4 with harmful indications if swallowed, suggesting the possible safe application. Furthermore, the compound vieloplain F did not exhibit any environmental toxicity violations.

CYP can act on aromatic or double bonds, probably leading to epoxy metabolites. Identifying a molecule's epoxidation site will help guide improvement to avoid epoxidation for safer drugs [35]. Besides, reactive metabolites can cause idiosyncratic adverse drug reactions and drug-induced liver injury [36]. Predicting epoxy and reactive metabolites can thus help predict potential adverse effects. The present study shows that vieloplain F has been doubtful to result in epoxides and reactive metabolites compared to control vemurafenib.

An endocrine disruptor is a substance that operates on nuclear receptors to disrupt the endocrine or hormonal systems [37]. Compounds are docked into eighteen integrated and well-validated crystal structures of fourteen different human nuclear receptors using Endocrine Disruptome's free web tool. As a result, achieving endocrine disruption potency for multiple targets at once is quite convenient. This eventually leads to rapid, well-informed judgments for in vitro and in vivo testing [38]. Vieloplain F was predicted to be non-endocrine disruptive, ensuring that it may be used safely. Among both compounds, including vieloplain F and vemurafenib, the highest profile for safety was achieved by vieloplain F.

To discover new anticancer drugs, sophisticated laboratory models are needed to predict their clinical behavior. This is mainly done from cell lines originating from human tumors [39]. In silico prediction will make the way more comfortable by offering useful predictions for making an informed decision before going into the In vitro experiments. The inhibitory effect of *Xylopia vielana* on cancer cell lines has been previously reported [13,15,16].

Vieloplain F has been studied as an anticancer agent in our recent study and has demonstrated potent activity against prostate cancer cell lines with an IC_{50} value of 9.5 μM [15]. However, particular attention to the deep mechanism has not yet been drawn. The present study demonstrated a specific targeted approach that can be useful in further anticancer drug development of B-Raf kinase.

B-Raf is a member of the Raf-kinase family of growth signal transduction protein kinases. This protein plays a role in regulating the MAP kinase/ERKs signaling pathway, affecting cell division, differentiation, and secretion. B-Raf kinase inhibitors are also considered one of the main treatments for malignant melanomas. Currently, only one B-Raf kinase inhibitor, vemurafenib, is approved by the FDA and is used in late-stage melanoma [39], but the main drawback of this drug is that patients start showing symptoms after receiving this therapy resistance towards this drug [8]. Therefore, research is underway to develop novel B-Raf kinase inhibitor compounds with maximum potency and minimum side effects. There is still scope to develop B-Raf kinase inhibitors with improved therapeutic efficacy and reduced side effects. As evident from our study, vieloplain F showed encouraging results in terms of B-Raf kinase inhibition compared to vemurafenib in the sense of molecular docking and simulation studies. Therefore, in these ways, special attention should be placed on investigating the therapeutic importance of this process.

Thus, the present study revealed the potential significant applications of vieloplain F and could be helpful against various diseases. The study relied on computational tools that reported pharmacological properties and bioactivities predictions. Besides, in this study, we reported the potentials of vieloplain F with comparison studies on vemurafenib, and in previous studies, the in vitro and in vivo potential of vieloplain F was displayed. Moreover, clinical studies are necessary to confirm the findings of the present work. Nonetheless, the results of this work will provide future guidance for the design and development of new lead compounds against Melanoma.

4. Materials and Methods

4.1. Isolation of Guaiane Dimer Vieloplain F

Isolation procedures for guaiane dimer, vieloplain F, and its structure elucidations based on spectroscopic methods (1D, 2DNMR, X-ray, and ESI-MS) along with a comparison of the relevant literature data reported for guaiane dimer, vieloplain F, were described in detail in our recent paper [15].

4.2. Prediction of Activity Spectra for Substances (PASS)

Prediction of activity spectra for substances (PASS) gives information about the bioactive compounds' plausible pharmacological activities. The free internet platform Pass online (http://www.pharmaexpert.ru/passonline, accessed on 10 August 2021) was used to make the PASS prediction. Only actions with Pa > Pi were considered feasible for a given compound. Pa > 0.7 indicated a high probability of experimental pharmacological effect, while Pa 0.5 to 0.7 indicated a moderate probability of experimental pharmacological action. If Pa was less than 0.5, the chances of pharmacological activity were negligible [40].

4.3. ADMET Analysis

ADMET (absorption, distribution, metabolism, excretion, and toxicity) are the essential measurement tools for any compound before being elected as a drug candidate. The online web tool swiss ADME (http://www.swissadme.ch/index.php, accessed on 10 August 2021) was used to obtain ADME properties of the vieloplain F [24], and the pharmacokinetic scores were predicted using the online web application pkCSM (http://biosig.unimelb.edu.au/pkcsm/prediction, accessed on 10 August 2021).

4.4. Prediction of Acute Rat Toxicity and Environmental Toxicity

The publicly accessible structure–activity relationship (GUSAR) software (http://www.way2drug.com/gusar/acutoxpredict.html, accessed on 10 August 2021) was used to predict median lethal dosage (LD_{50}) values for rats with oral administration [41]. The quantitative predictions of ecotoxicity were also assessed by GUSAR software (http://www.way2drug.com/gusar/environmental.html, accessed on 10 August 2021).

4.5. Prediction of Drug-Likeness

Swiss ADME (http://www.swissadme.ch, accessed on 10 August 2021) and Molinspiration Chemoinformatics tools (https://www.molinspiration.com/cgi-bin/properties, accessed on 10 August 2021) were used to predict drug-likeness. Lipinski's rule of five was considered a standard for accessing the drug-likeness [27].

4.6. Bioactivity Score Prediction

The bioactivity profile of a selected compound can be portrayed by the scoring system of G protein-coupled receptor (GPCR) ligand, ion channel modulator, nuclear receptor legend, a kinase inhibitor, protease inhibitor, and an enzyme inhibitor. These properties were determined by Molinspiration Chemoinformatics tools (https://www.molinspiration.com/cgi-bin/properties, accessed on 10 August 2021). According to studies by Roy, if the value was equal to or greater than 0.00 ($\geq$0), the compound was more active, while if the values were between −0.50 and 0.00, it was moderately active; nevertheless, if the values were less than −0.50 (<−5.0), it was thought to be inactive [42].

4.7. Prediction of Cardiac Toxicity

The blockage of the hERG K^+ channels has been linked to fatal cardiac arrhythmias. The pred-hERG 4.2 (http://predherg.labmol.com.br, accessed on 10 August 2021) web server, a web tool for early detection of putative hERG blockers and non-blockers, was used to predict cardiac toxicity [38].

4.8. Prediction of Epoxidation and Reactivity to Biological Macromolecules

Epoxides are metabolites produced by an enzyme cytochrome P450 operating aromatic or double bonds. Drug-metabolizing enzymes can bioactivate the drug into reactive metabolites, creating adducts when they bind to specific targets in DNA or proteins. The freely available web Xenosite (https://swami.wustl.edu/xenosite/submit, accessed on 10 August 2021) was used to predict epoxidation and reactivity to biological macromolecules [43].

4.9. Prediction of Endocrine Disruption Potential

Endocrine Disruptome is an unrestricted prediction tool for determining the potential for endocrine disruption via nuclear receptor binding. Fourteen human nuclear receptors and their eighteen validated structures that regulate reproduction, behavior, development, metabolism, and the immune system were utilized for molecular docking with the compounds in the freely accessible web platform Endocrine Disruptome (http://endocrinedisruptome.ki.si, accessed on 10 August 2021) [20].

4.10. Prediction of Cell Line Cytotoxicity

CLC-Pred (Cell Line Cytotoxicity Predictor) is a web-based program that predicts the cytotoxicity of chemical compounds in non-transformed and cancer cell lines depending on their structural formula. Prediction of cell line cytotoxicity of vieloplain F was made through CLC-Pred (http://www.way2drug.com/Cell-line/, accessed on 10 August 2021). The predicted output activity was represented in the probable activity (Pa) and probable inactivity (Pi) score. The scoring system was categorized into three portions according to activity. Pa > 0.5 was considered as the highest activity, Pa > 0.3 was considered as moderate activity, and Pa < 0.3 was considered as the lowest activity. [44].

4.11. Preparation, Analysis, Retrieval, and Visualization of Protein and Ligand Structures

ChemBioDraw (PerkinElmer Informatics, Waltham, MA, USA, v13.0) [45] was used to draw the compounds into .mol format. The control drug vemurafenib was downloaded from PubChem in .sdf file. The three-dimensional (3D) structures of B-Raf Kinase was accessed from Protein Data Bank (PDB) (www.rcsb.org, accessed on 10 August 2021) with PDBIDs 3OG7. The selected protein structure was minimized using Chiron portal (https://dokhlab.med.psu.edu/chiron/processManager.php, accessed on 10 August 2021) and visualized through UCSF Chimera 1.10.1 tool [46]. The Ramachandran plots of B-Raf kinase were accessed from the Discovery Studio 4.1 Client tool. The protein architecture and statistical percentage values of receptor proteins helices, beta-sheets, coils, and turn were predicted from online server VADAR 1.8 (http://vadar.wishartlab.com/, accessed on 10 August 2021) [47].

4.12. Molecular Docking

Docking studies of vieloplain F and vemurafenib were performed against B-Raf Kinase. To prepare the B-Raf Kinase structure, the unnecessary ligands and water molecules were removed to enhance docking results' efficacy. The ligands were sketched in the ACD/ChemSketch 2.1.2 tool (Advanced Chemistry Development, Toronto, Canada) and further minimized by UCSF Chimera 1.10.1. A docking experiment was used on all synthesized compounds against B-Raf Kinase using the PyRx docking tool 1.7 (https://pyrx.sourceforge.io/, accessed on 10 August 2021) [48]. To perform the docking experiment, grid box parametric dimension values were adjusted as $X = -1.3845$, $Y = -12.9405$, and $Z = -18.9916$, respectively. The default exhaustiveness = 8 value was used to obtain the finest binding conformational pose of protein-ligand docked complexes. All compounds were docked separately against the crystal structure of B-Raf kinase. The docked complexes were evaluated on lowest binding energy (kcal/mol) values, hydrogen and hydrophobic bond interaction pattern analysis using Discovery Studio (4.1) (Dassault Systemes BIOVIA, San Diego, CA, USA), and UCSF Chimera 1.10.1. The three-dimensional (3D) graphical depictions of all the docked complexes were accomplished by Discovery Studio (2.1.0) (Dassault Systemes BIOVIA, San Diego, CA, USA) and the UCSF Chimera 1.10.1 tool.

4.13. Molecular Dynamics Simulations

The MDS analysis was performed to find the interaction of ligand–protein stability. MD simulation studies also analyze the structure of the macromolecules transition to the functional significance of the complex. Simulation typically records atom movement concerning the time based on Newton's standard motion equation to predict the binding of the ligand in the biological environment. The MD simulation of the selected complex was undergone for 100 ns using Desmond v:3.6 New York, NY, USA module from Schrodinger suite [12]. The interaction of the complex obtained from MD was the initial structure of respective MD simulations followed by an established protocol where protein atoms were 10 Å away from the box. The ligand-receptor complex minimization and optimization were done through Wizard of Maestro (Schrödinger, New York, NY, USA). The systems were set up by applying the System Builder tool, a solvent standard TIP3P with an orthorhombic box was selected. The OPLS 2005 was used for simulation analysis. The Physiologic conditions of the model were minimized by adding 0.15 M NaC [49]. The models were rested before the start of the simulation. Lastly, simulations were run at 300 K temperature at 1 atm pressure, with an NPT ensemble was applied for all MDSs.

Moreover, the MDS trajectories were recorded after every 100 ps interval. The root means square deviation (RMSD) of both the protein and ligand were recorded to find the stabilities of simulation. The RMSF and RoG values were also calculated. MDSs were repeated thrice for each complex using the same parameters.

4.14. MMGBSA Calculations

The ligand-receptor complexes were minimized using the prime tool in maestro. After minimization, the molecular mechanics generalized born surface area (MMGBSA) was used to evaluate the binding free energies (ΔG) of complexes at both before (0 ns) and after (100 ns) simulation. All over the computation of binding free energies, an OPLS_2005 force field was utilized [50].

5. Conclusions

Natural products of herbal origin are well known to exhibit diverse biological activities compared to synthetic products. Here, in this study, one potent guaiane dimer, vieloplain F, was isolated from *Xylopia vielana* species and tested against B-Raf kinase to find a potent drug molecule against melanoma. The compound showed potent inhibitory activity against the B-Raf kinase protein receptor. The preliminary computational studies such as ADMET, bioactivities, and molecular docking studies proved that this guaiane dimer has a high binding affinity towards the targeted B-Raf kinase protein receptor. In addition, it revealed that our isolated natural compound is safer than the FDA-approved drug vemurafenib in cardiac and hepatotoxicity profile and has high binding energies towards targeted protein then vemurafenib. Overall, the present study acts as evidence to prove that this guaiane dimer isolated from the *Xylopia vielana* has the capacity to inhibit the B-Raf kinase protein receptor, which also opens the road for all the guaiane dimers that all these compounds should be screened for B-Raf kinase protein. The in silico studies can provide a platform for a potential compound against any specific disease but still, before any biomolecule needs to be selected, further studies must pass through deep in vivo and in vitro studies to confirm their results. The isolation methods, the quantity of the pure compound, and the complex structures of the natural products create a big question for future researchers to resolve this problem and do the wetlab assays independently.

In this study, the complete pharmacological profile encompassing PASS, bioactivity scores, ADMET, molecular docking, and molecular simulations will act as a foundation for other guaiane dimers to be investigated in the future for different types of cancers.

Supplementary Materials: The following are available online: Figure S1: Bioavailability radar chart for vemurafenib, Figure S2: Ramachandran graph of target protein, Table S1: Drug-likeness profile of Vieloplain F and Vemurafenib.

Author Contributions: S.S.u.H. performed all the experiments and paper writing. S.Q.A. and M.I. rechecked the manuscript for any errors. M.H. performed the docking studies. F.A. and S.G.B. checked the paper for simulations studies. I.B. language editing. H.-Z.J. designed the experiment. All authors have read and agreed to the published version of the manuscript.

Funding: The work was supported by NSFCs (81973191), Shanghai Natural Science Fund (19ZR1428100), Shanghai Engineering Research Center for the Preparation of Bioactive Natural Products (16DZ2280200), the Scientific Foundation of Shanghai China (13401900103, 13401900101), and the National Key Research and Development Program of China (2017YFC1700200).

Institutional Review Board Statement: Not applicable.

Informed Consent Statement: Not applicable.

Data Availability Statement: Not applicable.

Acknowledgments: The authors wish to thank to University of Oradea, Oradea, Romania for financial support in publishing this paper.

Conflicts of Interest: The authors declare no conflict of interest.

References

1. Hassan, S.S.u.; Muhammad, I.; Abbas, S.Q.; Hassan, M.; Majid, M.; Jin, H.Z.; Bungau, S. Stress driven discovery of natural products from actinobacteria with anti-oxidant and cytotoxic activities including docking and admet properties. *Int. J. Mol. Sci.* **2021**, *22*, 11432. [CrossRef] [PubMed]
2. Asmat, S.; Shah, A.; Shams, S.; Bungau, S.; Si, Y.; Xu, H.; Rahman, H.; Behl, T.; Gitea, D.; Pavel, F.; et al. Chemically Diverse and Biologically Active Secondary Metabolites from Marine *Phylum chlorophyta*. *Mar. Drugs* **2020**, *18*, 493. [CrossRef]
3. Khan, I.; Abbas, T.; Anjum, K.; Abbas, S.Q.; Shagufta, B.I.; Shah, S.A.A.; Akhter, N.; Hassan, S.S.u. Antimicrobial potential of aqueous extract of *Camellia sinensis* against representative microbes. *Pak. J. Pharm. Sci.* **2019**, *32*, 631–636. [PubMed]
4. Hassan, S.S.u.; Ishaq, M.; Zhang, W.; Jin, H.-Z. An overview of the mechanisms of marine fungi-derived antiinflammatory and anti-tumor agents and their novel role in drug targeting. *Curr. Pharm. Des.* **2021**, *27*, 2605–2614. [CrossRef]
5. Choi, W.K.; El-Gamal, M.I.; Choi, H.S.; Baek, D.; Oh, C.H. New diarylureas and diarylamides containing 1,3,4-triarylpyrazole scaffold: Synthesis, antiproliferative evaluation against melanoma cell lines, ERK kinase inhibition, and molecular docking studies. *Eur. J. Med. Chem.* **2011**, *46*, 5754–5762. [CrossRef]
6. Naves, L.B.; Almeida, L.; Ramakrishna, S. Understanding the Microenvironment of Melanoma Cells for the Development of Target Drug Delivery Systems. *Cit. EMJ Oncol.* **2017**, *5*, 85–92.
7. Morris, V.; Kopetz, S. BRAF inhibitors in clinical oncology. *F1000Prime Rep.* **2013**, *5*, 1–6. [CrossRef]
8. Holderfield, M.; Deuker, M.M.; McCormick, F.; McMahon, M. Targeting RAF kinases for cancer therapy: BRAF-mutated melanoma and beyond. *Nat. Rev. Cancer* **2014**, *14*, 455–467. [CrossRef]
9. Hassan, S.S.u.; Jin, H.Z.; Abu-Izneid, T.; Rauf, A.; Ishaq, M.; Suleria, H.A.R. Stress-driven discovery in the natural products: A gateway towards new drugs. *Biomed. Pharmacother.* **2019**, *109*, 459–467. [CrossRef]
10. Xie, Y.G.; Zhao, X.C.; Hassan, S.S.u.; Zhen, X.Y.; Muhammad, I.; Yan, S.K.; Yuan, X.; Li, H.L.; Jin, H. One new sesquiterpene and one new iridoid derivative from *Valeriana amurensis*. *Phytochem. Lett.* **2019**, *32*, 6–9. [CrossRef]
11. Xiao, Y.; Zhu, S.; Wu, G.; Hassan, S.S.u.; Xie, Y.; Ishaq, M.; Sun, Y.; Yan, S.K.; Qian, X.P.; Jin, H. Chemical Constituents of *Vernonia parishii*. *Chem. Nat. Compd.* **2020**, *56*, 134–136. [CrossRef]
12. Shams, S.; Zhang, W.; Jin, H.; Basha, S.H.; Priya, S.V.S.S. In-silico anti-inflammatory potential of guaiane dimers from *Xylopia vielana* targeting COX-2. *J. Biomol. Struct. Dyn.* **2020**. [CrossRef]
13. Xie, Y.G.; Yan, R.; Zhong, X.; Piao, H.; Muhammad, I.; Ke, X.; Yan, S.; Guo, Y.; Jin, H.Z.; Zhang, W.D. Xylopins A-F, six rare guaiane dimers with three different connecting modes from: *Xylopia vielana*. *RSC Adv.* **2019**, *9*, 9235–9242. [CrossRef]
14. Xie, Y.G.; Guo, Y.G.; Wu, G.J.; Zhu, S.L.; Cheng, T.F.; Zhang, Y.; Yan, S.K.; Jin, H.Z.; Zhang, W.D. Xylopsides A-D, four rare guaiane dimers with two unique bridged pentacyclic skeletons from: *Xylopia vielana*. *Org. Biomol. Chem.* **2018**, *16*, 7030–7034. [CrossRef]
15. Xie, Y.; Zhong, X.; Xiao, Y.; Zhu, S.; Muhammad, I.; Yan, S.; Jin, H.; Zhang, W. Vieloplains A-G, seven new guaiane-type sesquiterpenoid dimers from *Xylopia vielana*. *Bioorg. Chem.* **2019**, *88*, 102891. [CrossRef]

16. Hassan, S.S.u.; Shah, S.A.A.; Pan, C.; Fu, L.; Cao, X.; Shi, Y.; Wu, X.; Wang, K.; Wu, B. Production of an antibiotic enterocin from a marine actinobacteria strain H1003 by metal-stress technique with enhanced enrichment using response surface methodology. *Pak. J. Pharm. Sci.* **2017**, *30*, 313–324.
17. Marrero-Ponce, Y.; Siverio-Mota, D.; Gálvez-Llompart, M.; Recio, M.C.; Giner, R.M.; García-Domnech, R.; Torrens, F.; Arán, V.J.; Cordero-Maldonado, M.L.; Esguera, C.V.; et al. Discovery of novel anti-inflammatory drug-like compounds by aligning in silico and in vivo screening: The nitroindazolinone chemotype. *Eur. J. Med. Chem.* **2011**, *46*, 5736–5753. [CrossRef]
18. DiMasi, J.A. Success rates for new drugs entering clinical testing in the United States. *Clin. Pharmacol. Ther.* **1995**, *58*, 1–14. [CrossRef]
19. Lionta, E.; Spyrou, G.; Vassilatis, D.K.; Cournia, Z. Structure-based virtual screening for drug discovery: Principles, applications and recent advances. *Curr. Top. Med. Chem.* **2014**, *14*, 1923–1938. [CrossRef]
20. Kolšek, K.; Mavri, J.; Sollner Dolenc, M.; Gobec, S.; Turk, S. Endocrine disruptome—An open source prediction tool for assessing endocrine disruption potential through nuclear receptor binding. *J. Chem. Inf. Modeling* **2014**, *54*, 1254–1267. [CrossRef]
21. Chen, Y.; Tian, Y.; Gao, Y.; Wu, F.; Luo, X.; Ju, X.; Liu, G. In silico Design of Novel HIV-1 NNRTIs Based on Combined Modeling Studies of Dihydrofuro[3,4-d]pyrimidines. *Front. Chem.* **2020**, *8*, 164. [CrossRef] [PubMed]
22. Cruz, J.V.; Serafim, R.B.; da Silva, G.M.; Giuliatti, S.; Rosa, J.M.C.; Araújo Neto, M.F.; Leite, F.H.A.; Taft, C.A.; da Silva, C.H.T.P.; Santos, C.B.R. Computational design of new protein kinase 2 inhibitors for the treatment of inflammatory diseases using QSAR, pharmacophore-structure-based virtual screening, and molecular dynamics. *J. Mol. Modeling* **2018**, *24*, 1–16. [CrossRef] [PubMed]
23. Daina, A.; Zoete, V. A BOILED-Egg to Predict Gastrointestinal Absorption and Brain Penetration of Small Molecules. *ChemMedChem* **2016**, *11*, 1117. [CrossRef] [PubMed]
24. Daina, A.; Michielin, O.; Zoete, V. SwissADME: A free web tool to evaluate pharmacokinetics, drug-likeness and medicinal chemistry friendliness of small molecules. *Sci. Rep.* **2017**, *7*, 1–13. [CrossRef] [PubMed]
25. dos Santos, K.L.B.; Cruz, J.N.; Silva, L.B.; Ramos, R.S.; Neto, M.F.A.; Lobato, C.C.; Ota, S.S.B.; Leite, F.H.A.; Borges, R.S.; da Silva, C.H.T.P.; et al. Identification of novel chemical entities for adenosine receptor type 2a using molecular modeling approaches. *Molecules* **2020**, *25*, 1245. [CrossRef] [PubMed]
26. Veber, D.F.; Johnson, S.R.; Cheng, H.Y.; Smith, B.R.; Ward, K.W.; Kopple, K.D. Molecular properties that influence the oral bioavailability of drug candidates. *J. Med. Chem.* **2002**, *45*, 2615–2623. [CrossRef] [PubMed]
27. Lipinski, C.A.; Lombardo, F.; Dominy, B.W.; Feeney, P.J. Experimental and computational approaches to estimate solubility and permeability in drug discovery and development settings. *Adv. Drug Deliv. Rev.* **2012**, *43*, 3–26. [CrossRef]
28. Martin, Y.C. A bioavailability score. *J. Med. Chem.* **2005**, *46*, 3164–3170. [CrossRef]
29. Glass, C.K.; Ogawa, S. Combinatorial roles of nuclear receptors in inflammation and immunity. *Nat. Rev. Immunol.* **2006**, *6*, 44–55. [CrossRef]
30. Filimonov, D.A.; Druzhilovskiy, D.S.; Lagunin, A.A.; Gloriozova, T.A.; Rudik, A.V.; Dmitriev, A.V.; Pogodin, P.V.; Poroikov, V.V. Computer-aided prediction of biological activity spectra for chemical compounds: Opportunities and limitation. *Biomed. Chem. Res. Methods* **2018**, *1*, e00004. [CrossRef]
31. Spengler, E.K.; Kleiner, D.E.; Fontana, R.J. Vemurafenib-induced granulomatous hepatitis. *Hepatology* **2017**, *65*, 745–748. [CrossRef] [PubMed]
32. Truong, J.; Yan, A.T.; Cramarossa, G.; Chan, K.K.W. Chemotherapy-induced cardiotoxicity: Detection, prevention, and management. *Can. J. Cardiol.* **2014**, *30*, 869–878. [CrossRef] [PubMed]
33. Raschi, E.; Vasina, V.; Poluzzi, E.; De Ponti, F. The hERG K+ channel: Target and antitarget strategies in drug development. *Pharmacol. Res.* **2008**, *57*, 181–195. [CrossRef] [PubMed]
34. Nielsen, E.; Ostergaard, G.; Larsen, J.C. *Toxicological Risk Assessment of Chemicals: A Practical Guide*, 1st ed.; CRC Press: Boca Raton, FL, USA, 2008; ISBN 1420006940.
35. Hughes, T.B.; Miller, G.P.; Swamidass, S.J. Modeling epoxidation of drug-like molecules with a deep machine learning network. *ACS Cent. Sci.* **2015**, *1*, 168–180. [CrossRef] [PubMed]
36. Hughes, T.B.; Le Dang, N.; Miller, G.P.; Swamidass, S.J. Modeling reactivity to biological macromolecules with a deep multitask network. *ACS Cent. Sci.* **2016**, *2*, 529–537. [CrossRef] [PubMed]
37. De Coster, S.; van Larebeke, N. Endocrine-Disrupting Chemicals: Associated Disorders and Mechanisms of Action. *J. Environ. Public Health* **2012**, *2012*, 713696. [CrossRef]
38. Braga, R.C.; Alves, V.M.; Silva, M.F.B.; Muratov, E.; Fourches, D.; Lião, L.M.; Tropsha, A.; Andrade, C.H. Pred-hERG: A Novel web-Accessible Computational Tool for Predicting Cardiac Toxicity. *Mol. Inform.* **2015**, *34*, 698–701. [CrossRef]
39. Sharma, S.V.; Haber, D.A.; Settleman, J. Cell line-based platforms to evaluate the therapeutic efficacy of candidate anticancer agents. *Nat. Rev. Cancer* **2010**, *10*, 241–253. [CrossRef]
40. Filimonov, D.A.; Lagunin, A.A.; Gloriozova, T.A.; Rudik, A.V.; Druzhilovskii, D.S.; Pogodin, P.V.; Poroikov, V.V. Prediction of the biological activity spectra of organic compounds using the pass online web resource. *Chem. Heterocycl. Compd.* **2014**, *50*, 444–457. [CrossRef]
41. Lagunin, A.; Zakharov, A.; Filimonov, D.; Poroikov, V. QSAR modelling of rat acute toxicity on the basis of PASS prediction. *Mol. Inform.* **2011**, *30*, 241–250. [CrossRef]
42. Roy, S.; Samant, L.R.; Chowdhary, A. In silico pharmacokinetics analysis and ADMET of phytochemicals of Datura metel Linn. and Cynodon dactylon Linn. *J. Chem. Pharm. Res.* **2015**, *11*, 385–388.

43. Matlock, M.K.; Hughes, T.B.; Swamidass, S.J. XenoSite server: A web-available site of metabolism prediction tool. *Bioinformatics* **2015**, *31*, 1136–1137. [CrossRef] [PubMed]
44. Lagunin, A.A.; Dubovskaja, V.I.; Rudik, A.V.; Pogodin, P.V.; Druzhilovskiy, D.S.; Gloriozova, T.A.; Filimonov, D.A.; Sastry, N.G.; Poroikov, V.V. CLC-Pred: A freely available web-service for in silico prediction of human cell line cytotoxicity for drug-like compounds. *PLoS ONE* **2018**, *13*, e0191838. [CrossRef] [PubMed]
45. Li, Z.; Wan, H.; Shi, Y.; Ouyang, P. Personal experience with four kinds of chemical structure drawing software: Review on chemdraw, chemwindow, ISIS/draw, and chemsketch. *J. Chem. Inf. Comput. Sci.* **2004**, *44*, 1886–1890. [CrossRef]
46. Pettersen, E.F.; Goddard, T.D.; Huang, C.C.; Couch, G.S.; Greenblatt, D.M.; Meng, E.C.; Ferrin, T.E. UCSF Chimera—A visualization system for exploratory research and analysis. *J. Comput. Chem.* **2004**, *25*, 1605–1612. [CrossRef]
47. Willard, L.; Ranjan, A.; Zhang, H.; Monzavi, H.; Boyko, R.F.; Sykes, B.D.; Wishart, D.S. VADAR: A web server for quantitative evaluation of protein structure quality. *Nucleic Acids Res.* **2003**, *31*, 3316–3319. [CrossRef]
48. Dallakyan, S.; Olson, A.J. Small-molecule library screening by docking with PyRx. *Methods Mol. Biol.* **2015**, *1263*, 243–250. [CrossRef]
49. Ferreira, L.G.; Dos Santos, R.N.; Oliva, G.; Andricopulo, A.D. Molecular Docking and Structure-Based Drug Design Strategies. *Molecules* **2015**, *20*, 13384–13421. [CrossRef]
50. Jin, Z.; Wang, Y.; Yu, X.-F.; Tan, Q.-Q.; Liang, S.-S.; Li, T.; Zhang, H.; Shaw, P.-C.; Wang, J.; Hu, C. Structure-based virtual screening of influenza virus RNA polymerase inhibitors from natural compounds: Molecular dynamics simulation and MM-GBSA calculation. *Comput. Biol. Chem.* **2020**, *85*, 107241. [CrossRef]

molecules

MDPI

Article

Neuroprotective Effect of Natural Compounds in Paclitaxel-Induced Chronic Inflammatory Pain

Muhammad Faheem [1,*], Arif-ullah Khan [1,*], Muhammad Waqas Saleem [2], Fawad Ali Shah [1], Fawad Ali [3], Abdul Waheed Khan [4] and Shupeng Li [5,*]

1 Riphah Institute of Pharmaceutical Sciences, Riphah International University, Islamabad 45000, Pakistan; fawad.shah@riphah.edu.pk
2 Rural Health Centre, Head RajkanYazman, Bahawalpur 63236, Pakistan; waqassaleem2255@gmail.com
3 Department of Pharmacy, Kohat University of Science and Technology, Kohat 26000, Pakistan; fawad.alee@gmail.com
4 Department of Molecular Science and Technology, Ajou University, Suwon 16499, Korea; waheedmarwat31@gmail.com
5 State Key Laboratory of Oncogenomics, School of Chemical Biology and Biotechnology, Shenzhen Graduate School, Peking University, Shenzhen 518000, China
* Correspondence: fahimafridi345@gmail.com (M.F.); arif.ullah@riphah.edu.pk (A.-u.K.); lisp@pku.edu.cn (S.L.)

Abstract: The current study explored the effects of natural compounds, berbamine, bergapten, and carveol on paclitaxel-associated neuroinflammatory pain. Berbamine, an alkaloid obtained from *Berberis amurensisRuprhas* been previously researched for anticancer and anti-inflammatory potential. Bergapten is 5-methoxsalenpsoralen previously investigated in cancer, vitiligo, and psoriasis. Carveol obtained from caraway is a component of essential oil. The neuropathic pain model was induced by administering 2 mg/kg of paclitaxel (PTX) every other day for a week. After the final PTX injection, a behavioral analysis was conducted, and subsequently, tissue was collected for molecular analysis. Berbamine, bergapten, and carveol treatment attenuated thermal hypersensitivity, improved latency of falling, normalized the changes in body weight, and increased the threshold for pain sensation. The drugs increased the protective glutathione (GSH) and glutathione S-transferase (GST) levels in the sciatic nerve and spinal cord while lowering inducible nitric oxide synthase (iNOS) and lipid peroxidase (LPO). Hematoxylin and eosin (H and E) and immunohistochemistry (IHC) examinations confirmed that the medication reversed the abnormal alterations. The aforementioned natural substances inhibited cyclooxygenase-2 (COX-2), tumor necrosis factor-alpha (TNF-α), and nuclear factor kappa B (NF-κb) overexpression, as evidenced by enzyme-linked immunosorbant assay (ELISA) and Western blot and hence provide neuroprotection in chronic constriction damage.

Keywords: berbamine; bergapten; carveol; paclitaxel; ELISA; Western blot

Citation: Faheem, M.; Khan, A.-u.; Saleem, M.W.; Shah, F.A.; Ali, F.; Khan, A.W.; Li, S. Neuroprotective Effect of Natural Compounds in Paclitaxel-Induced Chronic Inflammatory Pain. *Molecules* **2022**, 27, 4926. https://doi.org/10.3390/molecules27154926

Academic Editors: Syed Shams ul Hassan, Mohamed M. Abdel-Daim, Tapan Behl, Simona Bungau and Luciana Mosca

Received: 22 June 2022
Accepted: 27 July 2022
Published: 2 August 2022

1. Introduction

Paclitaxel is well-known for its anti-tumor properties [1]. Chemotherapy is one of the most effective techniques for fighting cancer and enhancing the quality of life, resulting in fewer deaths throughout the globe. With appropriate use of chemotherapy, researchers predict a 35 percent increase in cancer survivors by 2022 [2]. One of the most prevalent dose-limiting side effects associated with the use of anticancer medicines is peripheral neuropathy [3].

Taxanes are a family of significant chemotherapeutic drugs that have been therapeutically used for decades to treat a variety of malignancies. Paclitaxel is a taxane-class medication that has been authorized by the FDA to treat lung cancer, breast cancer, prostate cancer, and ovarian cancer [4]. Its use has been linked to the development of dose-dependent neurotoxicity. The termination of treatment is due to the development of cold allodynia, mechanical hypersensitivity, and burning, shooting, and tingling sensations [5].

According to the research, the mechanism of PTX-induced neurotoxicity is thought to be initiated by the overexpression of inflammatory mediators in the spinal cord and the sciatic nerve. This overexpression leads to the disruption of transport in ion channels and improper intracellular signaling, which ultimately results in paclitaxel-induced peripheral neuropathy (PIPN) [6]. Recently published data show that because PTX can easily pass through the blood–brain barrier, it predominantly accumulates in the dorsal root ganglia (DRG). This, along with an alteration in the mitochondrial morphology and inflammation, leads to the development of PIPN, which causes cold allodynia as well as mechanical and thermal pain sensation [7,8].

The present study aims to look at the therapeutic potential of natural substances including berbamine, bargepten, and carveol in the treatment of paclitaxel-induced neuropathic pain. Herbal ingredients have been studied extensively for their ability to reduce paclitaxel-induced pain. Natural treatments such as curcumin [9], resveratrol [10], gallic acid [11], puerarin [12], and naringin [13] have been shown to reduce paclitaxel-induced discomfort. Chemically, berbamine (BBM) is a bisbenzylisoquinoline alkaloid obtained from the Chinese medicinal plant *Berberis amurensis Rupr*. It has anti-cancer, anti-inflammatory, and multidrug resistance properties, as well as a synergistic effect when combined with other medications [14].

Psoralen is a group of natural substances derived from the Ammi majus plant that are collectively known as Furocoumarin. Furocoumarin is a methoxsalen-based Furocoumarin. Bergapten (BRG) is a 5-methoxsalenpsoralen compound that has previously been studied in cancer, vitiligo, and psoriasis [15]. Carveol (CAR) is derived from the Caraway plant. It is a vital component of essential oil and is grown all over the globe. Caraway also contains pinene, thujene, phellandrene, camphene, limonene, and carvone as components [16].

To investigate the potential in PTX-induced PIPN, the above-mentioned three natural moieties (BBM, BRG, and CAR) listed were chosen. The available literature supports their potential for anticancer [14,15] and neuroprotective properties [17].

Hence, these compounds were examined in the current study for their therapeutic potential in PTX-induced peripheral and chronic inflammatory pain. The results reveal that treatment with natural compounds (BBM, BRG, and CAR) attenuated PTX-induced chronic inflammatory pain by downregulating NF-κB. The downregulation of NF-κB, a transcription factors which further attenuates inflammatory cytokines (COX-2 and TNF-α), is the proposed mechanistic pathway of the aforementioned natural compounds to hasten PTX-induced neuropathic pain. In addition, the treatment improved the antioxidant enzymes (GSH and GST) and diminished LPO and iNOS, which are the reasons of preventing oxidative stress and free radical generation to stop the PTX-induced progression of neurodegeneration and neuroinflammation.

2. Material and Methods

2.1. Chemicals

Proteinase K, PBS tablets, hydrogen peroxide (H_2O_2), formaldehyde, GSH, DTNB, CDNB, Mouse monoclonal anti-p-NF-κB, TNF-α, COX-2, Avidin-biotin complex kit, DAB, trichloroacetic acid (TCA), horseradish peroxidase-conjugated secondary antibodies, mounting media, COX-2, p-NF-κB, TNF-α ELISA and protein assay kit, skim milk and Bolt Mini Gels, X-ray film were used. The inducer paclitaxel was procured from the oncology pharmacy of Shifa International Hospital Islamabad, Islamabad, Pakistan. All the rest of the three compounds were procured from Sigma-Aldrich (Saint Louis, MO, USA).

2.2. Animals

Adult Sprague Dawley rats weighing 240–250 g (aged 10–12 weeks) were procured from Riphah International University's animal house (Islamabad, Pakistan) kept under in a controlled environment temperature (25–30 °C) and humidity. The experimental protocols for handling and dosing of animal were as per protocols set bythe Riphah Institute of Pharmaceutical Sciences (Ref. No. REC/RIPS/2019/28) (Islamabad, Pakistan).

The animals were randomly divided into five groups, each group with six rats as follows:

Group I: Control group, treated with saline 10 mL/kg

Group II: Disease group, treated with PTX 2 mg/kg IP on days 1, 3, 5, and 7 (induction phase)

Group III: Treated group, administered with PTX 2 mg/kg IP on days1, 3, 5, and 7 (induction phase) and then treated with compound (berbamine) for two weeks (7 to 21 days)

Group IV: Treated group, administered with PTX 2 mg/kg IP on days 1, 3, 5, and 7 (induction phase) and then treated with compound (bergapten) for two weeks (7 to 21 days)

Group V: Treated group, administered with PTX 2 mg/kg IP on days 1, 3, 5, and 7 (induction phase) and then treated with compound (carveol) for two weeks (7 to 21 days)

2.3. Paclitaxel-Induced Neuropathic Pain

Adult male Sprague Dawley rats were injected with paclitaxel to produce paclitaxel-induced peripheral neuropathic pain. The available pharmaceutical grade of paclitaxel was 6 mg/mL, which was further diluted (1:1 cremophor/ethanol) to 1 mg/mL and injected intraperitoneally (i.p) at a dose of 2 mg/kg every other day (1, 3, 5, 7) for a total of four injections and a final total dose of 8 mg/kg. On days 7, 14, and 21 (days after PTX's last injection), behavioral tests including temperature sensation, latency of falling, body weight, and mechanical pain threshold were performed. After behavioral tests on day 21, the sciatic nerve (SN) and spinal cord (SC) were removed for molecular investigation [18].

2.4. Thermal Pain Sensation

Acclimatized animal's paw sensitivity to heat was determined on a hot plate (54 ± 1 °C) with a cut-off time of time of 60 s. Thermal pain sensation was determined on days 7, 14, and 21, 30 min after treatment with the compound [19,20].

2.4.1. Latency of Falling

The latency of falling was recorded through the rotarod apparatus. Animals were first trained for 3 consecutive days to be able to remain on the rod for 60 s with 5 min cut-off time before starting the experiment. The animals were at a fixed rate (5–20 rpm). After recording the baseline reading (day 0), the latency of falling was recorded. Latency of falling was also determined on days 7, 14, and 21after treatment with compound [21].

2.4.2. Body Weight Analysis

The animals were first weighed at baseline before starting the experiments. The weight of each animal was recorded after therapy on days 7, 14, and 21, and the results were interpreted [22].

2.4.3. Pain Threshold (Mechanical Hypersensitivity)

A Conventional von-Frey filament apparatus was used to analyze the pressure-induced-pain threshold [23]. After its perpendicularly application to the subplanter region, the response of the animal was noticed on days 7, 14, and 21after treatment.

2.5. Oxidative Stress Markers

The samples (SN and SC) after homogenization in a phosphate buffer containing phenylmethylsulfonyl fluoride as a protease inhibitor were centrifuged at 4000× g for

10 min at 4 °C, and the supernatant was collected and processed for determination of GSH [24] and GST [25], lipid peroxidation [26], and nitric oxide (NO) [27].

2.6. Hematoxylin and Eosin Staining

De-paraffinization was performed through xylene (100%), ethanol (95 and 70%), and distilled water. There was washing with PBS and treatment with hematoxylin for 10 min. Each slide was dipped in 1% HCl, immersed in eosin for 5–10 min, washed, dried, and fixed in xylene. The images were acquired with the assistance of an Olympus (Model: CX31, Tokyo, Japan) microscope, and an automated self-quantification process was carried out with the assistance of ImageJ software (USA, version 1.46) [28].

2.7. Immuno-Histopathological Evaluation

Paraffinized slides were washed with xylene, ethanol, distilled water, and then in PBS. Proteinase K was applied for antigen retrieval. Endogenous peroxidases were blocked by hydrogen peroxide. Normal goat serum (5%), primary antibody secondary antibody, and ABC were then applied. Finally, the slides were exposed to a 0.1% DAB (diaminobenzidine peroxidase) solution and dehydrated by dipping in xylene/ethanol and dried in open air. The images were obtained through an Olympus (Model: CX31, Tokyo, Japan) microscope and an automated self-quantification method was applied utilizing ImageJ software (USA, version 1.46) [29].

2.8. ELISA

TNF-α, COX-2, and NF-κB were determined by using ELISA kits as explained by [30]. In the tissue, the designated antibodies' expression was determined by using an ELISA microplate reader.

2.9. Western Blot Assay

Equal amounts of protein (15–30 µg) underwent electrophoresis using 4–12% bolt Mini Gels 5% (*w*/*v*).Following the wet transfer, the PVDF membrane was blocked with skim milk to reduce nonspecific binding and incubated with primary antibodies at 4 °C for 12 to 16 h. After a reaction with a horse radish peroxidase (HRP)-conjugated secondary antibody, proteins were detected using enhanced chemiluminescence detection. The X-ray films were scanned, and the optical densities of the bands were analyzed through densitometry using the computer-based Sigma Gel program, version 1.0 (USA) [31].

2.10. Statistical Analysis

Data are presented as the mean ± SEM. H and E staining behavioral data and oxidative stress data were analyzed using one-way ANOVA, followed by post hoc Tukey's test using GraphPad Prism version 6.0 (San Diego, CA, USA). The *p*-value was calculated through GraphPad Instat software version 3.1 (San Diego, CA, USA). ImageJ software was used to analyze the morphological data. One-way ANOVA followed by post hoc Tukey's test was performed for ELISA and Western blot. Symbols $^{\#}$ or * represent significant difference values $p < 0.05$, $^{\#\#}$ or ** represent $p < 0.01$, and $^{\#\#\#}$ or *** represent $p < 0.001$ values.

3. Results

3.1. Effect on Thermal Pain Sensation

Berbamine elevated heat latency (HL) at 5 and 15 mg/kg on days 14 and 21, respectively, as compared to PTX. Bergapten at 50 and 100 mg/kg enhanced HL on days 14 and 21, respectively, vs. PTX. Carveol at 10 and 20 mL/kg increased HL on days 14 and 21, respectively, vs. PTX. PTX at 2 mg/kg decreased HL on days 7, 14, and 21 vs. control, presented in Figure 1.

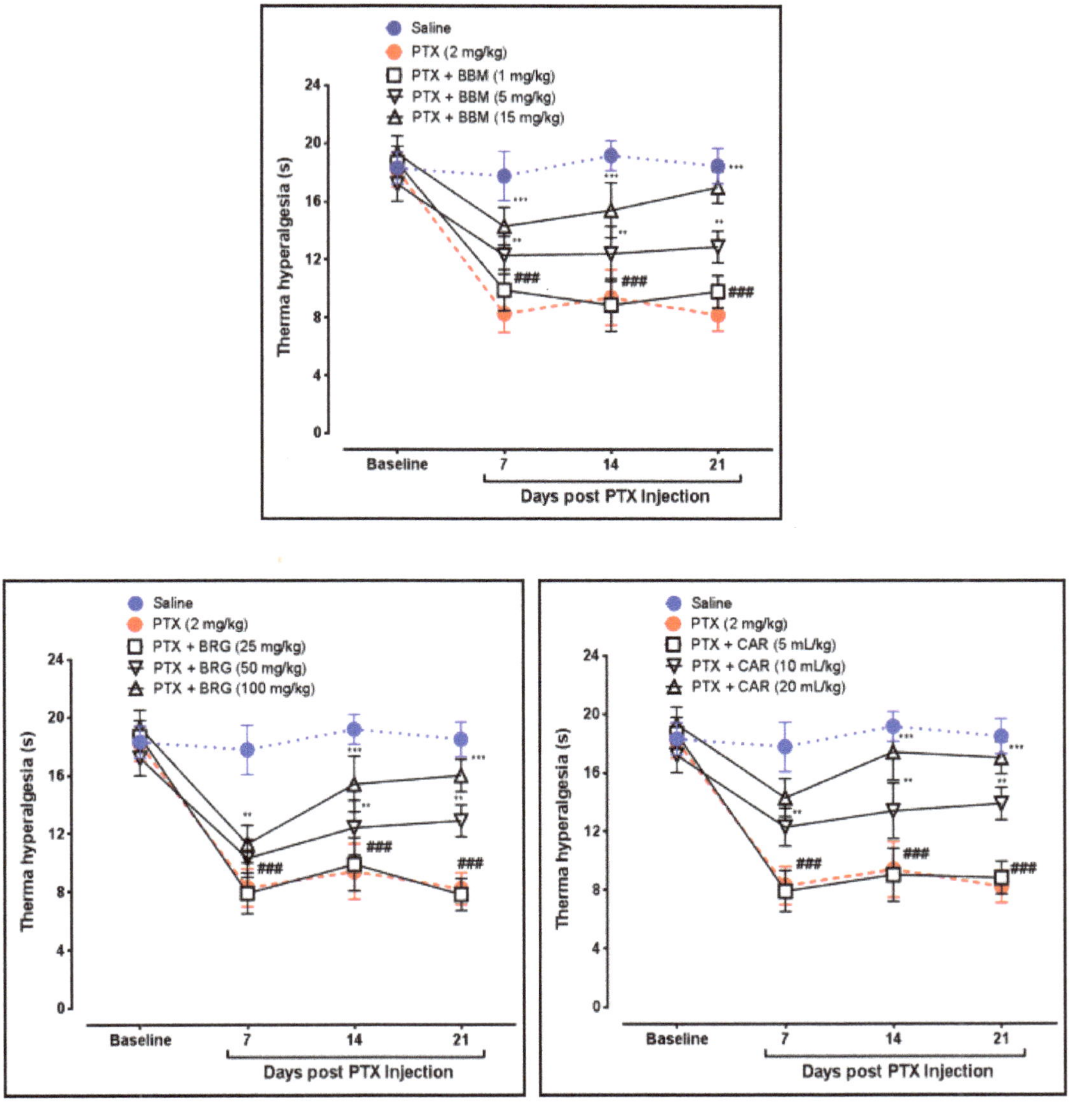

Figure 1. Line graph representing the effect of berbamine (1, 5, and 15 mg/kg), bargepten (25, 50, and 100 mg/kg), and carveol (5, 10, and 20 mL/kg) on thermal hyperalgesia on 7th, 14th, and 21st days. The data are expressed as the mean ± SEM, $n = 6$. One-way ANOVA with posthoc Tukey's test. ** $p < 0.01$ and *** $p < 0.001$ indicate a significant difference vs. PTX, and ### $p < 0.001$ indicates a significant difference vs. saline.

3.2. Effect on the Latency of Falling

Latency of falling (LF) or motor coordination was determined by the rotarod apparatus. Berbamine at 5 and 15 mg/kg augmented LF on days 14 and 21, respectively, vs. PTX. Bergapten at 100 mg/kg enhanced LF on days 14 and 21 vs. PTX. Carveol at 10 and 20 mL/kg raised LF on days 14 and 21, respectively, vs. PTX. PTX at 2 mg/kg decreased LF on days 7, 14, and 21 vs. that in the control group, presented in Figure 2.

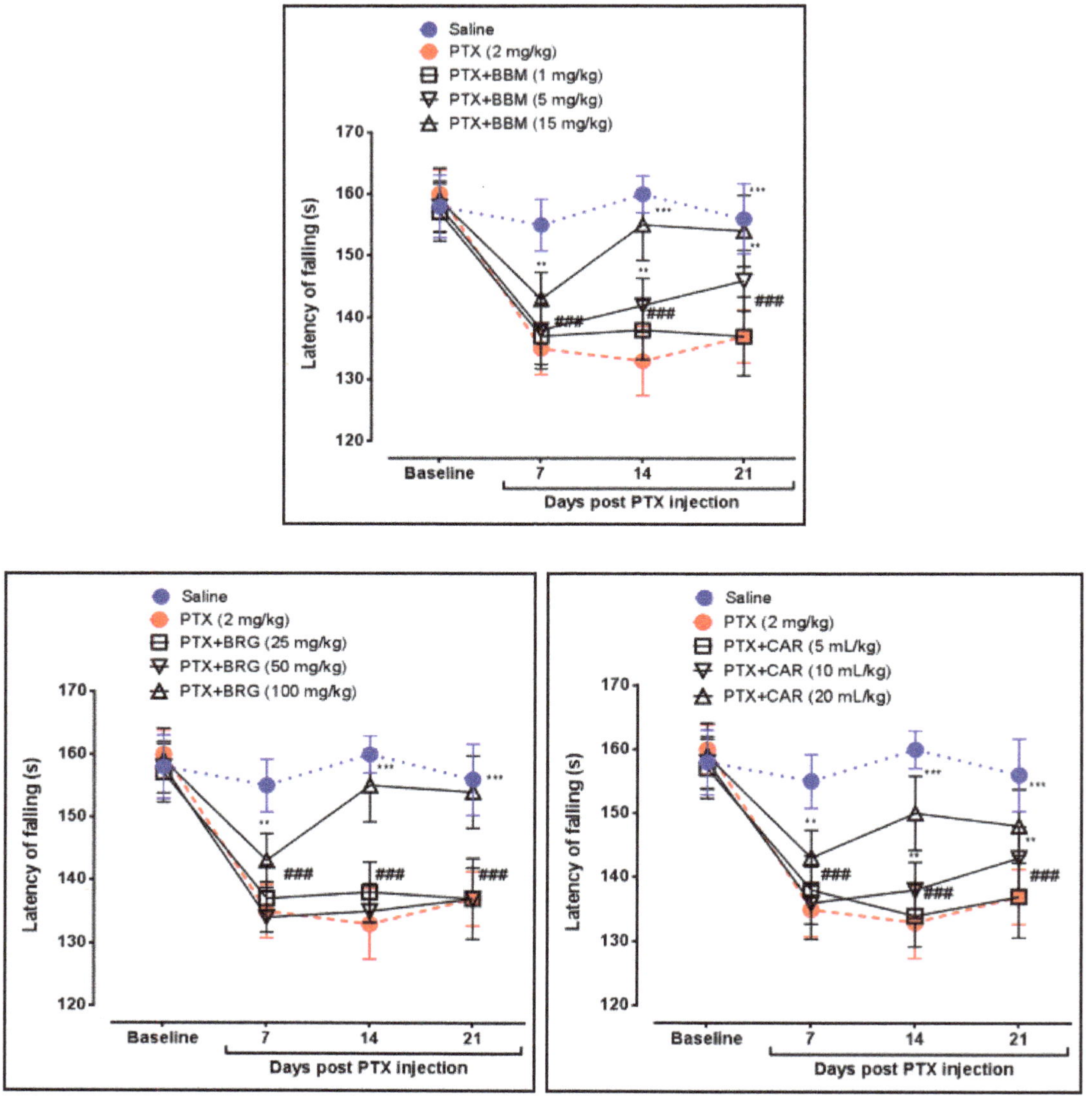

Figure 2. Line graph representing the effect of berbamine (1, 5, and 15 mg/kg), bargepten (25, 50, and 100 mg/kg), and carveol (5, 10, and 20 mL/kg) on latency of falling on 7th, 14th, and 21st days. The data are expressed as the mean ± SEM, $n = 6$. One-way ANOVA with posthoc Tukey's test. ** $p < 0.01$ and *** $p < 0.001$ indicate a significant difference vs. PTX, and ### $p < 0.001$ indicates a significant difference vs. saline.

3.3. *Effect on Changes in Body Weight*

Changes in body weight were also determined by utilizing weighing balance. Berbamine increased body weight at 5 and 15 mg/kg on days 14 and 21, respectively, vs. PTX. Bergapten increased body weight at 50 mg/kg on day 21 vs. PTX. Bergapten increased body weight at 100 mg/kg on days 14 and 21 vs. PTX. Carveol increased body weight at 10 and 20 mL/kg on days 14 and 21 vs. PTX. PTX at 2 mg/kg decreased the latency of falling on days 7, 14, and 21 vs. that in the control, presented in Figure 3.

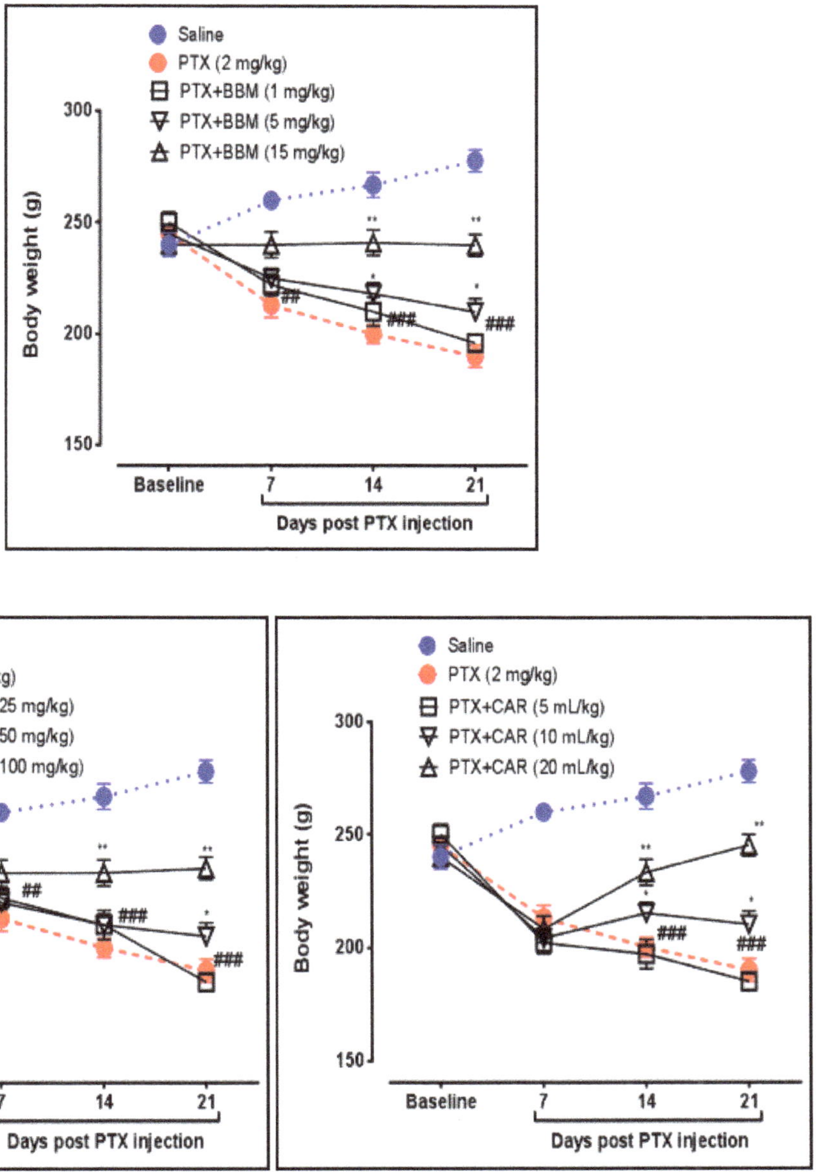

Figure 3. Line graph representing the effects of berbamine (1, 5, and 15 mg/kg), bargepten (25, 50, and 100 mg/kg), and carveol (5, 10, and 20 mL/kg) on body weight on 7th, 14th, and 21st days. The data are expressed as the mean ± SEM, $n = 6$. One-way ANOVA with post hoc Tukey's test. * $p < 0.05$ and ** $p < 0.01$ indicate a significant difference vs. PTX, and ### $p < 0.001$ indicates a significant difference vs. saline.

3.4. Effect on vonFrey-Induced Pain

Berbamine increased the paw withdrawal threshold (PWT) at 5 mg/kg on day 21 vs. PTX. At 15 mg/kg, berbamine improved PWT on day 14 vs. PTX. Berbamine at 15 mg/kg, augment PWT on day 21 vs. PTX. Bergapten at 50 mg/kg increased PWT on day 21 vs. PTX. Bergapten at 100 mg/kg elevated PWT on day 14 vs. PTX. Bergapten at 100 mg/kg, augmented PWT on day 21 vs. PTX. Carveol at 10 mL/kg raised paw PWT on days 14 and 21 vs. PTX. Carveol at 20 mL/kg increased PWT on days 14 and 21 vs. PTX. PTX at 2 mg/kg decreased PWT on days 7, 14, and 21 with $p < 0.001$ vs. that in the control, presented in Figure 4.

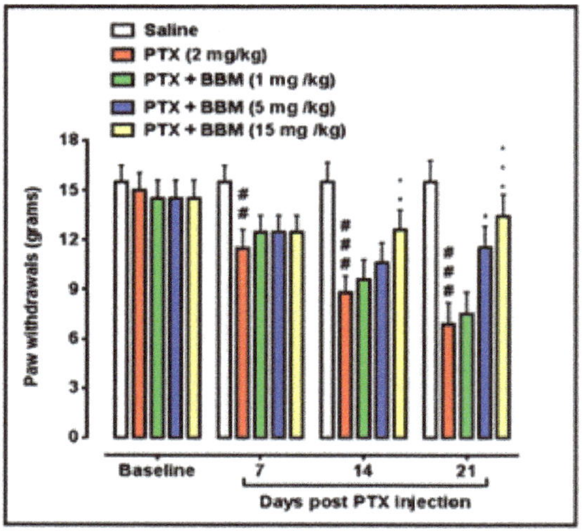

BBM

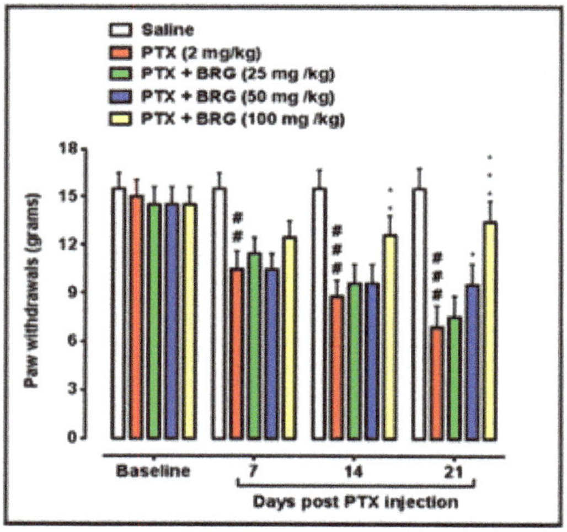

BRG

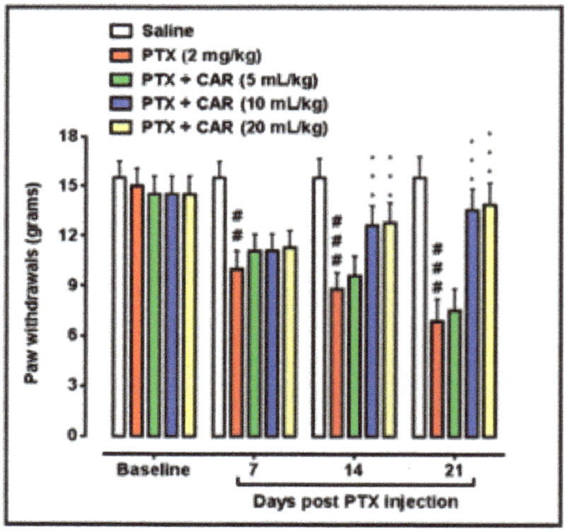

CAR

Figure 4. Bar graph representing the effects of berbamine (1, 5, and 15 mg/kg), bargepten (25, 50, and 100 mg/kg), and carveol (5, 10, and 20 mL/kg) on mechanical hypersensitivity on 7th, 14th, and 21st days. The data are expressed as the mean ± SEM, $n = 6$. One-way ANOVA with post hoc Tukey's test. * $p < 0.05$, ** $p < 0.01$, and *** $p < 0.001$ indicate a significant difference vs. PTX, and ## $p < 0.01$ and ### $p < 0.001$ indicate a significant difference vs. saline.

3.5. Effect on Oxidative Stress Enzyme

Berbamine, bergapten, and carveol were tested for their ability to inhibit oxidative-stress-causing enzymes. In the PTX-induced neuropathic pain group, PTX dramatically loweredthe levels of GSH and GST. In the SN and SC, berbamine dramatically boosted the protective markers GSH and GST. In both SN and SC, bergapten enhanced GSH and GST levels. Carveol reduced free radical production by increasing GSH and GST in the SN and SC, as seen in Tables 1 and 2. Berbamine, bergapten, and carveol were studied for their effects on iNOS and LPO. Destructive oxidative agents such as iNOS and LPO were observed to be elevated in the PTX group. In the SN and SC, berbamine significantly reduced iNOS and LPO. In the SN and SC, bergapten dramatically reduced iNOS and LPO. Carveol decreased iNOS and LPO in the SN and downregulated iNOS and LPO in the SC, as seen in Tables 1 and 2.

Table 1. Effects of berbamine (BBM), bargepten (BRG), and carveol (CAR) on the expression of GSH, GST, iNOS, and LPO in the sciatic nerve. The data expressed as the mean ± SEM, $n = 6$. One-way ANOVA with posthoc Tukey'stest.

Group	GSH (μmol/mg of Protein)	GST (μmol CDNB Conjugate/min/mg of Protein)	iNOS (μmol/mg of Protein)	LPO (nmol/TBARS/mg of Protein)
Saline (10 mL/kg)	48.22 ± 2.1	43.88 ± 1.5	34.22 ± 3.1	62.43 ± 1.8
PTX (2 mg/kg)	7.22 ± 1.7 ###	10.53 ± 2.6 ###	105.32 ± 3.2 ###	286.66 ± 2.2 ###
PTX+BBM (15 mg/kg)	30.29 ± 2.2 **	36.10 ± 3.4 **	66.21 ± 1.6 *	112.36 ± 2.8 *
PTX+BRG (100 mg/kg)	24.24 ± 2.2 **	29.14 ± 2.4 **	78.11 ± 2.2 *	128.10 ± 1.8 *
PTX+CAR (20 mL/kg)	28.14 ± 1.2 **	33.50 ± 1.4 **	56.11 ± 2.6 *	132.16 ± 3.8 *

* $p < 0.05$ and ** $p < 0.01$ indicate a significant difference vs. PTX, and ### $p < 0.001$ indicates a significant difference vs. saline.

Table 2. Effects of berbamine (BBM), bargepten (BRG), and carveol (CAR) on the expression of GSH, GST, iNOS, and LPO in the spinal cord. The data are expressed as the mean ± SEM, $n = 6$. One-wayANOVA with posthoc Tukey's test.

Group	GSH (μmol/mg of Protein)	GST (μmol CDNB Conjugate/min/mg of Protein)	iNOS (μmol/mg of Protein)	LPO (nmol/TBARS/mg of Protein)
Saline (10 mL/kg)	43.22 ± 1.8	35.71 ± 2.1	41.35 ± 1.2	62.33 ± 1.3
PTX (2 mg/kg)	9.41 ± 1.5 ###	7.53 ± 3.4 ###	98.32 ± 2.3 ###	195.68 ± 1.6 ###
PTX+BBM (15 mg/kg)	33.21 ± 1.4 **	26.14 ± 3.2 **	52.41 ± 2.5 **	76.66 ± 1.0 **
PTX+BRG (100 mg/kg)	25.21 ± 2.6 **	26.12 ± 3.2 **	73.41 ± 1.5 **	113.22 ± 1.8 ***
PTX+CAR (20 mL/kg)	29.14 ± 0.6 **	22.22 ± 1.2 **	63.11 ± 1.5 **	103.36 ± 2.0 ***

** $p < 0.01$ and *** $p < 0.001$ indicate a significant difference vs. PTX, and ### $p < 0.001$ indicates a significant difference vs. saline.

3.6. H and E Staining Examination

H and E staining demonstrated an organized cellular architecture, no infiltration, and intact intracellular spaces with no evidence of edema in the saline group. In the PTX group, compared to the saline group, PTX-induced pathological abnormalities, damage of the SN and SC with different types of injuries in the form of enlarged intracellular spaces, infiltration, and edema of a disorderly pattern were seen. As demonstrated in Figure 5, treatment with berbamine, bergapten, and carveol dramatically restored the PTX-induced damage and pathological development in the SN and SC compared to those in the PTX group.

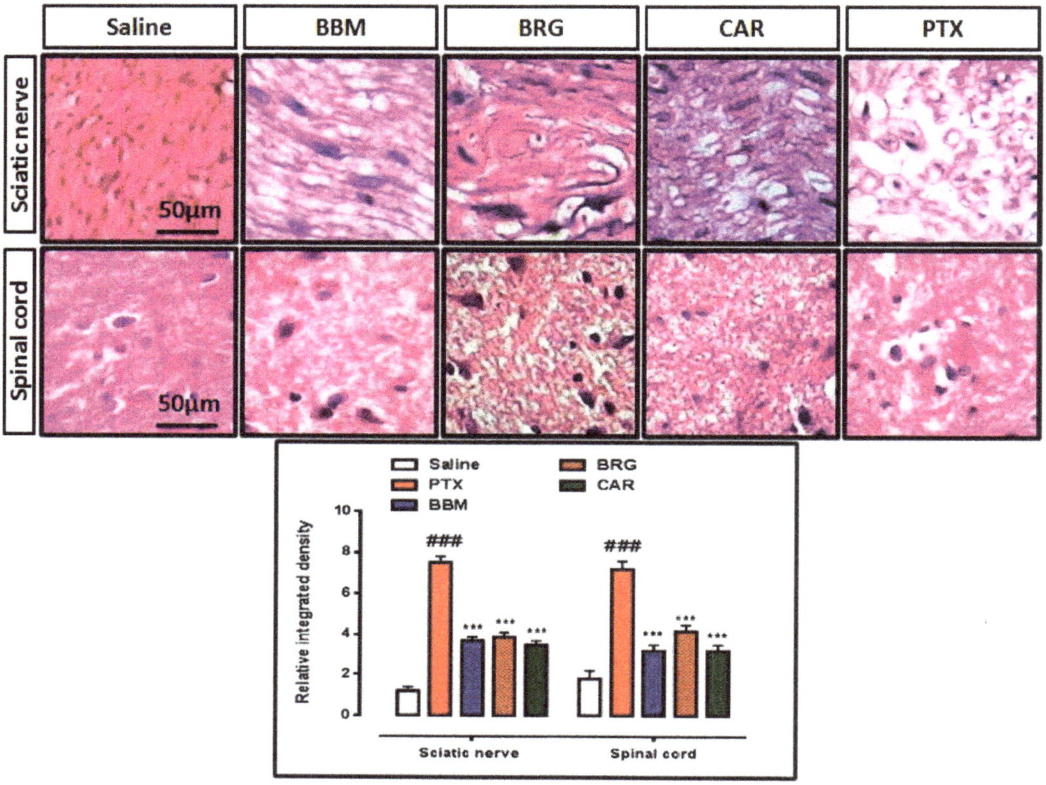

Figure 5. Representation of hematoxylin and eosin staining and the effects of berbamine, bargepten, and carveol on PTX-induced alteration in the sciatic nerve and spinal cord. The data are expressed as the mean ± SEM, $n = 6$. One-way ANOVA with posthoc Tukey's test. *** $p < 0.001$ indicates a significant difference vs. PTX, and ### $p < 0.001$ indicates a significant difference vs. saline. Morphological data were analyzed by ImageJ software. Bar 50 µm, magnification 40×.

3.7. IHC Analysis

The findings of IHC staining are presented in Figures 6 and 7. COX-2, TNF-α, and p-NF-κb were notably seen raised in the PTX group compared to those in the saline group in the SN and SC. Berbamine attenuated COX-2, TNF-α, and p-NF-κb significantly in the SN. Bergapten vanished COX-2, TNF-α, and p-NF-κb significantly in the SN. Carveol suppressed COX-2, TNF-α, and NF-κb significantly in the SN, presented in Figure 6A,B. Berbamine attenuated COX-2, TNF-α, and p-NF-κb significantly in the SC. Bergapten downregulated COX-2, TNF-α, and p-NF-κb significantly in the SC. Carveol reduced COX-2, TNF-α, and p-NF-κb significantly in the SC, presented in Figure 7A,B.

3.8. Effects on Inflammatory Marker (ELISA)

As shown in Figure 8, we studied the effects of berbamine, bergapten, and carveol on the expression of COX-2, TNF-α, and p-NF-κb. All three mediators were found raised in the PTX group vs. saline in the SN and SC.Berbamine at 1 mg/kg minimized expression of COX-2 in the SN and SC. At 5 mg/kg, it suppressed COX-2 expression in the SN and SC, at 15 mg/kg, it reduced COX-2 expression in the SN and SC. Bergapten at 25 mg/kg decreased COX-2 in the SN. At 50 mg/kg it downregulated COX-2 in the SN, and at 50 mg/kg it reduced COX-2 in the SC. Bergapten at 100 mg/kg minimized COX-2 SN and SC. Carveol at 10 mL/kg declined COX-2 SN and SC. At 20 mL/kg, it subsided COX-2 in the SN and

SC, as shown in Figure 8A. Berbamine at 1 mg/kg decreased TNF-α in the SN and SC. At 5 mg/kg, it downregulated TNF-α in the SN and SC, and at 15 mg/kg, it diminished TNF-α in the SN and SC. Bergapten at 25 mg/kg decreased TNF-α in the SN and SC. At 50 mg/kg, it downregulated TNF-α in the SN and SC. Bergapten at 100 mg/kg dropped TNF-α in the SN and SC. Carveol at 10 mL/kg decreased the expression of TNF-α in the sciatic nerve. At 20 mL/kg it downregulated TNF-α in the SN and SC as shown in Figure 8B. Berbamine at 1 mg/kg decreased p-NF-κb in the SN and SC. At 5 mg/kg it downregulated NF-κb in the SN and SC, and at 15 mg/kg it reduced NF-κbin the SN and SC. Bergapten at 25 mg/kg decreased p-NF-κb in the sciatic nerve. At 50 mg/kg it downregulated NF-κb in the SN and with $p < 0.05$ in the SC. Bergapten at 100 mg/kg decreased the expression of NF-κb in the SN and SC. Carveol at 10 mL/kg decreased NF-κb in the SN and SC. At 20 mg/kg it downregulated NF-κb in the SN and SC as shown in Figure 8C.

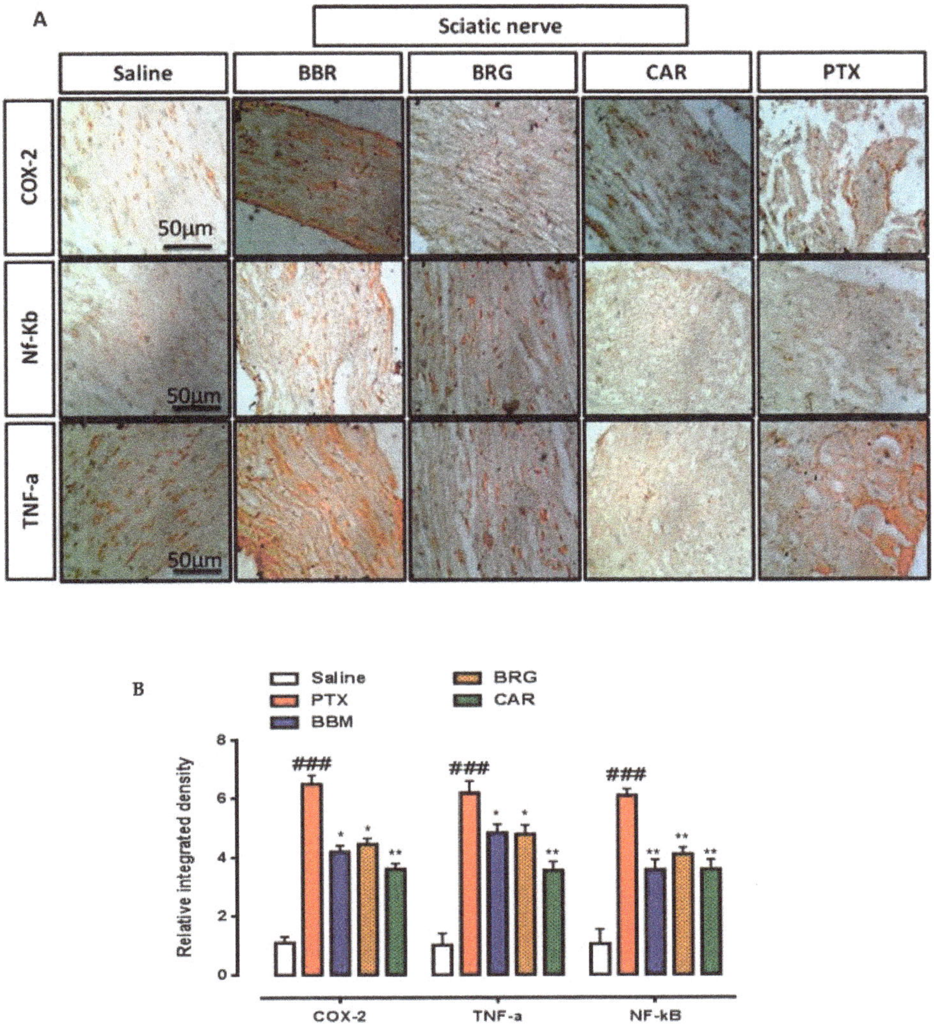

Figure 6. (**A**) Representation of immunohistochemistry results for COX-2, TNF-α, and p-NF-κbin the sciatic nerve of the rat. (**B**) Histograms showed comparatively higher expression of COX-2, TNF-α,

and p-NF-κbin the PTX group. The data are expressed as the mean ± SEM, $n = 6$. One-way ANOVA with posthoc Tukey's test. * $p < 0.05$ and ** $p < 0.01$ indicate a significant difference vs. PTX, and ### $p < 0.001$ indicates a significant difference vs. saline. Morphological data were analyzed by ImageJ software. Bar 50 µm, magnification 40×.

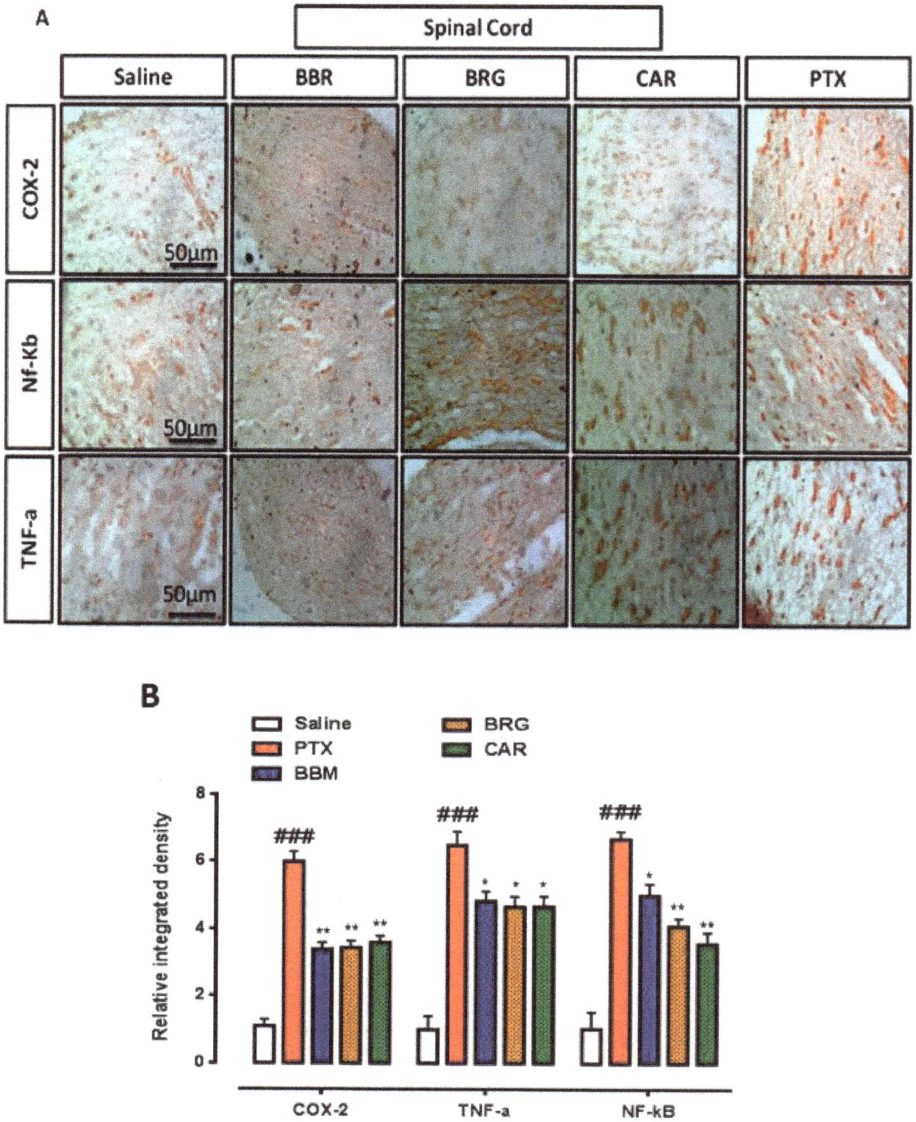

Figure 7. (**A**) Representation of immunohistochemistry results for COX-2, TNF-α, and p-NF-κbin the spinal cord of rat. Bar 50 µm, magnification 40x ($n = 6$/group). (**B**) Histograms showed comparatively higher expression of COX-2, TNF-α, and p-NF-κb in the PTX group. The data are expressed as the mean ± SEM, $n = 6$. One-way ANOVA with posthoc Tukey's test. * $p < 0.05$ and ** $p < 0.01$ indicate a significant difference vs. PTX, and ### $p < 0.001$ indicates a significant difference vs. saline. Morphological data were analyzed by ImageJ software. Bar 50 µm, magnification 40×.

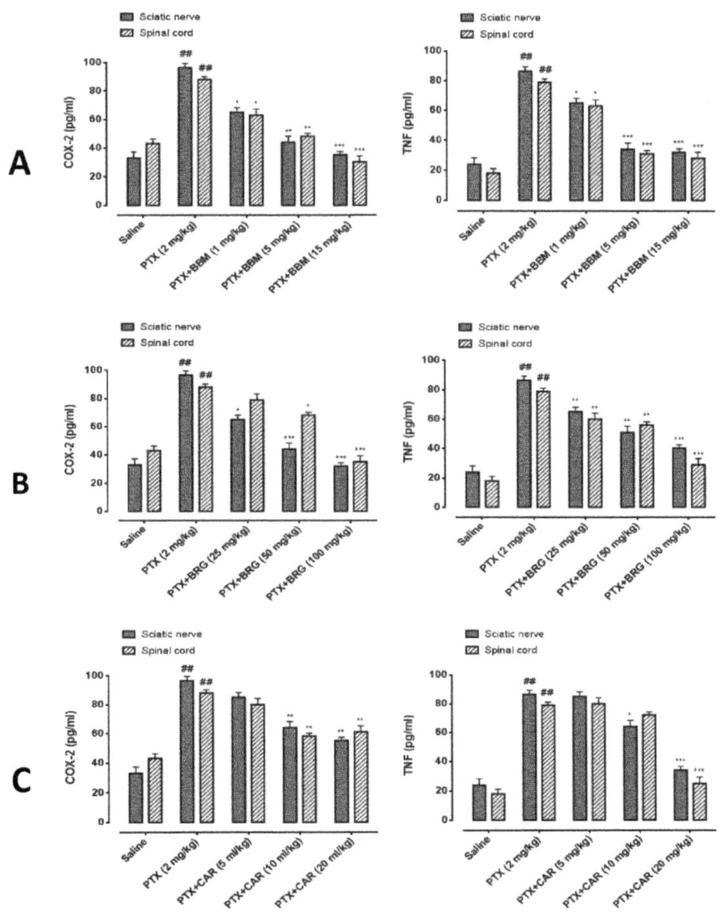

Figure 8. Representation of the effect of berbamine (**A**), bergapten (**B**), and carveol (**C**) on the expression of COX-2, TNF-α, and NF-κb in the sciatic nerve and spinal cord, quantified by using enzyme-linked immunosorbent assays. The data are expressed as the mean ± SEM, $n = 6$. One-way ANOVA with posthoc Tukey's test. * $p < 0.05$, ** $p < 0.01$, and *** $p < 0.001$ indicate a significant difference vs. PTX, and ## $p < 0.01$ indicates a significant difference vs. saline.

3.9. Western Blot Findings

The inflammatory markers were subsequently studied using Western blot analysis in both the sciatic nerve and the spinal cord, and the findings are shown in Figures 9 and 10. All inflammatory indicators were considerably elevated in the collected samples compared to saline, according to the findings. COX-2, TNF-, and NF-b were all inhibited by berbamine. COX-2, TNF-, and p-NF-b were all inhibited by bergapten. In comparison to the PTX group, carveol reduced COX-2, TNF-, and NF-b in the sciatic nerve sample. Berbamine inhibited COX-2, TNF-, and NF-b. Bergapten attenuated COX-2, TNF-α, and NF-κb. Carveol decreased the expression of COX-2, TNF-α, and NF-κb in the spinal cord sample vs. PTX.

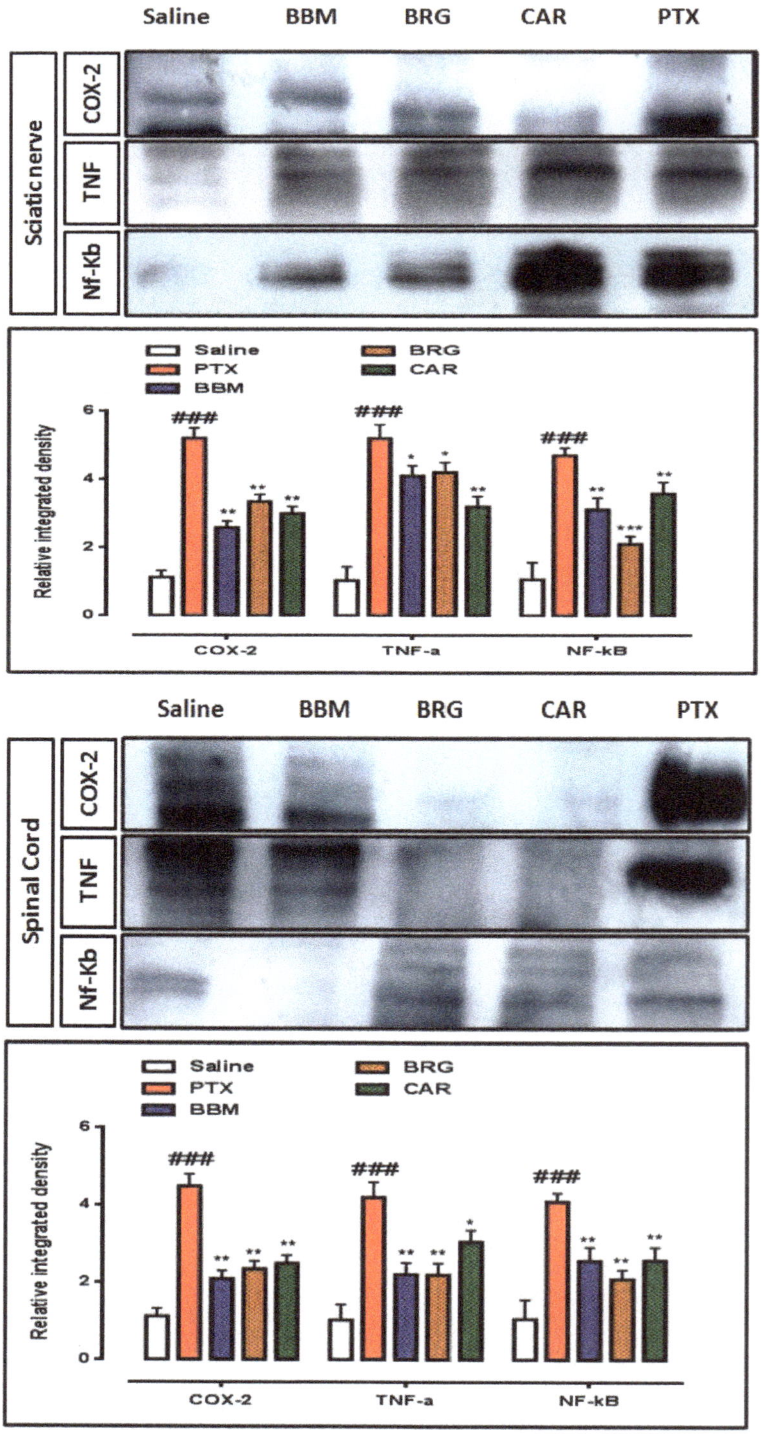

Figure 9. Representation of the effects of berbamine, bargepten, and carveol on the expression of COX-2, TNF-α, and NF-κb, quantified by using a Western blot sciatic nerve and spinal cord. The

data are expressed as the mean ± SEM, $n = 6$. One-way ANOVA with posthoc Tukey's test. * $p < 0.05$, ** $p < 0.01$, and *** $p < 0.001$ indicate a significant difference vs. PTX, and ### $p < 0.001$ indicate a significant difference vs. saline. Morphological data were analyzed by ImageJ software. Bar 50 μm, magnification 40×.

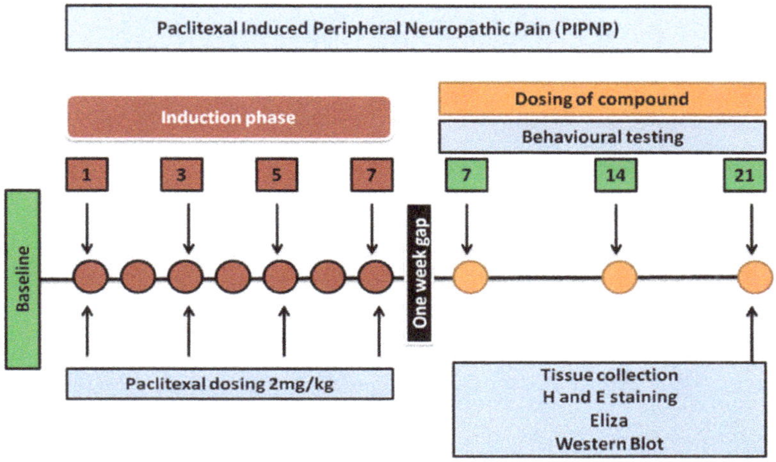

Figure 10. The scheme of the study. The baseline reading was collected on day0. The induction phase was started with administration of 2 mg/kg of PTX on day 1 and was continued on alternative days until day 7. After one week, the behavioral studies were conducted to confirm the development of neuropathy. At the same time, treatment was started and was continued until day 21.

4. Discussion

One of the most severe and disabling adverse medication reactions in more than 75% of cancer regimens is paclitaxel-induced peripheral neuropathy (PIPN) [32]. There are several ways to cause paclitaxel-induced peripheral neuropathy in rats, including a preventive technique (neuropathy is produced while the drug is present) and a pre-existing neuropathy method (neuropathy is established in the absence of drug). The potential of three natural compounds (BBM, BAR, and CAR) in chemotherapy-induced discomfort was secured using the pre-existing neuropathy method. To cause neuropathy, paclitaxel was administered at a dosage of 2 mg/kg on days 1, 3, and 5. Treatment was not started until neuropathy had developed and been verified by performing behavioral tests.

The pain threshold was supported by the following factors: body weight, mechanical hypersensitivity, latency before falling, and thermal pain feeling. As the number of cancer patients rises, there is an urgent need to create medications with a lower toxicity and higher efficacy profile in order to improve patient quality of life and reverse PIPN. Berberine [33], icariin [34], and melatonin [35] are only a few of the medications that have been studied for this purpose. Many natural products are renowned for their wide range of therapeutic potential, and researchers have worked hard to develop such substances as important nutrients to treat a variety of illnesses [36,37]. Moreover, as pain is the protective mechanism alarming us about the proper measure to be taken for the initiated problem, so people utilize different approaches to relieving pain associated with chronic illness [38]

In order to discover compounds for effective reversal of PIPN, we selected three natural compounds, BBM, BRG, and CAR. They were chosen to see how they may affect PIPN caused by PTX. The selection of these compounds was based on the fact that they are anti-inflammatory and neuroprotective [14,15,17] in nature and can be investigated in paclitaxel-associated chronic inflammation.

Several markers were investigated, including nuclear factor-2 and poly ADP-ribose polymerase, both of which are implicated in anticancer-induced neuropathic pain [39]. NF-kB, nitric oxide, COX-2, IL-1, IL-6, and TNF-α have also been shown to be overexpressed in PIPN [40]. Toll-like receptors (TLR-1, TLR-2, and TLR-4) and the transient receptor potential family of ion channels (TRPV1 and TRPV4) have been discovered, demonstrating their active role in chemotherapy-induced peripheral neuropathy [18,41,42]. It is generally established that oxidative stress is linked to all neuropathies. The induction of oxidative stress by paclitaxel is a well-known mechanism for the formation and maintenance of PIPN [42]. Its use causes the depletion of GSH and GST, which are implicated in blocking hazardous chemical metabolites in chemotherapy-induced pain [43–45]. In addition to GSH and GST, other harmful variables such as LPO and iNOS [42] are shown to be raised following PTX treatment and have a role in enhancing pain feeling. We tested the effects of three natural substances on paclitaxel-induced peripheral neuropathy. Figure 10 depicts the theme of the current research work's strategy.

The sciatic nerve and spinal cord were collected for further analysis. The disruption in latency of falling body weight and mechanical hypersensitivity are all key components of paclitaxel-induced neuropathy [46–48]. The behavioral study that we conducted on our drugs showed that berbamine, bergapten, and carveol were able to repair the behavioral deficit that was caused by paclitaxel-induced peripheral neuropathic pain. These findings are presented in Figures 1–4. The findings were determined to be in line with those of previous research. Numerous studies have concentrated their attention on the neuronal level of the sciatic nerve and spinal cord in paclitaxel-induced neuropathy and have found morphological abnormalities at both levels [49]. In addition, we investigated the morphology of two different tissues and came to the conclusion that PTX played a role in the progression of the disease by mediating neuronal degeneration, disrupting the development of myelin sheaths, and causing vacuolation. The fact that treatment with the natural substances mentioned above restored the damage caused by PTX suggests that these substances could be categorized as neuroprotective agents. This is demonstrated in Figures 5–7. Molecular research is primarily responsible for determining the expression of a wide variety of different types of inflammatory mediators and transcription factors. These molecular methods include the enzyme linked immunosorbent assay as well as the Western blot, both of which are utilized to evaluate the levels of endogenous marker expression [50–52]. The upregulation of inflammatory markers such as COX-2, TNF-a, and NF-kB, which are typically linked to chemotherapy-induced neuropathic pain, were also investigated as part of our research. Experiments using ELISA and Western blot demonstrated that PTX increased the expression of all of the pathogenic indicators that were previously mentioned. In an animal model of paclitaxel-induced neuropathic pain, treatment with the aforementioned treatments led to a reduction in the expression of COX-2, TNF-a, and NF-kb, which in turn led to a reduction in the severity of the pain.

5. Conclusions

This research explored that berbamine, bergapten, and carveol reversed paclitaxel-induced neuropathic pain by correcting behavioral deficits and increasing protective markers like GSH and GST, which are the key components involved in scavenging free radicals, thus protecting the body from their harmful consequences. The natural compounds decreased damaging factors like LPO and iNOS and downregulated overexpressed inflammatory mediators like COX-2, TNF-, and NF-kB, as evidenced by ELISA and Western blotting. This potential of the natural compounds (berbamine, bergapten, and carveol) makes them promising agents for neuroprotective effects. To broaden the list of medications useful in the therapy of neuropathic pain, further research is needed to clarify the pharmacokinetics and pharmacodynamics profile, as well as stability and dose form.

Author Contributions: M.F. collected the data and carried out the research, wrote the manuscript, and performed the bioinformatics part; A.-u.K., conceptualized and supervised the study; M.W.S. helped in proof reading and F.A.S. helped in performing the molecular study and acted as co-supervisors in the research. S.L. and A.W.K. and F.A. edited and revised the manuscript. All authors have read and agreed to the published version of the manuscript.

Funding: Shenzhen-Hong Kong Institute of Brain Sciences No: 2019SHIBS0004.

Institutional Review Board Statement: The experimental protocols for handling and dosing of animals were priorly approved by Riphah Institute of Pharmaceutical Sciences (Ref. No. REC/RIPS/2019/28).

Informed Consent Statement: Not applicable.

Data Availability Statement: All the data are incorporated in the manuscript.

Acknowledgments: The authors are grateful to the Institute of Pharmaceutical Science, Riphah International University Islamabad for providing a lab facility to conduct this research work.

Conflicts of Interest: The authors declare no competing interests.

Abbreviations

Berbamine	(BBM)
Bergapten	(BRG)
Carveol	(CAR)
Paclitaxel	(PTX)
Glutathione	(GSH)
Glutathione S-transferase	(GST)
Inducible nitric oxide synthase	(iNOS)
Lipid peroxidase	(LPO)
Hematoxylin and eosin	(H and E)
Immunohistochemistry	(IHC)
Cyclooxygenase	(COX-2)
Tumor necrosis factor-alpha	(TNF-α)
Nuclear factor kappa B	(NF-κb)
Enzyme-linked immunosorbant assay	(Elisa)
Phosphate buffer saline	(PBS)
Hydrogen peroxide	(H_2O_2)
5,5′-dithio-bis-2-nitrobenzoic acid	(DTNB)
2,4-Dinitrochlorobenzene	(CDNB)
Diaminobenzidine peroxidase	(DAB)
Trichloroacetic acid	(TCA)
Research and ethical committee	(REC)
Intraperitoneally	(i.p)
Sciatic nerve	(SN)
Spinal cord	(SC)
Nitric oxide	(NO)
Horseradish peroxidase	(HRP)
Heat latency	(HL)
Falling	(LF)
Paw withdrawal threshold	(PWT)
Paclitaxel induced peripheral neuropathic pain	(PIPNP)
Chemotherapy-induced peripheral neuropathic pain	(CIPNP)
Toll-like receptors	(TLR)
Transient receptor potential	(TRP)

References

1. Postma, T.; Vermorken, J.; Liefting, A.; Pinedo, H.; Heimans, J. Paclitaxel-induced neuropathy. *Ann. Oncol.* **1995**, *6*, 489–494. [CrossRef] [PubMed]
2. Zajączkowska, R.; Kocot-Kępska, M.; Leppert, W.; Wrzosek, A.; Mika, J.; Wordliczek, J. Mechanisms of chemotherapy-induced peripheral neuropathy. *Int. J. Mol. Sci.* **2019**, *20*, 1451. [CrossRef] [PubMed]
3. Augusto, C.; Pietro, M.; Cinzia, M.; Sergio, C.; Sara, C.; Luca, G.; Scaioli, V. Peripheral neuropathy due to paclitaxel: Study of the temporal relationships between the therapeutic schedule and the clinical quantitative score (QST) and comparison with neurophysiological findings. *J. Neurooncol.* **2008**, *86*, 89–99. [CrossRef] [PubMed]
4. Rowinsky, E.K.; Donehower, R.C. Paclitaxel (taxol). *N. Engl. J. Med.* **1995**, *332*, 1004–1014. [CrossRef] [PubMed]

5. Xue, X.; Liu, H.; Wang, S.; Hu, Y.; Huang, B.; Li, M.; Su, J. Neutrophil-erythrocyte hybrid membrane-coated hollow copper sulfide nanoparticles for targeted and photothermal/ anti-inflammatory therapy of osteoarthritis. *Compos. Part B Eng.* **2022**, *237*, 109855. [CrossRef]
6. Boyette-Davis, J.A.; Walters, E.T.; Dougherty, P.M. Mechanisms involved in the development of chemotherapy-induced neuropathy. *Pain Manag.* **2015**, *5*, 285–296. [CrossRef]
7. Klein, I.; Lehmann, H.C. Pathomechanisms of paclitaxel-induced peripheral neuropathy. *Toxics* **2021**, *9*, 229. [CrossRef] [PubMed]
8. Zhang, H.; Li, Y.; de Carvalho-Barbosa, M.; Kavelaars, A.; Heijnen, C.J.; Albrecht, P.J.; Dougherty, P.M. Dorsal root ganglion infiltration by macrophages contributes to paclitaxel chemotherapy-induced peripheral neuropathy. *J. Pain* **2016**, *17*, 775–786. [CrossRef]
9. Caillaud, M.; Thompson, D.; Toma, W.; White, A.; Mann, J.; Roberts, J.L.; Damaj, M.I. Formulated Curcumin Prevents Paclitaxel-Induced Peripheral Neuropathy through Reduction in Neuroinflammation by Modulation of α7 Nicotinic Acetylcholine Receptors. *Pharmaceutics* **2022**, *14*, 1296. [CrossRef]
10. Li, X.; Yang, S.; Wang, L.; Liu, P.; Zhao, S.; Li, H.; Jiang, Y.; Guo, Y.; Wang, X. Resveratrol inhibits paclitaxel-induced neuropathic pain by the activation of PI3K/Akt and SIRT1/PGC1α pathway. *J. Pain Res.* **2019**, *12*, 879. [CrossRef]
11. Kaur, S.; Muthuraman, A. Ameliorative effect of gallic acid in paclitaxel-induced neuropathic pain in mice. *Toxicol. Rep.* **2019**, *6*, 505–513. [CrossRef] [PubMed]
12. Zhang, X.-L.; Cao, X.-Y.; Lai, R.-C.; Xie, M.-X.; Zeng, W.-A. Puerarin relieves paclitaxel-induced neuropathic pain: The role of Nav1. 8 β1 subunit of sensory neurons. *Front. Pharmacol.* **2019**, *9*, 1510. [CrossRef] [PubMed]
13. Memariani, Z.; Abbas, S.Q.; ul Hassan, S.S.; Ahmadi, A.; Chabra, A. Naringin and naringenin as anticancer agents and adjuvants in cancer combination therapy: Efficacy and molecular mechanisms of action, a comprehensive narrative review. *Pharmacol. Res.* **2021**, *171*, 105264. [CrossRef] [PubMed]
14. Zhao, Y.; Lv, J.; Chen, J.; Jin, X.; Wang, M.; Su, Z.; Wang, L.; Zhang, H. Berbamine inhibited the growth of prostate cancer cells in vivo and in vitro via triggering intrinsic pathway of apoptosis. *Prostate Cancer Prostatic Dis.* **2016**, *19*, 358–366. [CrossRef] [PubMed]
15. Raquet, N.; Schrenk, D. Relative photomutagenicity of furocoumarins and limettin in the hypoxanthine phosphoribosyl transferase assay in V79 cells. *Chem. Res. Toxicol.* **2009**, *22*, 1639–1647. [CrossRef] [PubMed]
16. Srivastava, J.; Lambert, J.; Vietmeyer, N. *Medicinal Plants: An Expanding Role in Development*; World Bank Publications: Washinton, DC, USA, 1996; Volume 320.
17. Alvi, A.M.; Al Kury, L.T.; Alattar, A.; Ullah, I.; Muhammad, A.J.; Alshaman, R.; Li, S. Carveol attenuates seizure severity and neuroinflammation in pentylenetetrazole-kindled epileptic rats by regulating the Nrf2 signaling pathway. *Oxidative Med. Cell. Longev.* **2021**, *2021*, 9966663. [CrossRef] [PubMed]
18. Li, Y.; Zhang, T.; Wang, Z.; Liang, H.; Wu, Z.; Li, J.; Ou-Yang, J.; Zhu, B. Transcranial Focused Ultrasound Stimulation of Periaqueductal Gray for Analgesia. *IEEE Trans. Biomed. Eng.* **2022**, 1. [CrossRef]
19. Bardin, L.; Malfetes, N.; Newman-Tancredi, A.; Depoortere, R. Chronic restraint stress induces mechanical and cold allodynia, and enhances inflammatory pain in rat: Relevance to human stress-associated painful pathologies. *Behav. Brain Res.* **2009**, *205*, 360–366. [CrossRef] [PubMed]
20. Huang, C.; Hu, Z.-P.; Long, H.; Shi, Y.-S.; Han, J.-S.; Wan, Y. Attenuation of mechanical but not thermal hyperalgesia by electroacupuncture with the involvement of opioids in rat model of chronic inflammatory pain. *Brain Res. Bull.* **2004**, *63*, 99–103. [CrossRef] [PubMed]
21. Ba, X.; Wang, J.; Zhou, S.; Luo, X.; Peng, Y.; Yang, S.; Hao, Y.; Jin, G. Cinobufacini protects against paclitaxel-induced peripheral neuropathic pain and suppresses TRPV1 up-regulation and spinal astrocyte activation in rats. *Biomed. Pharmacother.* **2018**, *108*, 76–84. [CrossRef] [PubMed]
22. Joshi, R.P.; Negi, G.; Kumar, A.; Pawar, Y.B.; Munjal, B.; Bansal, A.K.; Sharma, S.S. SNEDDS curcumin formulation leads to enhanced protection from pain and functional deficits associated with diabetic neuropathy: An insight into its mechanism for neuroprotection. *Nanomed. Nanotechnol. Biol. Med.* **2013**, *9*, 776–785. [CrossRef]
23. Imran, M.; Al Kury, L.T.; Nadeem, H.; Shah, F.A.; Abbas, M.; Naz, S.; Khan, A.-U.; Li, S. Benzimidazole containing acetamide derivatives attenuate neuroinflammation and oxidative stress in ethanol-induced neurodegeneration. *Biomolecules* **2020**, *10*, 108. [CrossRef] [PubMed]
24. Guedes, R.P.; Dal Bosco, L.; da Rosa Araújo, A.S.; Belló-Klein, A.; Ribeiro, M.F.M.; Partata, W.A. Sciatic nerve transection increases gluthatione antioxidant system activity and neuronal nitric oxide synthase expression in the spinal cord. *Brain Res. Bull.* **2009**, *80*, 422–427. [CrossRef] [PubMed]
25. Iqbal, S.; Shah, F.A.; Naeem, K.; Nadeem, H.; Sarwar, S.; Ashraf, Z.; Imran, M.; Khan, T.; Anwar, T.; Li, S. Succinamide derivatives ameliorate neuroinflammation and oxidative stress in scopolamine-induced neurodegeneration. *Biomolecules* **2020**, *10*, 443. [CrossRef] [PubMed]
26. Kumar, K.S.; Hsieh, H.W.; Wang, S.-Y. Anti-inflammatory effect of lucidone in mice via inhibition of NF-κB/MAP kinase pathway. *Int. Immunopharmacol.* **2010**, *10*, 385–392. [CrossRef] [PubMed]
27. Mohsin Alvi, A.; Tariq Al Kury, L.; Umar Ijaz, M.; Ali Shah, F.; Tariq Khan, M.; Sadiq Sheikh, A.; Nadeem, H.; Khan, A.-U.; Zeb, A.; Li, S. Post-treatment of synthetic polyphenolic 1,3,4 oxadiazole compound A3, attenuated ischemic stroke-induced neuroinflammation and neurodegeneration. *Biomolecules* **2020**, *10*, 816. [CrossRef] [PubMed]

28. Ali, A.; Shah, F.A.; Zeb, A.; Malik, I.; Alvi, A.M.; Alkury, L.T.; Rashid, S.; Hussain, I.; Ullah, N.; Khan, A.U. NF-κB inhibitors attenuate MCAO induced neurodegeneration and oxidative stress—A reprofiling approach. *Front. Mol. Neurosci.* **2020**, *13*, 33. [CrossRef]
29. Hassan, S.S.U.; Muhammad, I.; Abbas, S.Q.; Hassan, M.; Majid, M.; Jin, H.Z.; Bungau, S. Stress driven discovery of natural products from actinobacteria with anti-oxidant and cytotoxic activities including docking and admet properties. *Int. J. Mol. Sci.* **2021**, *22*, 11432. [CrossRef] [PubMed]
30. Al Kury, L.T.; Zeb, A.; Abidin, Z.U.; Irshad, N.; Malik, I.; Alvi, A.M.; Khalil, A.A.K.; Ahmad, S.; Faheem, M.; Khan, A.-U. Neuroprotective effects of melatonin and celecoxib against ethanol-induced neurodegeneration: A computational and pharmacological approach. *Drug Des. Devel. Ther.* **2019**, *13*, 2715. [CrossRef] [PubMed]
31. Majid, M.; Farhan, A.; Asad, M.I.; Khan, M.R.; Hassan, S.S.U.; Haq, I.U.; Bungau, S. An Extensive Pharmacological Evaluation of New Anti-Cancer Triterpenoid (Nummularic Acid) from Ipomoea batatas through In Vitro, In Silico, and In Vivo Studies. *Molecules* **2022**, *27*, 2474. [CrossRef]
32. Sisignano, M.; Baron, R.; Scholich, K.; Geisslinger, G. Mechanism-based treatment for chemotherapy-induced peripheral neuropathic pain. *Nat. Rev. Neurol.* **2014**, *10*, 694–707. [CrossRef] [PubMed]
33. Singh, J.; Saha, L.; Singh, N.; Kumari, P.; Bhatia, A.; Chakrabarti, A. Study of nuclear factor-2 erythroid related factor-2 activator, berberine, in paclitaxel induced peripheral neuropathy pain model in rats. *J. Pharm. Pharmacol.* **2019**, *71*, 797–805. [CrossRef] [PubMed]
34. Gui, Y.; Zhang, J.; Chen, L.; Duan, S.; Tang, J.; Xu, W.; Li, A. Icariin, a flavonoid with anti-cancer effects, alleviated paclitaxel-induced neuropathic pain in a SIRT1-dependent manner. *Mol. Pain* **2018**, *14*, 1744806918768970. [CrossRef]
35. Ambriz-Tututi, M.; Rocha-González, H.I.; Cruz, S.L.; Granados-Soto, V. Melatonin: A hormone that modulates pain. *Life Sci.* **2009**, *84*, 489–498. [CrossRef] [PubMed]
36. Bano, I.; Horky, P.; Abbas, S.Q.; Majid, M.; Bilal, A.H.M.; Ali, F.; Bungau, S. Ferroptosis: A New Road towards Cancer Management. *Molecules* **2022**, *27*, 2129. [CrossRef]
37. Xie, Y.G.; Zhao, X.C.; ul Hassan, S.S.; Zhen, X.Y.; Muhammad, I.; Yan, S.K.; Jin, H.Z. One new sesquiterpene and one new iridoid derivative from Valeriana amurensis. *Phytochem. Lett.* **2019**, *32*, 6–9. [CrossRef]
38. Laktasic Zerjavic, N.; Hrkic, E.; Zagar, I.; Delimar, V.; Kovac Durmis, K.; Spoljarić Carevic, S.; Peric, P. Local cryotherapy, comparison of cold air and ice massage on pain and handgrip strength in patients with rheumatoid arthritis. *Psychiatr. Danub.* **2021**, *33*, 757–761.
39. Komirishetty, P.; Areti, A.; Yerra, V.G.; Ruby, P.; Sharma, S.S.; Gogoi, R.; Sistla, R.; Kumar, A. PARP inhibition attenuates neuroinflammation and oxidative stress in chronic constriction injury induced peripheral neuropathy. *Life Sci.* **2016**, *150*, 50–60. [CrossRef]
40. Yerra, V.G.; Negi, G.; Sharma, S.S.; Kumar, A. Potential therapeutic effects of the simultaneous targeting of the Nrf2 and NF-κB pathways in diabetic neuropathy. *Redox Biol.* **2013**, *1*, 394–397. [CrossRef] [PubMed]
41. Pascual, M.; Baliño, P.; Aragón, C.M.; Guerri, C. Cytokines and chemokines as biomarkers of ethanol-induced neuroinflammation and anxiety-related behavior: Role of TLR4 and TLR2. *Neuropharmacology* **2015**, *89*, 352–359. [CrossRef]
42. Hara, T.; Chiba, T.; Abe, K.; Makabe, A.; Ikeno, S.; Kawakami, K.; Utsunomiya, I.; Hama, T.; Taguchi, K. Effect of paclitaxel on transient receptor potential vanilloid 1 in rat dorsal root ganglion. *PAIN®* **2013**, *154*, 882–889. [CrossRef]
43. Ishii, N.; Tsubouchi, H.; Miura, A.; Yanagi, S.; Ueno, H.; Shiomi, K.; Nakazato, M. Ghrelin alleviates paclitaxel-induced peripheral neuropathy by reducing oxidative stress and enhancing mitochondrial anti-oxidant functions in mice. *Eur. J. Pharmacol.* **2018**, *819*, 35–42. [CrossRef] [PubMed]
44. Mir, O.; Alexandre, J.; Tran, A.; Durand, J.-P.; Pons, G.; Treluyer, J.-M.; Goldwasser, F. Relationship between GSTP1 Ile105Val polymorphism and docetaxel-induced peripheral neuropathy: Clinical evidence of a role of oxidative stress in taxane toxicity. *Ann. Oncol.* **2009**, *20*, 736–740. [CrossRef] [PubMed]
45. Mccormick, B.; Lowes, D.; Colvin, L.; Torsney, C.; Galley, H. MitoVitE, a mitochondria-targeted antioxidant, limits paclitaxel-induced oxidative stress and mitochondrial damage in vitro, and paclitaxel-induced mechanical hypersensitivity in a rat pain model. *BJA Br. J. Anaesth.* **2016**, *117*, 659–666. [CrossRef]
46. Naveed, M.; Ullah, R.; Khan, A.; Shal, B.; Khan, A.U.; Khan, S.Z.; Khan, S. Anti-neuropathic pain activity of a cationic palladium (II) dithiocarbamate by suppressing the inflammatory mediators in paclitaxel-induced neuropathic pain model. *Mol. Biol. Rep.* **2021**, *48*, 7647–7656. [CrossRef] [PubMed]
47. Pawar, S.H.; Upaganlawar, A.B.; Upasani, C.D. Attenuation of Hyperalgesia and Allodynia by some Phenolic Acids in Paclitaxel Induced Neuropathy. *bioRxiv* **2021**, *1*, 427–445.
48. Sullivan, K.A.; Grant, C.V.; Jordan, K.R.; Vickery, S.S.; Pyter, L.M. Voluntary wheel running ameliorates select paclitaxel chemotherapy-induced sickness behaviors and associated melanocortin signaling. *Behav. Brain Res.* **2021**, *399*, 113041. [CrossRef] [PubMed]
49. Son, D.B.; Choi, W.; Kim, M.; Go, E.J.; Jeong, D.; Park, C.-K.; Kim, Y.H.; Lee, H.; Suh, J.-W. Decursin alleviates mechanical allodynia in a paclitaxel-induced neuropathic pain mouse model. *Cells* **2021**, *10*, 547. [CrossRef] [PubMed]
50. Yardım, A.; Kandemir, F.M.; Çomaklı, S.; Özdemir, S.; Caglayan, C.; Kucukler, S.; Çelik, H. Protective effects of curcumin against paclitaxel-induced spinal cord and sciatic nerve injuries in rats. *Neurochem. Res.* **2021**, *46*, 379–395. [CrossRef] [PubMed]

51. Kashem, M.A.; Li, H.; Toledo, N.P.; Omange, R.W.; Liang, B.; Liu, L.R.; Li, L.; Yang, X.; Yuan, X.-Y.; Kindrachuk, J. Toll-like interleukin 1 receptor regulator is an important modulator of inflammation responsive genes. *Front. Immunol.* **2019**, *10*, 272. [CrossRef]
52. Mahmood, F.; Khan, J.A.; Mahnashi, M.H.; Jan, M.S.; Javed, M.A.; Rashid, U.; Bungau, S. Anti-Inflammatory, Analgesic and Antioxidant Potential of New (2 S, 3 S)-2-(4-isopropylbenzyl)-2-methyl-4-nitro-3-phenylbutanals and Their Corresponding Carboxylic Acids through In Vitro, In Silico and In Vivo Studies. *Molecules* **2022**, *27*, 4068. [CrossRef] [PubMed]

molecules

MDPI

Article

Comprehensive Phytochemical Profiling, Biological Activities, and Molecular Docking Studies of *Pleurospermum candollei*: An Insight into Potential for Natural Products Development

Maqsood Ahmed [1], Kashif-ur-Rehman Khan [1,*], Saeed Ahmad [1], Hanan Y. Aati [2,*], Chitchamai Ovatlarnporn [3], Muhammad Sajid-ur Rehman [4], Tariq Javed [5], Anjum Khursheed [1], Bilal Ahmad Ghalloo [1], Rizwana Dilshad [1] and Maryam Anwar [1]

1 Department of Pharmaceutical Chemistry, Faculty of Pharmacy, The Islamia University of Bahawalpur, Bahawalpur 63100, Pakistan; dr.maqsoodrao@gmail.com (M.A.); rsahmed_iub@yahoo.com (S.A.); anjumkhursheedrana@gmail.com (A.K.); drbilal29@hotmail.com (B.A.G.); rizwanadilshad8@gmail.com (R.D.); mariyama2014@hotmail.com (M.A.)
2 Department of Pharmacognosy, College of Pharmacy, King Saud University, Riyadh 11495, Saudi Arabia
3 Department of Pharmaceutical Chemistry, Faculty of Pharmaceutical Sciences, Prince of Songkla University, Hat Yai 90110, Thailand; chitchamai.oo@psu.ac.th
4 Department of Pharmacognosy, Faculty of Pharmacy, The Islamia University of Bahawalpur, Bahawalpur 63100, Pakistan; sj_pharmacist@iub.edu.pk
5 Lahore Pharmacy College (LMDC), Lahore 53400, Pakistan; tjavedpk@gmail.com
* Correspondence: kashifur.rahman@iub.edu.pk (K.-u.-R.K.); hati@ksu.edu.sa (H.Y.A.)

Citation: Ahmed, M.; Khan, K.-u.-R.; Ahmad, S.; Aati, H.Y.; Ovatlarnporn, C.; Rehman, M.S.-u.; Javed, T.; Khursheed, A.; Ghalloo, B.A.; Dilshad, R.; et al. Comprehensive Phytochemical Profiling, Biological Activities, and Molecular Docking Studies of *Pleurospermum candollei*: An Insight into Potential for Natural Products Development. *Molecules* **2022**, *27*, 4113. https://doi.org/10.3390/molecules27134113

Academic Editors: Syed Shams ul Hassan, Mohamed M. Abdel-Daim, Tapan Behl and Simona Bungau

Received: 27 May 2022
Accepted: 24 June 2022
Published: 26 June 2022

Abstract: The purpose of this study was to find the biological propensities of the vegetable plant *Pleurospermum candollei* by investigating its phytochemical profile and biological activities. Phytochemical analysis was done by spectroscopic methods to investigate the amount of total polyphenols, and biological evaluation was done by the different antioxidant, enzyme inhibitory (tyrosinase, α-amylase, and α-glucosidase), thrombolytic, and antibacterial activities. The highest amount of total phenolic and flavonoid contents was observed in methanolic extract (240.69 ± 2.94 mg GAE/g and 167.59 ± 3.47 mg QE/g); the fractions showed comparatively less quantity (57.02 ± 1.31 to 144.02 ± 2.11 mg GAE/g, and 48.21 ± 0.75 to 96.58 ± 2.30 mg QE/g). The effect of these bioactive contents was also related to biological activities. GCMS analysis led to the identification of bioactive compounds with different biological effects from methanolic extract (antioxidant; 55.07%, antimicrobial; 56.41%), while the identified compounds from the *n*-hexane fraction with antioxidant properties constituted 67.86%, and those with antimicrobial effects constituted 82.95%; however, the synergetic effect of polyphenols may also have contributed to the highest value of biological activities of methanolic extract. Molecular docking was also performed to understand the relationship of identified secondary metabolites with enzyme-inhibitory activities. The thrombolytic activity was also significant (40.18 ± 1.80 to 57.15 ± 1.10 % clot lysis) in comparison with streptokinase (78.5 ± 1.53 to 82.34 ± 1.25% clot lysis). Methanolic extract also showed good activity against Gram-positive strains of bacteria, and the highest activity was observed against *Bacillus subtilis*. The findings of this study will improve our knowledge of phytochemistry, and biological activities of *P. candollei*, which seems to be a ray of hope to design formulations of natural products for the improvement of health and prevention of chronic diseases; however, further research may address the development of novel drugs for use in pharmaceuticals.

Keywords: *Pleurospermum candollei*; natural compounds; antioxidant; antidiabetic; thrombolytic; antibacterial; pharmaceuticals; molecular docking

1. Introduction

Recently, the research on food plants and their bioactive ingredients is increasing due to the increased awareness by human consumers about the nutritional and functional properties of food ingredients, which have beneficial effects related to antioxidants and other biological activities and play an extensive role in maintaining and improving the human health [1,2]. The lower prevalence of metabolic disorders and other chronic diseases in the people of rural areas may also be attributed to the high consumption of plant foods, which are rich sources of bioactive compounds [3,4]. The plants with a high content of bioactive phytochemicals are loved as medicines due to their anti-inflammatory, antibacterial, anticancer sedative, antidepressant, anxiolytic, anticonvulsant, antispasmodic, and anti-HIV properties [5]. Laboratory-based work not only increases the knowledge of phytochemical composition but also finds the relationship between disease and the bioactive compounds [6], which leads to the development of formulations and functional foods with improved health benefits and safety from the plants [7]. The thrombotic events are also growing these days with a possible relation with the pandemic of COVID-19 and other infections in addition to numerous factors [8,9]. These pieces of evidence inspired us to select a plant for investigation that has shreds of evidence of use in traditional medicine but has not been evaluated scientifically for biological propensities nor for its bioactive compounds' usefulness as a functional food to provide health-promoting effects and prevent human disease.

Pleurospermum candollei Benth. ex C.B.Clarke. is commonly known as Braq Shandun/Shoogroon/Shabdun in the Karakoram and Himalayan zones and belongs to the family Umbelliferae/Apiaceae. It is a 30–40 cm-long herb consumed as a vegetable by local tribes [10]. People of the Karakoram area use it for taste and nutritional benefits. This specie has also been used for different ailments and is available for commercial sale also in these regions. The whole plant is used to cure abdominal problems and stomach disorders. It also decreases cholesterol and blood pressure and provides relief from heart problems and gastric troubles [11,12]. One teaspoonful of dried plant powder is used with milk once a day for a week to treat headache and fever and can also be used by cooking with leafy vegetables for this purpose [13]. It is employed for the treatment of respiratory disorders, and the evidence also shows its good effects in pain, unconsciousness, and cerebral disorders. Many other ailments are also getting cured by the people of the Himalayan zone using this plant [14]. Stem powder of this plant has been used in joints problem and back pain in the area of Gilgit-Baltistan. Male and female infertility have also been treated using preparations from this plant [11]. It is also used to treat diarrhea in animals [15]. Ali et al., separated some compounds also to correlate the anti-inflammatory properties of *P. candollei* [14].

The Apiaceae family and *Pleurospermum* genus have been reported as good sources of natural antioxidants and used for their medicinal, pharmaceutical, nutraceutical, cosmaceutical, and food value due to the presence of many bioactive phytochemicals and their versatile biological activities [16,17]. *P. candollei*, despite the shreds of evidence of traditional uses for health benefits, has not been thoroughly investigated for its phytochemical potential and biological activities, therefore, its use as a source of antioxidants and other bioactive compounds for nutraceuticals and pharmaceutical formulations as well as functional foods, is still unknown. So the current study was designed to explore the chemical profile via total bioactive contents and GCMS analysis; a comprehensive biological screening was performed through antioxidant, enzyme-inhibitory (along with in silico studies), antibacterial, and thrombolytic activities to disclose ingredients for nutraceuticals, functional foods, and pharmaceutical possibilities. This study may also help to provide a rationale for the use of *P. candollei* in traditional medicinal system.

2. Results and Discussion

2.1. Determination of Bioactive Phytochemicals

2.1.1. Quantification of Bioactive Contents (TPC and TFC)

Many researchers are interested in investigating herbs because of their phenolic phytochemicals with unique carbon scaffolds. It substantiates the importance of measuring total bioactive compounds (phenolic contents) to authenticate the use of various plants in formulating new nutraceuticals/pharmaceuticals and functional foods. It substantiates the importance of measuring total bioactive compounds (phenolic contents) to authenticate the use of various plants in formulating new nutraceuticals/pharmaceuticals and functional foods [18].

So the bioactive contents of *P. candollei* were determined by quantitative methods of total phenolic contents (TPC) and total flavonoid contents (TFC). Fractions of increasing polarity were used in this experiment and the results are shown in Table 1. In the present work, we used the solvents of increasing polarity (*n*-hexane, chloroform, and *n*-butanol) to make fractions of different polarities from crude extract (methanolic extract). We used the methanol (80%) to prepare the crude extract because phenolic and flavonoid compounds can efficiently be extracted using a methanol solvent, as proved by the literature [19,20]. However, separation of the extract based on polarity results in the clustering of different bioactive phytochemicals. It gives a better understanding of different behaviors of biological effects in different extracts and facilitates the bio-guided isolation of compounds also for use in pharmaceuticals and biomedicine [21]. According to the results of our study, the quantity of phenolic and flavonoid contents was highest in PCME (240.69 ± 2.94 mg GAE/g dried wt. and 167.59 ± 3.47 mg QE/g dried wt.), and their quantity decreased with the polarity of the fraction. PCHF (non-polar fraction) had the least phenolic and flavonoid contents (TPC = 57.02 ± 1.31 mg GAE/g dried wt., and TFC = 48.21 ± 0.75 mg QE/g dried wt.), while PCBF had the highest contents (TPC = 144.02 ± 2.11 mg GAE/g dry wt. and TFC = 96.58 ± 2.30 mg QE/g dried wt.). The pieces of evidence also support our results by the findings that phenolic compounds more effectively recovered with solvents of higher polarity [22,23]. Herein, it is clear from obtained data that the methanolic extract and *n*-butanol fraction had higher bioactive contents in the form of TPC and TFC, and the bioactive content's potential of the least polar fraction was least but significant. To the best of our knowledge, no previous research has been done on the tested species of Pleurospermum to obtain data regarding its bioactive phytochemicals. Our findings were also in agreement with a recent study in which Al-Dalahmeh et al., reported that the highly polar fractions have higher levels of flavonoid and phenolic contents [24].

Table 1. Total phenolic and total flavonoid contents (TPC and TFC) of *P. candollei* (sample conc. 1 mg/mL).

Sample Codes	TPC (mg GAE/g Dried wt.)	TFC (mg QE/g Dried wt.)
PCME	240.69 ± 2.94	167.59 ± 3.47
PCHF	57.02 ± 1.31	48.21 ± 0.75
PCCF	97.02 ± 1.83	88.32 ± 1.45
PCBF	144.02 ± 2.11	96.58 ± 2.30

Values were taken in triplicates and reported as mean ± SD. PCME: *P. candollei* methanolic extract; PCHF: *P. candollei n*- hexane fraction; PCCF: *P. candollei* chloroform fraction; PCBF: *P. candollei n*-butanol fraction; TPC: total phenolic content; TFC: total flavonoid content; GAE: gallic acid equivalent; QE: Quercetin equivalent.

2.1.2. Characterization of Bioactive Compounds by GCMS Analysis

GC-MS (gas chromatography-mass spectrometry) analysis is preferable for non-polar as well as for volatile compounds, and methanolic extract contained polar and non-polar compounds; moreover, the *n*-hexane fraction mainly contained volatile and non-polar compounds [25]. So methanolic extract (PCME) and *n*-hexane fraction (PCHF) of *P. candollei* were subjected to gas chromatography-mass spectrometry (GC-MS) in the search for non-polar and volatile bioactive compounds (Figures 1 and 2, respectively). For the identification of natural compounds of the plant, the NIST library was used and 22 natural compounds were tentatively identified from PCME and presented in Table 2, whereas 51 natural compounds were identified from PCHF and shown in Table 3, along with their retention time in minutes, compound name, molecular formula, molecular weight, chemical class, and the percent peak area. Major constituents of PCME extract were the terpenoids, which constitute 82.61% of the non-polar components of the extract; hopanoids were the main subclass of the terpenoids; while the other compounds were steroids, alcohols, and flavonoids. The PCHF fraction mainly contained the saturated and unsaturated fatty acids (16.62%), and their esters (40.73%). Other classes of bioactive compounds were phenolic compounds (15.69%), coumarins (7.9%), steroids (4.64%), ester of fatty alcohols (2.72%), and terpenoids (1.55%), flavon (1.26%), and alkenes and alkynes (1.06%). Whereas the class of compounds present in minor quantity (less than 1% each) was alcohols (0.52%), phenolic acids (0.09%), and pyridine derivatives (0.49%). Similar results have been reported from the essential oil of a closely related species *Pleurospermum amabile*, in which Wangchuk et al. revealed many phenolics, terpenoids, and other bioactive compounds of pharmacological interest [26]. Major pharmacological activities of identified phytochemicals from *P. candollei* extract were given in Table 2. Those of the *n*-hexane fraction were presented in Table 3, along with their reference from the literature.

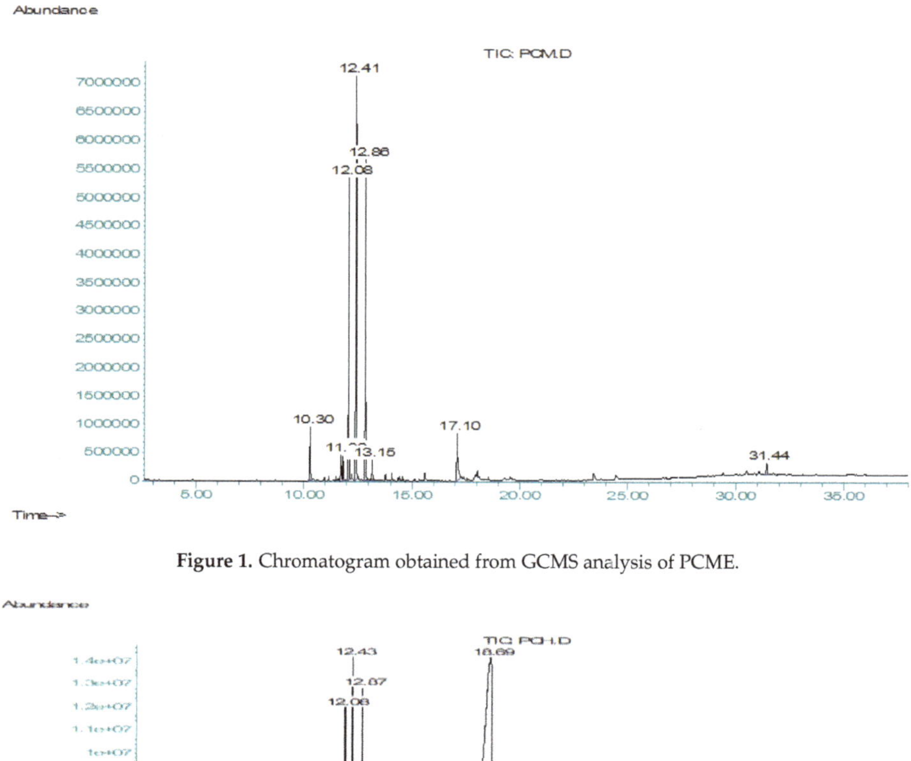

Figure 1. Chromatogram obtained from GCMS analysis of PCME.

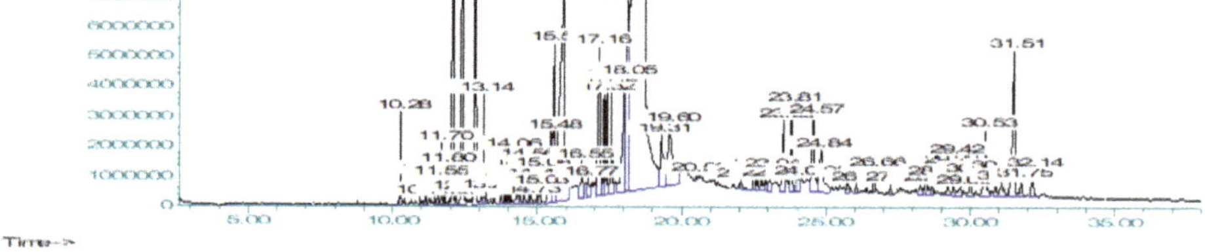

Figure 2. Chromatogram obtained from GCMS analysis of PCHF.

Table 2. Phytochemical profiling of methanolic extract (PCME) of *P. candollei* through GCMS analysis.

Sr.no.	RT	Tentative Identification of Compounds	Molecular Formula	Molecular Weight	Chemical Class	Area %	Reported Activities from Literature
1	15.13	Hexadecanoic acid, methyl ester	$C_{17}H_{34}O_2$	270.5	Fatty acid ester	0.05	Antioxidant, antimicrobial [27]
2	17.34	β-Amyrin	$C_{30}H_{50}O$	426.70	Terpenoid	2.77	Antioxidant, anti-inflammatory, antibacterial, antiulcer, antiarthritic, antidiahreal [28]
3	17.42	2(1H)Naphthalenone, 3,5,6,7,8,8a-hexahydro-4,8a-dimethyl-6-(1-methylethenyl)-	$C_{15}H_{22}O$	218.33	Terpenoid	0.45	Anticancer, antioxidant ani-inflammatory, analgesic, sedative [29]
4	18.46	4,6,6-Trimethyl-2-(3-methylbuta-1,3-dienyl)-3-oxatricyclo[5.1.0.0(2,4)]octane	$C_{15}H_{22}O$	218.33	Terpenoid	9.20	Antioxidant [27]
5	19.21	A-Neooleana-3(5),12-diene	$C_{30}H_{48}$	408.7	Terpenoid	9.92	Anti-inflammatory, antimicrobial [30], antidiabetic [31]
6	19.31	3-Epimoretenol	$C_{30}H_{50}O$	426.7	Terpenoid	3.37	Anti-inflammatory, analgesic [32]
7	19.77	9,19-Cyclolanost-24-en-3-ol, acetate, (3beta)-	$C_{32}H_{52}O_2$	468.8	Steroid ester	8.00	Antibacterial, antioxidant [33]
8	20.27	Lanosterol	$C_{30}H_{50}O$	426.7	Steroid	1.97	Antioxidant [34], cytoprotective, neuroprotective, anti-inflammatory [35]
9	23.27	Phenanthrene, 7-ethenyl-1,2,3,4,4a,4b,5,6,7,8,8a,9-dodecahydro-1,1,4b,7-tetramethyl-	$C_{20}H_{32}$	272.5	Terpenoid	0.58	Antioxidant, antibacterial [36]
10	25.21	D:C-Friedours-7-en-3-one	$C_{30}H_{48}O$	424.7	Terpenoid	1.93	Antioxidant, anti-inflammatory, antibacterial [37]
11	25.33	9,19-Cycloergost-24(28)-en-3-ol, 4,14-dimethyl-, acetate, (3β,4α,5α)-	$C_{32}H_{52}O_2$	468.75	Steroid ester	0.66	Anti-inflammatory, antibacterial [32]
12	25.49	2′-Hydroxy-3,4,4′,6′-tetramethoxychalcone	$C_{19}H_{20}O_6$	344.4	Flavonoid	1.93	Antioxidant, antibacterial, antidiabetic [38]
13	25.63	A′-Neogammacer-22(29)-en-3-one	$C_{30}H_{48}O$	424.7	Terpenoid	0.47	Antibacterial, antioxidant [39]
14	26.22	Cedran-diol, 8S,14-	$C_{15}H_{26}O_2$	238.37	Terpenoid	9.22	Anti-inflammatory, anticancer [40]-
15	26.64	Taraxasterol	$C_{30}H_{50}O$	426.7	Terpenoid	19.00	Antidiabetic [31], anti-inflammatory, analgesic [41]
16	26.93	Hop-22(29)-en-3.beta.-ol	$C_{30}H_{50}O$	426.7	Terpenoid	12.13	Antibacterial, antioxidant [42]
17	27.18	Lupeol	$C_{30}H_{50}O$	426.7	Terpenoid	10.40	Antimicrobial, antioxidant, anticancer, anti-inflammatory [43]
18	27.55	Urs-12-en-3-ol, acetate, (3beta)-	$C_{32}H_{52}O_2$	468.8	Ester	3.31	Antioxidant, antimicrobial, anticancer [44]
19	28.21	Friedelan-3-one	$C_{30}H_{50}O$	426.7	Terpenoid	0.79	Antioxidant, antidiabetic, antimicrobial [45]
20	29.40	2,4,6-Cycloheptatrien-1-one, 3-hydroxy-	$C_7H_6O_2$	300.31	Alcohol	0.38	-
21	30.69	Lup-20(29)-en-3-ol, acetate, (3β)-	$C_{32}H_{52}O_2$	468.8	Terpenoid	2.38	Anti-inflammatory, analgesic, antibacterial [32], tyrosinase-inhibitory effect [46]
22	31.12	Olean-18-en-28-oic acid, 3-oxo-, methyl ester	$C_{31}H_{48}O_3$	468.7	Fatty acid ester	1.09	Antimicrobial, antioxidant [47]

RT; Retention time in minutes.

Table 3. Phytochemical profiling of *n*-hexane (PCHF) fraction of *P. candollei* through GCMS analysis.

Sr.no.	RT	Tentative Identification of Compounds	Molecular Formula	Molecular Weight	Chemical Class	Area %	Reported Activities from Literature
1	10.28	Methyl iso-eugenol 2	$C_{11}H_{14}O_2$	178.22	Phenolic	0.55	Antibacterial, antioxidant. [48]
2	10.94	Benzene, 1,2,3-trimethoxy-5-(2-propenyl)-	$C_{12}H_{16}O_3$	208.25	Phenolic	0.06	Anti-inflammatory, antioxidant [49] antifungal, antimicrobial [50]
3	11.15	Bicyclo[6.3.0]undec-1(8)-en-3-ol, 2,2,5,5-tetramethyl-	$C_{15}H_{26}O$	222.19	Terpenoid	0.17	Cytotoxic, antiplasmodial, antiviral, anti-inflammatory [51]
4	11.55	3-Hydroxy-4-methoxycinnamic acid	$C_{10}H_{10}O_4$	194.18	Phenolic	0.28	Antioxidant, anti-inflammatory [52]
5	11.80	9-Anthracenecarboxylic acid	$C_{15}H_{10}O_2$	222.24	Aromatic carboxylic acid	0.24	Antimicrobial, antifungal [53], antibacterial, anticancer [54]
6	11.97	Diepi-.alpha.-cedrene epoxide	$C_{15}H_{24}O$	220.35	Terpenoid	0.06	Cytotoxic, antibacterial [54]
7	12.08	Isoelemicin	$C_{12}H_{16}O_3$	208.25	Phenolic	3.35	Antimicrobial [55]
8	12.20	1,3-Benzodioxole, 4,5-dimethoxy-7-(2-propenyl)-	$C_{12}H_{14}O_4$	222.24	Phenolic	0.14	Antimicrobial, antioxidant, anticancer [56]
9	12.43	Asarone	$C_{12}H_{16}O_3$	208.25	Phenolic	6.30	Hypoglycemic, antimicrobial, anti-Alzheimer's disease, anticonvulsive, antiepileptic and antioxidant properties [57,58]
10	12.87	Apiol	$C_{12}H_{14}O_4$	222.23	Phenolic	3.97	Antioxidant, antimicrobial [59]
11	13.07	2H-1-Benzopyran-2-one, 7-methoxy-	$C_{10}H_8O_3$	176.17	Coumarin	0.12	Antioxidant, analgesic, anticoagulant, anti-inflammatory, antimicrobial [60].
12	13.14	Aspidinol	$C_{13}H_{18}O_4$	238.28	Phenolic	0.59	Antibacterial [61]
13	13.22	Tetradecanoic acid	$C_{14}H_{28}O_2$	228.37	Fatty acid	0.14	-
14	13.88	Phenol,2-[[(4-methylphenyl)imino]methyl]-	$C_{14}H_{13}NO$	211.26	Phenolic	0.15	–
15	13.97	2,4,6-Trimethoxyacetophenone	$C_{11}H_{14}O_4$	210.23	Phenolic	0.14	Antibacterial and synergistic effect with antibiotics [62]
16	14.06	1-Methoxy-3-(2-hydroxyethyl)nonane	$C_{12}H_{26}O_2$	202.33	Alcohol	0.31	Antifungal, antioxidant [63]
17	14.34	9-Octadecyne	$C_{18}H_{34}$	250.5	Alkyne	0.34	Larvicidal, antioxidant [64]
18	14.42	Tricyclo[7.2.0.0(2,6)]undecan-5-ol	$C_{15}H_{26}O$	222.37	Terpenoid	0.18	-
19	14.79	7,10,13-Hexadecatrienoic acid, methyl ester	$C_{17}H_{28}O_2$	264.40	Fatty acid ester	0.17	Antioxidant, anti-inflammatory, antimicrobial [65]
20	15.03	3,4-dihydrocoumarin	$C_9H_8O_2$	148.16	Coumarin	0.12	Anticoagulant, antifungal, anticancer, antibacterial [66]
21	15.08	Hexadecanoic acid, methyl ester	$C_{17}H_{34}O_2$	270.45	Fatty acid ester	0.20	Antibacterial, antifungal [67]
22	15.48	11,14,17-Eicosatrienoic acid, methyl ester	$C_{20}H_{34}O_2$	306.5	Fatty acid	1.52	Anti-microbial, anti-cancer, anti-hair fall, CNS depressant, analgesic, anti-inflammatory, antipyretic, anti-arthritic, anti-coronary [68]
23	15.92	n-Hexadecanoic acid	$C_{16}H_{32}O_2$	256.42	Fatty acid	9.40	Antioxidant, antibacterial, anti-inflammatory [69]
24	17.16	7H-Furo[3,2-g][1]benzopyran-7-one	$C_{11}H_6O_5$	218.16	Coumarin	2.42	-
25	17.32	9,12-Octadecadienoic acid, methyl ester	$C_{19}H_{34}O_2$	294.5	Fatty acid ester	0.84	Antioxidant, anti-inflammatory, antimicrobial [65]
26	17.42	9,12,15-Octadecatrienoic acid, methyl ester	$C_{19}H_{32}O_2$	278.4	Fatty acid ester	0.92	Antimicrobial [70] Antioxidant, anti-inflammatory, antimicrobial [65]
27	17.58	Phytol	$C_{20}H_{40}O$	296.5	Terpenoid	1.14	Antioxidant, anticancer [71]
28	18.05	Z,Z-10,12-Hexadecadien-1-ol acetate	$C_{18}H_{32}O_2$	280.4	Ester of fatty alcohol	2.55	Antimicrobial [72]
29	18.15	9,12-Octadecadienoic acid (Z,Z)	$C_{18}H_{32}O_2$	280.4	Fatty acid	3.55	Antibacterial, antifungal, anti-inflammatory, antineoplastic [73]
30	18.69	2-Chloroethyl linoleate	$C_{20}H_{35}ClO_2$	342.9	Fatty acid ester	39.69	Cytotoxic, antioxidant, antimicrobial [74]

Table 3. *Cont.*

Sr.no.	RT	Tentative Identification of Compounds	Molecular Formula	Molecular Weight	Chemical Class	Area %	Reported Activities from Literature
31	19.31	Flavone	$C_{15}H_{10}O_2$	222.24	Flavonoid	1.26	Antibacterial, antiviral, antifungal, antioxidant, anti-inflammatory [75]
32	19.59	Pimpinellin	$C_{13}H_{10}O_5$	246.21	Furocoumarin	2.57	Strong antibacterial [76]
34	22.05	Phenol, 2,2′-methylenebis[6-(1,1-dimethylethyl)-4-methyl-	$C_{23}H_{32}O_2$	340.49	Phenolic	0.16	α-amylase inhibitor [77]
35	22.52	1,3,14,16-Nonadecatetraene	$C_{19}H_{32}$	260.45	Alkene	0.23	-
36	22.77	(R)-(-)-14-Methyl-8-hexadecyn-1-ol	$C_{17}H_{32}O$	252.4	Alcohol	0.21	-
37	22.92	1,5,9,13-Tetradecatetraene	$C_{14}H_{22}$	190.32	Alkene	0.14	-
38	23.04	9-Tricosene, (Z)-	$C_{23}H_{46}$	322.6	Alkene	0.35	-
39	23.55	4-(3-Methyl-2-oxobutoxy)-7H-furo[3,2-g][1]benzopyran-7-one	$C_{16}H_{14}O_5$	286.28	Coumarin	1.20	Antibacterial [76]
40	24.57	7H-Furo(3,2-g)(1)benzopyran-7-one, 4,9-dihydroxy-	$C_{11}H_6O_5$	218.16	Coumarin	1.47	Antibacterial, antiacetyl, and butyrylcholinesterase [76]
41	24.84	6-Acetylchrysene	$C_{19}H_{14}$	242.3	Phenanthrene	1.27	-
42	25.76	13-Tetradecen-1-ol acetate	$C_{16}H_{30}O_2$	254.41	Ester fatty alcohol	0.17	Antibacterial, antioxidant [78]
43	27.22	3,4-Dimethoxycinnamic acid	$C_{11}H_{12}O_4$	208.21	*Cinnamic acid*	0.09	Neuroprotactive, antioxidant, anticancer [79]
44	28.40	N-hydroxy-N′-[2-(trifluoromethyl)phenyl]pyridine-3-carboximidamide	$C_{13}H_{10}F_3N_3O$	281.23	Pyridine derivative	0.49	Antioxidant, anti-inflammatory, antimicrobial [80]
45	28.69	Stigmastan-6,22-dien, 3,5-dedihydro-	$C_{29}H_{46}$	394.7	Steroid	0.18	Antifungal [81]
46	29.66	Stigmastane-3,6-dione	$C_{29}H_{48}O_2$	428.7	Steroid	0.23	-
47	30.53	Stigmasta-5,22-dien-3-ol, acetate	$C_{31}H_{50}O_2$	454.7	Steroid	1.16	Antimicrobial, antioxidant [82]
48	30.88	Ergosta-4,6,22-trien-3.beta.-ol	$C_{28}H_{44}O$	396.6	Steroid	0.29	-
49	31.51	Clionasterol acetate	$C_{31}H_{52}O_2$	456.7	Steroid	2.51	-
50	31.75	3β-acetoxy-pregna-5,16-dien-20-one	$C_{23}H_{32}O_3$	298.5	Steroid	0.27	Anti-inflammatory, antibacterial [83]
51	32.14	Vitamin E	$C_{29}H_{50}O_2$	430.7	Chromanol derivative	0.46	Antioxidant, anticancer, anti-inflammatory [84].

RT; retention time in minutes.

2.2. Biological Activities

2.2.1. Antioxidant Activities

The antioxidant quality of a herbal drug or any plant represents its value as a food/drug supplement, which may prevent the oxidative damage of biological molecules in the human body [85]. As phenolic contents have strong antioxidant properties [86], so they are important components of natural products, nutraceuticals, pharmaceuticals, and functional foods [87], and the formulations rich in phenolic compounds reduce the risk of disease by decreasing oxidative stress [88] through a mechanism where the hydroxy group on the aromatic ring acts as an electron donor and is directly involved in the quenching of free radicals [89].

Data of our experiment revealed a good quantity of phenolic and flavonoid contents (Table 1), so we investigated the antioxidant potential of *P. candollei* from its methanolic extract and different fractions. Results of DPPH and ABTS assays showed (Figure 3 and Table 4) that the methanolic extract and polar fractions have more radical quenching effects compared to a non-polar fraction. The antioxidant values of the PCME were highest (165.06 ± 3.03 and 253.09 ± 2.11 mg TE/g dry weight for DPPH and ABTS assays respectively), while the antioxidant property of *n*-butanol fraction (PCBF) was found to be 101.95 ± 1.04 and 181.29 ± 1.91 mg TE/ g dry weight for DPPH and ABTS assays, respectively. The *n*-hexane fraction (PCHF) had the lowest value (DPPH = 58.09 ± 1.54, and ABTS = 113.41 ± 1.06). The higher values of antioxidant activities of methanolic extract also correlated with the antioxidant compounds (1, 2, 4, 8, 9, 10, 11, 16, 17, 19, 23, 25, 26, 27, 30, 31, 42, 43, 44, 47, 51, from Table 2), which constituted 67.86% of non-polar compounds of the total extract (Table 2); however, the highest phenolic contents also contributed to the higher antioxidant activities of PCME [90]. Moreover, the radical scavenging effect of PCHF was also significant, which may be due to the presence of many compounds (1, 2, 4, 5, 8, 9, 10, 11, 17, 19, 22, 26, 27, 29, 30, 42, 43, 44, 47, and 51, from Table 3) possessing antioxidant activities, as proven by many pieces of researches (Table 3). Those bioactive natural compounds (with antioxidant properties) contributed to 63.46% of the total PCHF fraction. So it depicted the significant antioxidant results in the PCHF fraction. Our findings were also consistent with the results of a previous study in which radical scavenging effects decreased in fractions of lesser polarity [91], this might be due to the synergistic effect of phenolic compounds in more polar fractions and they were collectively present in PCME.

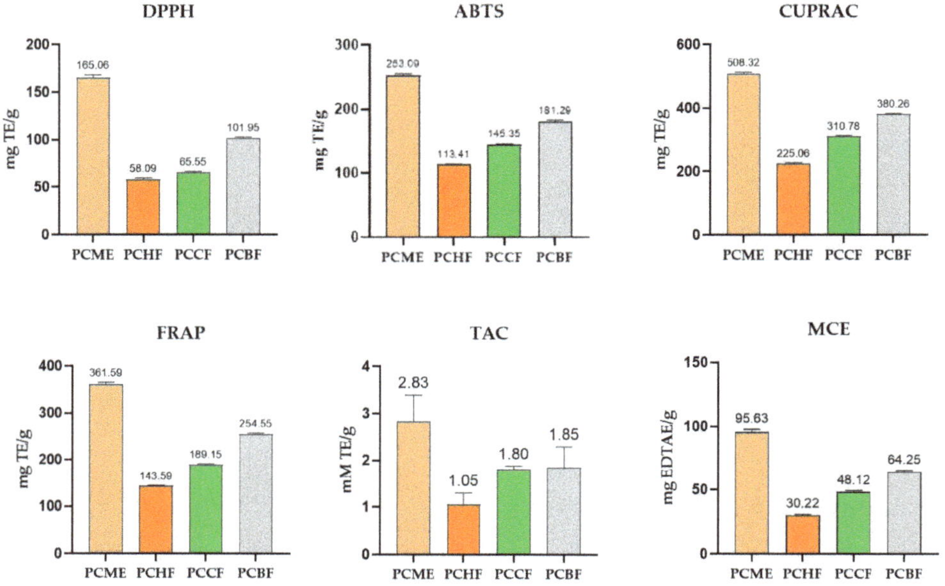

Figure 3. Antioxidant activities (mg and mM equivalents of standard antioxidants/g extract or fraction) of *P. candollei*.

Table 4. IC_{50} values of antioxidant activities of methanolic extract and different fractions of *P. candollei* (sample conc. 1 mg/mL).

Sample Codes	Radical Scavenging Assay		Reducing Power Assay		Reducing/Metal-Chelating Assay	
	DPPH (IC_{50} mg/mL)	ABTS (IC_{50} mg/mL)	CUPRAC (EC_{50} mg/mL)	FRAP (EC_{50} mg/mL)	TAC (EC_{50} mM/mL)	MCE (IC_{50} mg/mL)
PCME	3.64 ± 0.73^{a}	2.49 ± 0.84^{a}	1.20 ± 0.45^{a}	1.67 ± 0.68^{a}	0.45 ± 0.06^{a}	2.16 ± 0.48^{a}
PCHF	10.34 ± 1.41^{d}	5.31 ± 1.04^{d}	2.67 ± 0.80^{d}	4.20 ± 1.06^{d}	0.24 ± 0.02^{d}	7.00 ± 0.86^{d}
PCCF	9.23 ± 1.95^{c}	4.14 ± 1.21^{c}	1.93 ± 0.79^{c}	3.17 ± 1.10^{c}	0.29 ± 0.08^{c}	4.38 ± 0.62^{c}
PCBF	5.90 ± 1.06^{b}	3.32 ± 0.95^{b}	1.59 ± 0.16^{b}	2.36 ± 0.75^{b}	0.30 ± 0.04^{b}	3.28 ± 0.35^{b}

Values were taken in triplicates ($n = 3$) and reported as mean ± SD PCME: *P. candollei* methanolic extract, PCHF: *P. candollei n*-hexane fraction, PCCF: *P. candollei* chloroform fraction, PCBF: *P. candollei n*-butanol fraction, DPPH: 2,2-diphenyl-1-picrylhydrazyl assay, ABTS: 2,2′-azino-bis(3-ethylbenzothiazoline-6-sulfonic acid) assay, CUPRAC: cupric reducing antioxidant capacity, FRAP: Ferric reducing antioxidant power, TAC: Phosphomolybdenum assay, MCE: metal-chelating effect. [a, b, c, d] Letters in one column indicate significant differences in the activities of tested extract and fractions ($p < 0.05$).

The antioxidant effect was also determined by using reducing power assays like CUPRAC and FRAP. Again, the PCME showed the highest antioxidant activity (for CUPRAC; 508.32 ± 4.05, FRAP; and 361.59 = 3.52 mg TE/g dry weight, respectively), while the activity decreased with polarity and the following pattern in the results was observed as PCBF > PCCF > PCHF. Similarly, the antioxidant activity by the FRAP method was also strongest in PCME and weakest in PCHF extract. The high reducing power may also be attributed to the presence of a higher quantity of antioxidant polyphenols and other bioactive compounds in more polar fractions [90] of *P. candollei*. However, the identified antioxidant compounds presented in Table 3 might be contributing to the significant reducing power of PCHF, while the highest antioxidant activity of PCME may be due to the synergetic effects of polyphenols in addition to the identified bioactive secondary metabolites presented in Table 2.

Total antioxidant capacity was measured by phosphomolybdenum assay and it was found that PCME has the highest antioxidant capacity (2.83 ± 0.56 mmol TE/g dry extract). However, fractions also showed significant TAC; *n*-butanol (PCBF), chloroform (PCCF), and *n*-hexane (PCHF), when compared to each other; they were: 1.85 ± 0.45, 1.80 ± 0.08, and 1.50 ± 0.25 respectively. Although, the results of PCHF showed lesser total antioxidant capacity, they were not far behind the other fractions in this regard. In addition, evidence is also present in favor of mid to non-polar volatile organic compounds, which have also shown therapeutic benefits of antioxidants relevant to traditional medicine and pharmaceuticals [92]. The findings of our study are further supported by the high number of polyphenolic compounds (Table 1) and antioxidant compounds identified by GCMS (Tables 2 and 3).

Chelation therapy is gaining importance with the increase in pollution with heavy metals. Herbal drugs have phytochemicals to detoxify those metallic ions [93]. Although iron is essential for life it may produce free radicals, leading to oxidative damage, and its chelation prevents oxidation [94]. The metal-chelating antioxidant effect of the PCME was highest (95.63 ± 2.18 mg EDTAE/g dry weight), and the metal-chelating effect of fractions was found as PCHF = 30.22 ± 0.51, PCCF = 81.22 ± 1.01, and PCBF = 84.22 ± 1.01 mg EDTAE/g of dry weight of the fraction. The metal-chelating activity of polar fractions might be due to the presence of polyphenolic compounds [95]. The behavior of the PCHF fraction as a metal-chelating agent may be due to the presence of terpenoids [96] and other bioactive compounds (identified by GCMS and shown in Table 3), and there may be a synergetic antioxidant effect among the polyphenols (Table 1) and terpenoids, along with other bioactive natural compounds (Table 2), identified from GCMS in this study. Human health can be maintained by exploring herbs and developing formulations rich in phenolic phytochemicals, showing antioxidant properties [97,98], and *P. candollei* revealed bioactive phytochemicals and antioxidant activities in the current investigation, which provide a rational for the health benefits of this plant.

2.2.2. In Vitro Enzyme Inhibition Activities

Enzyme inhibition may be an important technique in therapeutics, as various enzymes have a vital role in the physiology of human organs, and they are also involved in the etiology of different diseases. The bioactivity of tested enzymes in this study was linked with different diseases. Enzymes like α-amylase and α-glucosidase are associated with type 2 diabetes, and tyrosinase is involved in skin hyperpigmentation [99]. We presented the results of inhibition studies for these enzymes in Table 5.

Table 5. Enzyme-inhibitory activities of methanolic extract and different fractions of *P. candollei* (sample conc. 1 mg/mL).

Sample Codes	Tyrosinase (mg KAE/g Dried wt.)	α-Amylase (mmol ACAE/g Dried wt.)	α-Glucosidase (mmol ACAE/g Dried wt.)
PCME	112.29 ± 2.79	0.93 ± 0.07	1.88 ± 0.15
PCHF	52.61 ± 1.26	0.53 ± 0.08	0.46 ± 0.01
PCCF	82.91 ± 1.79	0.69 ± 0.06	0.78 ± 0.04
PCBF	90.15 ± 2.10	0.81 ± 0.05	0.95 ± 0.09

Values were taken in triplicates and reported as mean ± SD. PCME, *P. candollei* methanolic extract; PCHF, *P. candollei* -hexane fraction; PCCF, *P. candollei* chloroform fraction; PCBF, *P. candollei n*-butanol fraction; KAE, kojic acid equivalent; ACAE, acarbose equivalent.

Plant-derived antioxidants and tyrosinase inhibitors have gained prime importance in the cosmaceutical industry and natural products, as they have functional ingredients protecting skin from pigmentation, aging, and other skin disorders [100]. Various side effects of chemically synthetic skin remedies, like hydroquinone, made herbal skin remedies of prime importance [101]. In the current study, we already reported the good antioxidant potential of *P. candollei* that is reflectde in the tyrosinase inhibitory effect also. The highest tyrosinase inhibitory activity was revealed in methanolic extract (112.29 ± 2.79 mg KAE/g of dry weight), and tyrosinase inhibition of fractions was observed in the range of 52.61 ± 1.26 to 90.15 ± 2.10 mg KAE/g of dry weight (Table 5). A correlation between phenolic compounds, antioxidant effects, and tyrosinase inhibition was found in an Indian study [102], which also substantiates the findings of our current work. Herein, we observed that the PCME has the highest tyrosinase inhibition followed by the PCBF, PCCF, and PCHF, respectively, which may be due to the presence of different antioxidant and anti-inflammatory compounds present in high quantities (Table 2) in PCME and the synergetic effect of phenolic and flavonoid groups of compounds with varying biological potential [103]. Lup-20(29)-en-3-ol, acetate, (3β)- has been reported for tyrosinase inhibition [46], and we identified this compound in significant quantity (2.38%) as presented in Table 2, which further strengthened our findings of tyrosinase inhibition by PCME. Moreover, the *n*-hexane fraction substantially suppressed the activity of the tyrosinase enzyme due, to the identified antioxidant phytochemicals (Table 3), and evidence from the literature also favors the effect of antioxidants as melanin inhibitors [104]. A correlation was also found between DPPH antioxidation and the tyrosinase-inhibitory effect of plants. In addition to the radical scavenging mechanism, the identified compounds with anti-inflammatory (Tables 2 and 3) effects may also be the contributors to skin-whitening effects that are much desired in Asian countries [105]. So, our data favor the *P. candollei* as a potential source to provide the bioactive and antioxidant ingredients for cosmeceuticals or the pharmaceutical industry.

Currently, modernization, aging, and lifestyle changes are the main factors, which increase oxidative stress and lead to the development of metabolic disorders like type 2 diabetes mellitus (characterized by hyperglycemia) [106], ultimately resulting in impaired quality of life [107]. The digestive enzymes (α-amylase and α-glucosidase) are responsible for converting the ingested carbohydrates into glucose and lead to the systemic hyperglycemia. So, inhibitors of these enzymes are the antihyperglycemic agents available among the best treatment options for hyperglycemia, which control carbohydrate digestion in the intestine. However, these drugs are associated with many unwanted effects. So, there is a need for a more selective and safer agent for the induction of satisfactory therapeutic effects [108]. In an attempt to find that effect, we evaluated the fractions of *P. candollei* against α-glucosidase and α-amylase activity. Herein, the PCME showed good results for the inhibition of α-amylase and α-glucosidase (0.93 ± 0.13 mmol ACAE/g) as shown in Table 5. Among the studied fractions, PCBF had the highest activity of α-amylase inhibition (0.81 ± 0.05 mmol ACAE/g), while the other two fractions have lesser but significant activities (PCCF = 0.69 ± 0.06 and PCHF = 0.53 ± 0.08 mmol ACAE/g) compared with PCBF. For α-glucosidase-inhibitory activity, again the PCME was more effective with an inhibition value of 1.88 ± 0.15 mmol ACAE/g, respectively, while the results of other fractions showed that PCBF was more effective (0.95 ± 0.09 mmol ACAE/g), and a decrease in activity was observed with the decrease in polarity of fraction (PCCF = 0.78 ± 0.04 > 0.46 ± 0.01 mmol ACAE/g). Our finding was consistent with the reported evidence in literature in which an intimate positive correlation was found between TPC and TFC and in vitro antioxidant activity and various enzyme inhibition assays [109]. Moreover, some compounds were identified with antioxidant and anti-inflammatory effects by GCMS analysis of PCME and PCHF, which also contribute to the antidiabetic activity of the plant [110]. Evidence from the literature proved that many plant extracts and fractions show antidiabetic effects, and some plants may show antidiabetic activity even more so than the standard drug (acarbose) [111]. Although PCME showed moderate inhibition and fractions of *P. candollei* possess mild inhibition of digestive enzymes (α-glucosidase and α-amylase) as compared to equivalents of acarbose, however, our results correlate with the study on similar specie *P. benthamii* [112]. Higher antidiabetic activity of PCME was also attributed to the identified compounds A-Neooleana-3(5),12-diene, 2′-Hydroxy-3,4,4′,6′-tetra methoxy chalcone, Friedelan-3-one, and Taraxasterol, which had been reported for antidiabetic property, and collectively they constituted 31.64% of non-polar compounds of the PCME extract (Table 2, with references). These shreds of evidence widened the scope of our work on this plant species for inhibiting the activity of tested enzymes to tackle hyperglycemia and provide health benefits.

Overall, we found significant results from enzyme-inhibition assays, which establish the effectiveness and health benefits of the methanolic extract and different fractions of *P. candollei* as a potential candidate for the natural products, nutraceuticals, and pharmaceutical industry, and it may be due to the presence of a good quantity of compounds with antioxidant and anti-inflammatory property (identified molecules), and their synergic effects may also be contributing to different biological activities.

2.2.3. Thrombolytic Activity

The current pandemic of COVID-19 has caused a drastic effect on the healthcare system and global economy, and the vaccination of COVID-19, although reducing the death rate, may cause thrombosis (which is also a complication of many other infections) [9,113]. Thrombosis has gained attention as a deadly complication of respiratory viral infections also (like influenza and coronavirus) [8]. Thrombosis may lead to myocardial infarction, acute ischemic stroke, venous thromboembolism, and other cardiovascular complications [114]. Experts often recommended thromboproprophylactic medications to combat thrombotic disorders of various etiologies [115], and remarkable attempts have been done in the discovery and development of safer remedies from natural constituents, so various plant sources have been investigated for antiplatelet, anticoagulant, antithrombotic, and thrombolytic effects [116]. That is why we also investigated the clot-dissolving property of our plant.

According to the results of our study, PCME was found to have maximum thrombolytic activity in a range of 55.38 ± 1.51 to 59.85 ± 1.51% of clot lysis, which was declared good when compared with streptokinase activity (78.5 ± 1.53 to 82.34 ± 1.25% clot lysis). The following order of thrombolytic activity was observed in the fractions of *P. candollei*, PCBF > PCCF > PCHF. However, results from all fractions seemed to be significant as shown in Table 6, and the results were in agreement with previous findings reported in the literature, in which the researchers found that more polar extract showed the highest thrombolytic activity in comparison with less polar fractions [117]. It was also proven by various studies that flavonoids, tannins, alkaloids, and saponins from polar organic extracts may be responsible for the thrombolytic activity of plant extracts, and developing pharmaceutical formulations or consumption of such plants as food can prevent coronary events and stroke due to thrombolytic properties [118]. So, our plant may be a potential source for developing a formulation to reduce pill burden for the patients with high thrombotic risk if further investigated by in-vivo studies.

Table 6. Thrombolytic activity (% clot lysis) of fractions of *P. candollei* from different blood samples (subject A–E).

Sample Codes	Subject A	Subject B	Subject C	Subject D	Subject E
PCME	55.38 ± 1.51	58.16 ± 1.9	55.45 ± 1.18	58.65 ± 1.25	59.85 ± 1.51
PCHF	40.18 ± 1.80	43.1 ± 1.69	42.51 ± 0.98	43.8 ± 0.82	42.63 ± 1.35
PCCF	41.54 ± 0.95	48.15 ± 1.41	47.15 ± 1.11	50.14 ± 1.61	47.85 ± 1.80
PCBF	52.15 ± 0.68	57.25 ± 0.94	55.10 ± 1.12	56.95 ± 1.70	57.15 ± 1.10
Streptokinase	78.5 ± 1.53	80.14 ± 0.91	81.43 ± 1.39	82.34 ± 1.25	79.12 ± 2.3

Values were taken in triplicates and reported as mean ± SD. PCME; *P. candollei* methanolic extract, PCHF; *P. candollei* n-hexane fraction, PCCF; *P. candollei* chloroform fraction, PCBF; *P. candollei* n-butanol fraction.

2.2.4. Antibacterial Activity

Traditional antibiotics become less effective against pathogenic bacteria due to the potential of these pathogens to develop drug resistance with extensive use of antibacterial drugs, which leads to global health threats. So, there is a need to develop new antibacterials to combat the growth of evolving bacteria [119]. Plants are the factories of nature to synthesize the secondary metabolites of varying functions, including the defense against microbial infection and parasitic infestation [120]. Plant-derived molecules may prove to be effective as antibacterial agents or may act synergistically to enhance the efficacy of older antibiotics, consequently restoring their clinical use [121], and the triterpenoids identified by GCMS may also be contributing in the antibacterial activity of methanolic extract [122]. In search of antibacterial effects, we found a significant activity of different fractions as shown in Table 7. The significant activity of PCME and PCHF can be correlated with the presence of various metabolites identified by GCMS, having antibacterial, antifungal, antiviral, and antioxidant activities (reported in the literature) as shown in Tables 2 and 3. It was found that 56.41% of phytochemicals were identified from PCME with antimicrobial properties and in PCHF these compounds were 82.95%, as presented in Tables 2 and 3. Our findings were close to the observation of an analysis of *P. amabile*, which describes the activity of essential oil, against *Bacillus subtilis*, and also identified some bioactive molecules by GCMS of oil. However, they also found that polar extracts were more active compared with non-polar extract [26], which correlates with our finding that PCME shows slightly greater activity, which may be due to the synergistic effect of phenolic compounds in the PCME, in addition to the identified antibacterial and antimicrobial natural compounds of the studied plant.

Table 7. Antibacterial activity of methanolic extract and different fractions of *P. candollei*.

Strain Name	Zone of Inhibition (mm) of Standard (Co-Amoxiclav) (Concentration = 1 mg/mL)	Concentration (mg/mL)	Zone of Inhibition of PCME Extract (mm)	Zone of Inhibition of PCHF Fraction (mm)	Zone of Inhibition of PCCF Extract (mm)	Zone of Inhibition of PCBF Extract (mm)
Bacillus subtilis	23	10	7	-	-	7
		20	13	10	12	12
		40	18	16	16.5	18
Micrococcus luteus	20	10	7	6	6	6
		20	13	10	11	13
		40	17	16	17	17
Staphylococcus epidermidis	24	10	5	-	-	-
		20	10	8	8	10
		40	12	10	13	15
Bacillus pumilus	22	10	6	-	-	-
		20	11	7	9	9
		40	15	12	14	14.5
Staphylococcus aureus	23	10	7	-	6	6
		20	13	10	11	12
		40	15.5	12	12.5	14
Escherichia coli	25	10	-	-	-	-
		20	8	-	7	7.5
		40	10	6	8	9
Bordetella bronchiseptica	26	10	-	-	-	-
		20	8	-	6	8
		40	9	10	11	11.5
Pseudomonas aeruginosa	18	10	-	-	-	-
		20	-	-	-	-
		40	7	-	-	6

PCME, *P. candollei* methanolic extract; PCHF, *P. candollei* *n*-hexane fraction; PCCF, *P. candollei* chloroform fraction; PCBF, *P. candollei* *n*-butanol fraction.

Although PCHF showed a higher percentage of identified antibacterial and antimicrobial compounds (82.95%), its activity was weaker than polar fractions due to less quantity of polyphenols (Table 1), which may potentiate the antibacterial activity if present. Later, in 2014, some phenolic compounds also isolated from *P. amabile*, which showed antibacterial activity, further attributed the greater antimicrobial activity of more polar fraction due to the presence of phenolic compounds [123,124], and phenolic compounds were already detected in higher quantity from PCME, as shown in Table 1. The highest activity was observed against *Bacillus subtilis* and the lowest against *Pseudomonas aeruginosa*, and the antibacterial effect was also increased with the increase in the concentration of tested samples. The first five strains of Table 7 are Gram-positive, which showed greater inhibition by PCME, and polar fractions (PCBF and PCCF). However, the last three strains (*Escherichia coli, Bor-*

detella bronchiseptica, and *Pseudomonas aeruginosa*) are Gram-negative, which showed the least activities. Herein, the lesser zone of inhibition may be due to the high resistance of Gram-negative bacteria due to the cell membrane made of lipopolysaccharide, which is impermeable to the non-polar (lipophilic) metabolites. Peptidoglycan is the external layer in Gram-positive strains of bacteria, which is permeable to lipophilic metabolites [125]. These findings were required to compare the antibacterial activity of the plant of study with its bioactive phytochemicals and strong antioxidants potential; however, more investigation and purification of phenolic components may lead to the development of natural antibacterial agents for functional foods, nutraceuticals, and pharmaceuticals, to improve the shelf life and to extend health benefits in those systems.

2.3. Molecular Docking Studies

Computational chemistry provided the opportunity to the researchers to investigate the interactions between the residues of the receptor protein and ligands. The docking technique is loaded with the tools that facilitate the understanding of interaction of a presumed rigid active site with the ligand molecule and helps in drug discovery [126]. Molecular docking studies were performed for tyrosinase, α-amylase, and α-glucosidase receptors. Binding affinities and binding interactions of compounds (identified from GCMS) with major peak area were estimated for their interaction with the active sites of tested enzymes.

2.3.1. Molecular Docking against Tyrosinase Enzyme

Ten identified compounds were selected from methanolic extract (Table 2) and docked against the tyrosinase enzyme. All the docked compounds showed lower binding energies (−7.2 to −8.6 kcal/mol) with tyrosinase active site as compared to kojic acid (−6.0 kcal/mol), which predict the higher binding affinity of these ligands as compared to kojic acid (Table 8). Molecular docking of the tyrosinase receptor was also performed by selecting six identified compounds from n-hexane fraction, and results were calculated as binding energies in kcal/mol (Table 8).

Table 8. Binding energies of docked compounds against tyrosinase, α-amylase, and α-glucosidase.

Name of Compound	Binding Energy of Ligand with Tyrosinase (kcal/mol)	Binding Energy of Ligand with α-Amylase (kcal/mol)	Binding Energy of Ligand with α-Glucosidase (kcal/mol)
Taraxasterol	−8.6	−9.5	−8.5
beta-Amyrin	−8.3	−9.1	−8.8
Hopa-22(29)-ene-3alpha-ol	−8.0	−8.9	−9.0
A-Neooleana-3(5),12-dien	−7.9	−10.9	−8.7
Urs-12-en-3-ol, acetate, (3beta)-	−7.7	−8.8	−8.4
Lupeol	−7.4	−8.8	−9.1
Lanosterol	−7.3	−9.7	−8.7
Lup-20(29)-en-3-ol, (3beta)-	−7.3	−8.7	−9.0
3-Epimoretenol	−7.3	−8.9	−8.6
9,19-Cyclolanost-24-en-3-ol, acetate, (3beta)-	−7.2	−7.9	−8.5
Standad	−6.0 [1]	−7.7 [2]	−7.0 [2]

[1] Kojic acid and [2] acarbose.

The following compounds from n-hexane fraction show greater affinities (lower binding energy); 2-chloroethyl linoleate (−6.6 kcal/mol); 9, 12-octadecadienoic acid (Z, Z) (−6.6 kcal/mol); and apiol (−6.1 kcal/mol) as compared to kojic acid (−6.0 kcal/mol). The higher negative value of binding energy showed the greater affinity of the ligand with tyrosinase active sites. For the further validation of the results, the ligands were docked again with the tyrosinase by using Autodock-1.5.6, and the same results were obtained in terms of the binding energy values. This study revealed that the good inhibition of tyrosinase by methanolic extract (as shown by experimental work) may be due to the higher binding affinities of docked ligands compared with kojic acid. Binding interactions of these compounds also revealed the binding forces with residues of tyrosinase active site. Different bonding interactions were involved, including hydrogen bonding, van der Waals interactions, and π bonding interactions, and some diagrammatic presentations are given in Figure 4.

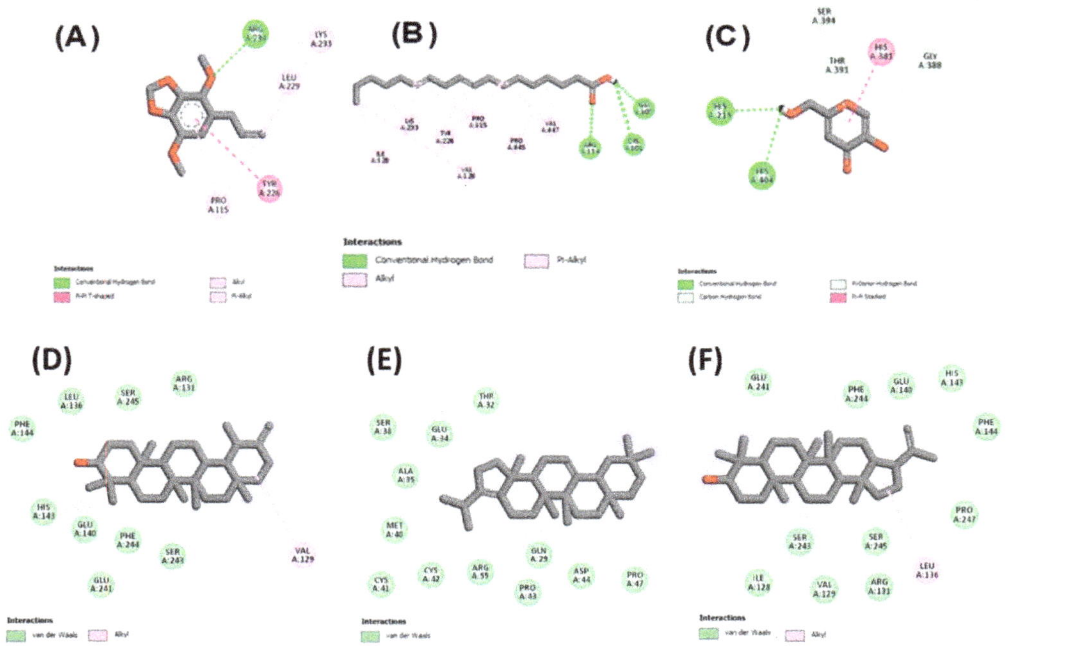

Figure 4. Interactions of tyrosinase active site residues with (**A**) apiol, (**B**) octadecadienoic acid, (**C**) kojic acid, (**D**) taraxasterol, (**E**) a-neooleana-3(5), 12-dien, (**F**) hopa-22(29)-ene-3alpha-ol.

2.3.2. Molecular Docking against α-Amylase, and α-Glucosidase

The highest anti-diabetic activity was observed from experimental work on methanolic extract of our plant. We docked the major identified compounds from GCMS, and it revealed that all the compounds had a greater negative value of binding energy (−7.9 to −10.9 kcal/mol), which depicted higher binding affinities than acarbose (−7.7 kcal/mol) for α-amylase inhibition. Binding energies were presented in Table 8. The docking of ligands from *n*-hexane fraction resulted in lower binding affinities (binding energy; −5.0 to −6.1 kcal/mol) compared to acarbose (−7.7 kcal/mol) for α-amylase inhibition as shown in Table 8. In the case of α-glucosidase, we observed the same pattern as in the case of α-amylase that the selected ligands from methanolic extract (Table 2) showed greater affinities (binding energy; −8.4 to −9.1 kcal/mol) than standard ligand acarbose (binding energy; −7.0 kcal/mol), so the binding energies of docked ligands were lower than the standard drug, which showed higher affinity of docked ligands with active sites of α-glucosidase, and stronger antidiabetic activity was also shown by the methanolic extract in our experimental work. Herein, the compounds selected from *n*-hexane extract showed lesser binding affinities due to their higher binding energies (−5.1 to −5.9 kcal/mol) compared to the standard drug acarbose (−7.0), as shown in Table 8, and similar results were also found by experimental work. The results of molecular docking against the digestive enzymes of hyperglycemic responsibility helped us to understand the mechanism of mild to moderate results of antidiabetic activity from our experimental work for the inhibition of those enzymes. Figure 5 showed the binding interactions of α-amylase with the selected ligands, identified from GCMS, while Figure 6 showed the interaction of selected ligands with α-glucosidase enzyme.

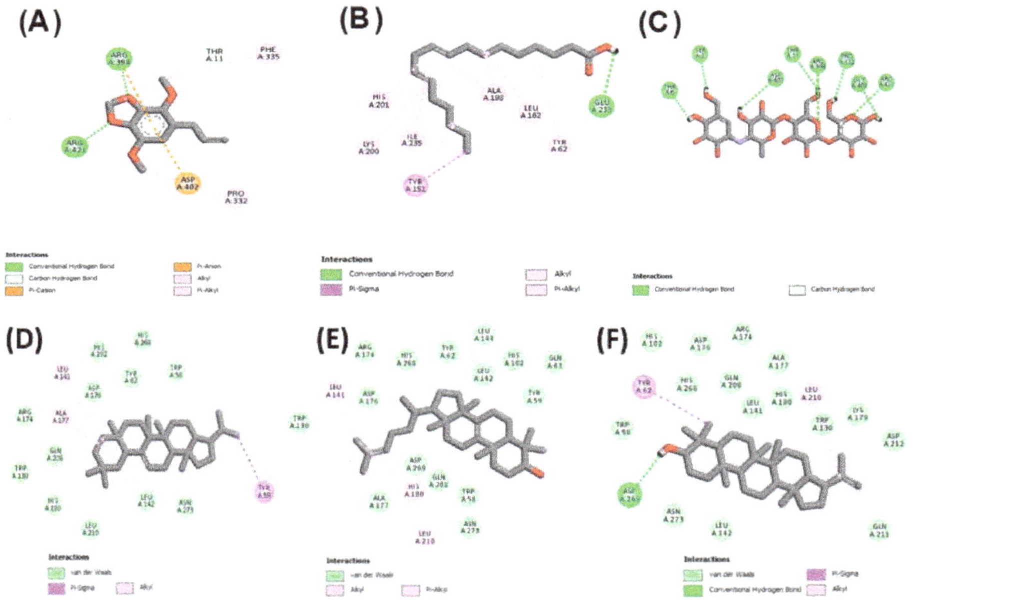

Figure 5. Interactions of residues of α-amylase active site (2D presentation) with (**A**) apiol, (**B**) octadecadienoic acid, (**C**) acarbose, (**D**) a-neooleana-3(5), 12-dien, (**E**) lanosterol, (**F**) 3-epimoretenol.

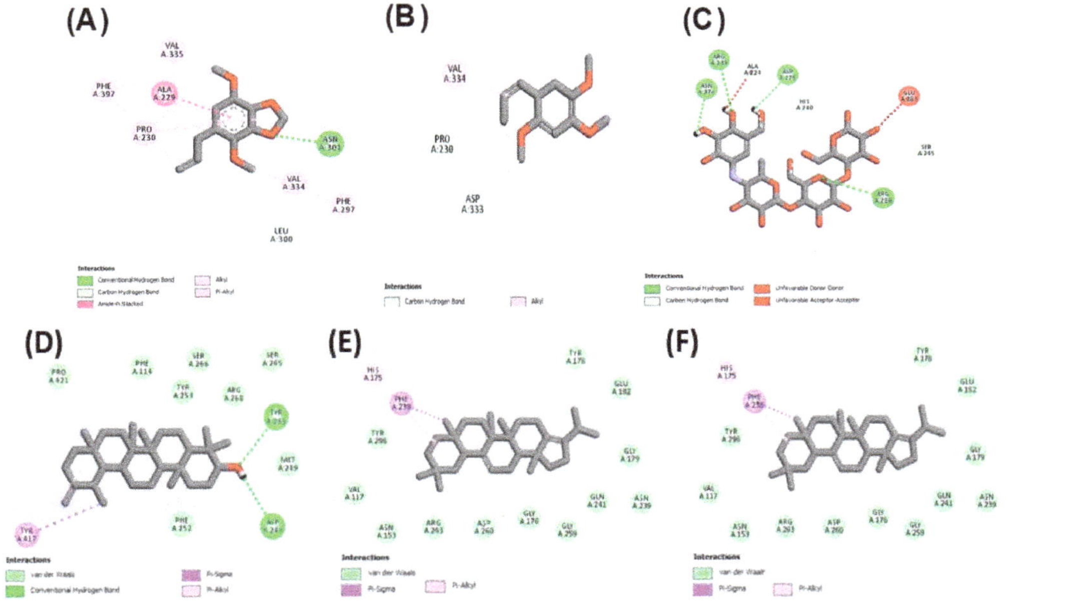

Figure 6. Interactions of residues of α-glucosidase active site (2D presentation) with (**A**) apiol, (**B**) asarone, (**C**) acarbose, (**D**) taraxasterol, (**E**) hopa-22(29)-ene-3alpha-ol, (**F**) a-neooleana-3(5), 12-dien.

3. Materials and Methods

3.1. Plant Collection and Authentication

The aerial parts of *P. candollei* were collected in July 2021 from Astore, Gilgit Baltistan, Pakistan. The plant was authenticated by the Herbarium of Hazara University Mansehra, Pakistan, with collection number 516 and voucher number 1885.

3.2. Extraction and Fractionation

The plant was air-dried in shade and macerated in a hydroalcoholic solvent. The solvent system contained aqueous methanol (80%). This solvent system was employed as it has the property of efficient extraction of phenolic and flavonoid phytochemicals. Powdered plant material (2.5 kg) was macerated in a volume of 8 L solvent. The extract was decanted and marc squeezed. Then the obtained liquid was filtered and dried by a rotary evaporator (Heidolph, Schwabach, Germany) at 42 °C temperature and at reduced pressure. Finally, a semi-solid residue (408 g) was obtained with subsequent drying. Furthermore, fractionation of the extract was done with the help of a separating funnel and using the solvents of increasing polarity (n-hexane, chloroform, and n-butanol). Fractions were further consolidated by employing a rotary evaporator and air-dried to obtain dry material for further analysis [127].

3.3. Determination of Bioactive Phytochemicals

3.3.1. Quantification of Bioactive Contents (TPC and TFC)

The total phenol content (TPC) of fractions was estimated quantitatively by Folin-Ciocalteau (FC reagent) method as already mentioned in the literature, with some changeovers [128]. Three readings were taken for the total phenolic contents and recorded as mg of gallic acid equivalents/gram of dried weight of fraction (mg GAE/g wt.). The total flavonoid contents (TFCs) of fractions were measured with the previously defined method with some adaptations [129]. The experiment was performed in triplicates, and total flavonoid contents were noted in mg of quercetin equivalents/g of the dried weight of fraction (mg QE/g wt.).

3.3.2. Characterization of Bioactive Compounds by GCMS Analysis

The PCME extract and PCHF fraction of *P. candollei* were subjected to GC-MS analysis by following a procedure mentioned in the literature. The Percent peak area for each compound was calculated from the whole area of peaks [130].

3.4. Biological Activities

3.4.1. Antioxidant Activities

The antioxidant potential of *P. candollei* was tested by six different methods, which include DPPH, ABTS, FRAP, CUPRAC, and TAC (phosphomolybdenum assay) by previously described methods [131,132]. The results were expressed as equivalent to Trolox, whereas the metal-chelating property was expressed as equivalent to ethylene diamine tetraacetic acid (EDTA), and all the readings were taken in triplicates. Moreover, the IC_{50} and EC_{50} values were also calculated due to the good antioxidant results of the extract and fractions (equivalent to the standard antioxidants).

3.4.2. In Vitro Enzyme Inhibition Activities

The enzyme-inhibitory potential of all fractions was tested against tyrosinase, α-amylase, and α-glucosidase by using the methods reported in the literature [132]. Tyrosinase inhibition was measured as kojic acid equivalent, and inhibition of enzymes, responsible for hyperglycemia (α-amylase and α-glucosidase) was observed by acarbose equivalent potential of the extract, and its fractions and three readings were taken for each assay.

3.4.3. Thrombolytic Activity

For thrombolytic activity, the blood samples were obtained from healthy volunteers by following the guidelines of the ethical committee of the Islamia University of Bahawalpur. Five volunteers were included in the study who had not been using antidepressants, oral contraceptives, or anticoagulants. Analysis was performed by taking 5.0 mL of water (sterile) and dropped into the commercially available lyophilized streptokinase (SK) injection (15,000,000 i.u.). Then it was used as a positive control for thrombolytic activity. After that, 100 µL of solution of the SK (30,000 I.U) was for thrombolysis. In the Eppendorf tubes (after weighing the tubes), 500 µL of the venous blood (from a volunteer) was taken to form a clot. The liquid (serum) was completely removed by aspiration without the disruption of the clot. The weight of the clot was measured by using the formula, weight of thrombus = weight of tube with thrombus, which is the weight of the empty tube. Finally, a 100 µL volume of sample solution (1 mg/mL) was added to each tube, and 100 µL of streptokinase in the control tube. Negative control was distilled water, which was added to each tube in a volume of 100 µL. The test tubes were incubated at 37 °C for a period of 90 minutes and weighed again after removing the fluid (to observe clot lysis). The difference in weight was calculated before and after clot lysis by the given formula and expressed as a percent of thrombolytic activity of fractions, and the experiment was performed in triplicates [133].

3.4.4. Antibacterial Activity

The antibacterial effect of the plant was determined by using five Gram-positive (*Bacillus subtilis* ATCC1692, *Micrococcus luteus* ATCC 4925, *Staphylococcus epidermidis* ATCC 8724, *Bacillus pumilus* ATCC 13835, and *Staphylococcus aureus* ATCC 6538) and three Gram-negative strains (*Escherichia coli* ATCC 25922, *Bordetella bronchiseptica* ATCC 7319, and *Pseudomonas aeruginosa* ATCC 9027). These bacterial strains were procured from drug testing laboratory (DTL) Bahawalpur. Disc diffusion technique was employed to check the inhibition zones in mm. The procedure of the activity was described by previous research [134]. In disc diffusion assay, the sample of the extract/fraction or the standard (antibiotic) was diffused from a disk (loaded with sample or antibiotic) over the medium (nutrient agar), thereby creating a concentration gradient. Discs were made up of filter paper and were 5 mm in diameter. A known concentration of the sample (10, 20, and 40 mg per mL in DMSO) was impregnated on each disc, and the discs were placed on nutrient agar media (inoculated with test strains of bacteria). Co-amoxiclav in the concentration of 1 mg/mL was used as a standard antibacterial drug, and a disc without sample/standard was placed as a negative control (loaded with DMSO). The petri dishes were incubated at 37 °C for 24 h to provide optimum conditions for the growth of bacteria. A clear/distinct zone showing no bacterial growth is called "Zone of Inhibition", and the diameter of this zone was measured in mm to determine the antibacterial activity of the sample or standard.

3.5. Molecular Docking Studies

In computer-aided drug designing, molecular docking is a useful technique to understand the mechanism of interaction of protein receptors with ligands. Structures of the standard ligands (kojic acid for tyrosinase and acarbose for amylase and glucosidase) were downloaded from the PubChem database in the form of SDF (structured data format) files. Biomacromolecules (tyrosinase; 10.2210/pdb3NM8/pdb, amylase; 10.2210/pdb3VX1/pdb, and glucosidase; 10.2210/pdb5ZCB/pdb) were obtained in protein data bank (PDB) format. The preparation of ligands as PDB files was done for molecular retrieval and different software tools were used, such as auto Dock vina software, Discovery Studio, PyRx, and Babel. Discovery Studio (Discovery Studio 2021 client) was used to prepare the enzyme molecules that were obtained from Protein Data Bank. Ligand molecules were identified from GC-MS and selected based on major peak areas. Babel was used for preparing ligand molecules. Vina embedded in PyRx was run by uploading prepared receptors and ligands. These structures of ligands were placed in the area of the active site of the enzyme with AutoDock vina. Outcomes of the interaction of enzyme-active site residues and ligand molecules were evaluated using the Discovery Studio Visualizer, and 2D interactions were presented in Figures 4–6 [133,135].

4. Conclusions

The purpose of current work was to quantify and characterize the phytochemical composition and biological activities from the methanolic extract and different extraction fractions of *P. candollei*. Polyphenols and other bioactive phytochemicals were revealed by TPC, TFC, and GCMS analysis, and the identified molecules had antioxidant potential, antibacterial/antimicrobial potential, and other biological properties; moreover, biological activities of the plant were also found to be in relation to the quantity and quality of bioactive natural compounds. High values of antioxidant activities encourage the use of this plant to develop natural products for various health benefits and make it a potential source for use in natural products, pharmaceuticals, and nutraceuticals. Significant results of inhibition against tyrosinase, α-amylase, and α-glucosidase; good thrombolytic property; and antibacterial activity may also help to establish the nutritional and phyto-therapeutical role of the plant. A molecular docking study was used to explain the relationship of enzyme inhibition with the identified bioactive natural compounds. The findings of this study will not only improve our understanding of phytochemistry and biological effects of *P. candollei*, but also provide basic scientific knowledge to understand the health-improving and bioactive properties for developing the nutraceutical possibilities and natural products. However, future research may address the development of novel drugs for use in pharmaceuticals.

Author Contributions: Conceptualization, M.A. (Maqsood Ahmed) and K.-u.-R.K.; methodology, B.A.G.; software, B.A.G. and A.K.; validation, H.Y.A.; formal analysis, M.S.-u.R.; investigation, R.D. and M.A. (Mariyam Anwar); resources, H.Y.A.; data curation, C.O.; writing—original draft preparation, M.A. (Maqsood Ahmed) and H.Y.A.; writing—review and editing, M.A. (Maqsood Ahmed) and H.Y.A.; visualization, T.J.; supervision, K.-u.-R.K. and S.A.; project administration, K.-u.-R.K.; funding acquisition, H.Y.A. All authors have read and agreed to the published version of the manuscript.

Funding: The authors are grateful to the King Saud University, Riyadh, Saudi Arabia for funding this study through Project number RSP2022R504.

Institutional Review Board Statement: All the trials were carried out in accordance with the NIH guidelines and were approved by the Department of Pharmaceutical chemistry's concerned committee (1009/AS & RB/12/07/2021).

Informed Consent Statement: Not applicable.

Data Availability Statement: Not applicable.

Acknowledgments: The authors are thankful to Researchers Supporting Project number (RSP2022R504), King Saud University, Riyadh, Saudi Arabia.

Conflicts of Interest: The authors declare no conflict of interest.

Sample Availability: Not applicable.

References

1. Paz, M.; Gúllon, P.; Barroso, M.F.; Carvalho, A.P.; Domingues, V.F.; Gomes, A.M.; Becker, H.; Longhinotti, E.; Delerue-Matos, C. Brazilian fruit pulps as functional foods and additives: Evaluation of bioactive compounds. *Food Chem.* **2015**, *172*, 462–468. [CrossRef] [PubMed]
2. Hassan, W.; Noreen, H.; Rehman, S.; Gul, S.; Amjad Kamal, M.; Paul Kamdem, J.; Zaman, B.; BT da Rocha, J. Oxidative stress and antioxidant potential of one hundred medicinal plants. *Curr. Top. Med. Chem.* **2017**, *17*, 1336–1370. [CrossRef] [PubMed]
3. Uuh-Narváez, J.J.; González-Tamayo, M.A.; Segura-Campos, M.R. A study on nutritional and functional study properties of Mayan plant foods as a new proposal for type 2 diabetes prevention. *Food Chem.* **2021**, *341*, 128247. [CrossRef] [PubMed]
4. Satil, F.; Selvi, S. Ethnobotanical features of *Ziziphora* L.(*Lamiaceae*) taxa in Turkey. *Int. J. Nat. Life Sci.* **2020**, *4*, 56–65.
5. Xie, Y.-G.; Zhao, X.-C.; ul Hassan, S.S.; Zhen, X.-Y.; Muhammad, I.; Yan, S.-K.; Yuan, X.; Li, H.-L.; Jin, H.-Z.J.P.L. One new sesquiterpene and one new iridoid derivative from *Valeriana amurensis*. *Phytochem. Lett.* **2019**, *32*, 6–9. [CrossRef]

6. ul Hassan, S.S.; Ishaq, M.; Zhang, W.D.; Jin, H.Z. An Overview of the Mechanisms of Marine Fungi-Derived Anti-Inflammatory and Anti-Tumor Agents and their Novel Role in Drug Targeting. *Curr. Pharm. Des.* **2021**, *27*, 2605–2614. [CrossRef]
7. Jideani, A.I.; Silungwe, H.; Takalani, T.; Omolola, A.O.; Udeh, H.O.; Anyasi, T.A. Antioxidant-rich natural fruit and vegetable products and human health. *Int. J. Food Prop.* **2021**, *24*, 41–67. [CrossRef]
8. Veizades, S.; Tso, A.; Nguyen, P.K. Infection, inflammation and thrombosis: A review of potential mechanisms mediating arterial thrombosis associated with influenza and severe acute respiratory syndrome coronavirus 2. *Biol. Chem.* **2021**, *403*, 231–241. [CrossRef]
9. Elegbede, J.; Lateef, A.; Azeez, M.; Asafa, T.; Yekeen, T.; Oladipo, I.; Aina, D.; Beukes, L.; Gueguim-Kana, E.J.W.; Valorization, B. Biofabrication of gold nanoparticles using xylanases through valorization of corncob by *Aspergillus niger* and *Trichoderma longibrachiatum*: Antimicrobial, antioxidant, anticoagulant and thrombolytic activities. *Waste Biomass Valorization* **2020**, *11*, 781–791. [CrossRef]
10. Ghosh, S.K.; Sur, P.K. Sanjeevani and Bishalyakarani plants-myth or real! *Int. J. Curr. Res.* **2015**, *7*, 19964–19969. Available online: https://scholar.google.com.hk/scholar?hl=zh-CN&as_sdt=0%2C5&q=Ghosh%2C+Swapan+Kr.+%22Pradip+Kr+Sur+Current+Research%2C+7%2C%289%29%2C+19964-19969&btnG= (accessed on 25 May 2022).
11. Abbas, Q.; Hussain, A.; Khan, S.W.; Hussain, A.; Shinwari, S.; Hussain, A.; Ullah, A.; Zafar, M.; Ali, K. Floristic Diversity, Ethnobotany and Traditional Recipes of Medicinal Plants of Maruk Nallah, Haramosh Valley, District Gilgit, Gilgit Baltistan: Traditional recipes of Maruk Nallah, Haramosh Valley, District Gilgit. *Proc. Pak. Acad. Sci. B Life Environ. Sci.* **2019**, *56*, 97–112.
12. Abbas, Z.; Khan, S.M.; Abbasi, A.M.; Pieroni, A.; Ullah, Z.; Iqbal, M.; Ahmad, Z. Ethnobotany of the balti community, tormik valley, karakorum range, Baltistan, Pakistan. *J. Ethnobiol. Ethnomed.* **2016**, *12*, 1–16. [CrossRef] [PubMed]
13. Shah, A.; Bharati, K.A.; Ahmad, J.; Sharma, M. New ethnomedicinal claims from Gujjar and Bakerwals tribes of Rajouri and Poonch districts of Jammu and Kashmir, India. *J. Ethnopharmacol.* **2015**, *166*, 119–128. [CrossRef] [PubMed]
14. Ali, I.; Mu, Y.; Atif, M.; Hussain, H.; Li, J.; Li, D.; Shabbir, M.; Bankeu, J.J.K.; Cui, L.; Sajjad, S. Separation and anti-inflammatory evaluation of phytochemical constituents from *Pleurospermum candollei* (Apiaceae) by high-speed countercurrent chromatography with continuous sample load. *J. Sep. Sci.* **2021**, *44*, 2663–2673. [CrossRef]
15. Khan, K.U.; Shah, M.; Ahmad, H.; Ashraf, M.; Rahman, I.U.; Iqbal, Z.; Khan, S.M.; Majid, A. Investigation of traditional veterinary phytomedicines used in Deosai Plateau, Pakistan. *Glob. Vet.* **2015**, *15*, 381–388. [CrossRef]
16. Thiviya, P.; Gamage, A.; Piumali, D.; Merah, O.; Madhujith, T. Apiaceae as an Important Source of Antioxidants and Their Applications. *Cosmetics* **2021**, *8*, 111. [CrossRef]
17. Kawarty, A.M.A.; Behçet, L.; Çakilcioğlu, U. An ethnobotanical survey of medicinal plants in Ballakayati (Erbil, North Iraq). *Turk. J. Bot.* **2020**, *44*, 345–357. [CrossRef]
18. Aryal, S.; Baniya, M.K.; Danekhu, K.; Kunwar, P.; Gurung, R.; Koirala, N. Total phenolic content, flavonoid content and antioxidant potential of wild vegetables from Western Nepal. *Plants* **2019**, *8*, 96. [CrossRef] [PubMed]
19. El Aanachi, S.; Gali, L.; Nacer, S.N.; Bensouici, C.; Dari, K.; Aassila, H. Phenolic contents and in vitro investigation of the antioxidant, enzyme inhibitory, photoprotective, and antimicrobial effects of the organic extracts of *Pelargonium graveolens* growing in Morocco. *Biocatal. Agric. Biotechnol.* **2020**, *29*, 101819. [CrossRef]
20. Fettach, S.; Mrabti, H.; Sayah, K.; Bouyahya, A.; Salhi, N.; Cherrah, Y.; El Abbes, F.M. Phenolic content, acute toxicity of Ajuga iva extracts and assessment of their antioxidant and carbohydrate digestive enzyme inhibitory effects. *S. Afr. J. Bot.* **2019**, *125*, 381–385. [CrossRef]
21. Pieters, L.; Vlietinck, A.J. Bioguided isolation of pharmacologically active plant components, still a valuable strategy for the finding of new lead compounds? *J. Ethnopharmacol.* **2005**, *100*, 57–60. [CrossRef]
22. Herrera-Pool, E.; Ramos-Díaz, A.L.; Lizardi-Jiménez, M.A.; Pech-Cohuo, S.; Ayora-Talavera, T.; Cuevas-Bernardino, J.C.; García-Cruz, U.; Pacheco, N. Effect of solvent polarity on the Ultrasound Assisted extraction and antioxidant activity of phenolic compounds from habanero pepper leaves (*Capsicum chinense*) and its identification by UPLC-PDA-ESI-MS/MS. *Ultrason. Sonochem.* **2021**, *76*, 105658. [CrossRef] [PubMed]
23. Hapsari, S.; Yohed, I.; Kristianita, R.A.; Jadid, N.; Aparamarta, H.W.; Gunawan, S. Phenolic and flavonoid compounds extraction from *Calophyllum inophyllum* leaves. *Arab. J. Chem.* **2022**, *15*, 103666. [CrossRef]
24. Al-Dalahmeh, Y.; Al-Bataineh, N.; Al-Balawi, S.S.; Lahham, J.N.; Al-Momani, I.F.; Al-Sheraideh, M.S.; Mayyas, A.S.; Abu Orabi, S.T.; Al-Qudah, M.A. LC-MS/MS Screening, Total Phenolic, Flavonoid and Antioxidant Contents of Crude Extracts from Three Asclepiadaceae Species Growing in Jordan. *Molecules* **2022**, *27*, 859. [CrossRef] [PubMed]
25. Sharma, D.K.; Dave, R.; Shah, K. Proximate analysis, preliminary phytochemical screening and characterization of compounds by GC–MS from "Cycas revoluta". *Vegetos* **2022**, 1–7. [CrossRef]
26. Wangchuk, P.; Keller, P.A.; Pyne, S.G.; Taweechotipatr, M.; Kamchonwongpaisan, S. GC/GC-MS analysis, isolation and identification of bioactive essential oil components from the Bhutanese medicinal plant, *Pleurospermum amabile*. *Nat. Prod. Commun.* **2013**, *8*, 1934578X1300800930. [CrossRef]
27. Wang, T.; Xi, M.; Guo, Q.; Wang, L.; Shen, Z. Chemical components and antioxidant activity of volatile oil of a Compositae tea (*Coreopsis tinctoria* Nutt.) from Mt. Kunlun. *Ind. Crops Prod.* **2015**, *67*, 318–323. [CrossRef]
28. Henneh, I.T.; Huang, B.; Musayev, F.N.; Al Hashimi, R.; Safo, M.K.; Armah, F.A.; Ameyaw, E.O.; Adokoh, C.K.; Ekor, M.; Zhang, Y. Structural elucidation and in vivo anti-arthritic activity of β-amyrin and polpunonic acid isolated from the root bark of *Ziziphus abyssinica* HochstEx. A Rich (*Rhamnaceae*). *Bioorg. Chem.* **2020**, *98*, 103744. [CrossRef]

29. Devi, J.A.I.; Muthu, A.K. Gas Chromatography-Mass Spectrometry Analysis of Phytocomponents in the Ethanolic Extract from Whole Plant of *Lactuca runcinata* DC. *GAS* **2015**, *8*, 202–206.
30. de Castro Jorge, N.; Guedes, L.M.; Aguilera, N.; Becerra, J.; dos Santos Isaias, R.M. Allelopathic potential of the extracts of non-galled stems and globoid stem galls of *Eremanthus erythropappus* (DC) McLeish (*Asteraceae*). *Biochem. Syst. Ecol.* **2022**, *100*, 104379. [CrossRef]
31. Adinortey, C.A.; Kwarko, G.B.; Koranteng, R.; Boison, D.; Obuaba, I.; Wilson, M.D.; Kwofie, S.K. Molecular Structure-Based Screening of the Constituents of *Calotropis procera* Identifies Potential Inhibitors of Diabetes Mellitus Target Alpha Glucosidase. *Curr. Issues Mol. Biol.* **2022**, *44*, 963–987. [CrossRef]
32. Gao, R.; Su, Z.; Yin, Y.; Sun, L.; Li, S. Germplasm, chemical constituents, biological activities, utilization, and control of Chinese tallow (*Triadica sebifera* (L.) Small). *Biol. Invasions* **2016**, *18*, 809–829. [CrossRef]
33. Aziz, S.S.; El-Zayat, M.M.; El-Khateeb, A.Y. Biological Activity and Composition of the Essential Oil and Fatty Constituents of Petroleum ether Extract of *Brassica juncea* (L.). *J. Plant Prod.* **2020**, *11*, 57–59. [CrossRef]
34. Jin, J.; Sheraliev, G.; Xie, D.; Zhang, W.; Jin, Q.; Wang, X. Characteristics of Specialty Natural Micronutrients in Certain Oilseeds and Oils: Plastochromanol-8, Resveratrol, 5-Hydroxytryptamine Phenylpropanoid Amides, Lanosterol, Ergosterol and Cyclolinopeptides. *J. Am. Oil Chem. Soc.* **2016**, *93*, 155–170. [CrossRef]
35. Upadhyay, A.; Amanullah, A.; Mishra, R.; Kumar, A.; Mishra, A. Lanosterol suppresses the aggregation and cytotoxicity of misfolded proteins linked with neurodegenerative diseases. *Mol. Neurobiol.* **2018**, *55*, 1169–1182. [CrossRef]
36. Wintola, O.A.; Afolayan, A.J. Chemical constituents and biological activities of essential oils of *Hydnora africana* thumb used to treat associated infections and diseases in South Africa. *Appl. Sci.* **2017**, *7*, 443. [CrossRef]
37. Zhang, J.-P.; Tian, X.-H.; Yang, Y.-X.; Liu, Q.-X.; Wang, Q.; Chen, L.-P.; Li, H.-L.; Zhang, W.-D. *Gleditsia* species: An ethnomedical, phytochemical and pharmacological review. *J. Ethnopharmacol.* **2016**, *178*, 155–171. [CrossRef]
38. Salleh, W.M.N.H.W.; Kassim, H.; Tawang, A. Traditional uses, chemical profile and biological activities of piper hispidum Sw.: A review. *Biointerface Res. Appl. Chem.* **2021**, *11*, 13115–13129.
39. Akbar, M.; Ali, U.; Khalil, T.; Iqbal, M.S.; Amin, A.; Naeem, R.; Nazir, A.; Waqas, H.M.; Aslam, Z.; Jafri, F.I. *Cornus macrophylla*, the antibacterial activity of organic leaf extracts and the characterization of the more lipophilic components by GC/MS. *Molecules* **2020**, *25*, 2395. [CrossRef]
40. Mahadev, R.; Ramakrishnaiah, H.; Krishna, V.; Kumar, N.N. Chemical Composition of the Essential Oil from the Fruits of *Solanum erianthum* D. Don. *J. Essent. Oil Bear. Plants* **2012**, *15*, 387–391. [CrossRef]
41. Zhang, X.; Xiong, H.; Liu, L. Effects of taraxasterol on inflammatory responses in lipopolysaccharide-induced RAW 264.7 macrophages. *J. Ethnopharmacol.* **2012**, *141*, 206–211. [CrossRef]
42. Shah, R.; Alabri, S.J.; Ashehi, A.S.A.; Asiyabi, N.S.A.; AlMamari, W.K.A.; AlSabahi, J.N.A.; Al-Ruqaishi, H. Antibacterial Activity and Chemical Composition of Crude Extract and Oil of *Zygophyllum* (Fagonia) *luntii* (Baker) 1894 (Family *Zygophyllaceae*). *J. Agric. Mar. Sci.* **2020**, *25*, 58–66. [CrossRef]
43. Liu, K.; Zhang, X.; Xie, L.; Deng, M.; Chen, H.; Song, J.; Long, J.; Li, X.; Luo, J. Lupeol and its derivatives as anticancer and anti-inflammatory agents: Molecular mechanisms and therapeutic efficacy. *Pharmacol. Res.* **2021**, *164*, 105373. [CrossRef] [PubMed]
44. PRABHA, M.; Brintha, M. Molecular Profiling and Antioxidant Potential of *Citrus limon* (L.) Burm. F Fruits. *Nveo-Nat. Volatiles Essent. Oils J.* **2021**, *8*, 8360–8373.
45. Harley, B.K.; Amponsah, I.K.; Ben, I.O.; Adongo, D.W.; Mireku-Gyimah, N.A.; Baah, M.K.; Mensah, A.Y.; Fleischer, T.C. *Myrianthus libericus*: Possible mechanisms of hypoglycaemic action and in silico prediction of pharmacokinetics and toxicity profile of its bioactive metabolite, friedelan-3-one. *Biomed. Pharmacother.* **2021**, *137*, 111379. [CrossRef] [PubMed]
46. Malik, W.; Ahmed, D.; Izhar, S. Tyrosinase inhibitory activities of Carissa opaca Stapf ex haines roots extracts and their phytochemical analysis. *Pharmacogn. Mag.* **2017**, *13*, S544. [PubMed]
47. Monção, N.B.N.; Araújo, B.Q.; Silva, J.D.N.; Lima, D.J.B.; Ferreira, P.M.P.; Airoldi, F.P.d.S.; Pessoa, C.; Citó, A.M.d.G.L. Assessing chemical constituents of *Mimosa caesalpiniifolia* stem bark: Possible bioactive components accountable for the cytotoxic effect of *M. caesalpiniifolia* on human tumour cell lines. *Molecules* **2015**, *20*, 4204–4224. [CrossRef]
48. Siva, S.; Li, C.; Cui, H.; Lin, L. Encompassment of isoeugenol in 2-hydroxypropyl-β-cyclodextrin using ultrasonication: Characterization, antioxidant and antibacterial activities. *J. Mol. Liq.* **2019**, *296*, 111777. [CrossRef]
49. Peng, W.; Chen, J.; Lou, J.; Ni, C.; Ashraf, M.A. Molecules and Functions of Rosewood: Dalbergia Cearensis. *Caribb. J. Sci.* **2018**, *51*, 458–465.
50. Prabha, S.; Kumar, J. Gas Chromatographic and Mass Spectroscopic (GC-MS) Analysis of Rhizome of *Acorus calamus* Linn. for Identification of Potent Antimicrobial Bio-active Compounds. *J. Sci. Res.* **2021**, *13*, 263–273. [CrossRef]
51. Qin, G.-F.; Liang, H.-B.; Liu, W.-X.; Zhu, F.; Li, P.-L.; Li, G.-Q.; Yao, J.-C. Bicyclo [6.3.0] undecane sesquiterpenoids: Structures, biological activities, and syntheses. *Molecules* **2019**, *24*, 3912. [CrossRef]
52. Merlani, M.; Barbakadze, V.; Amiranashvili, L.; Gogilashvili, L. Synthesis of new dihydroxylated derivatives of ferulic and isoferulic acids. *Bull. Georg. Natl. Acad. Sci* **2018**, *12*, 119–124.
53. McCann, M.; Curran, R.; Ben-Shoshan, M.; McKee, V.; Tahir, A.A.; Devereux, M.; Kavanagh, K.; Creaven, B.S.; Kellett, A. Silver (I) complexes of 9-anthracenecarboxylic acid and imidazoles: Synthesis, structure and antimicrobial activity. *Dalton Trans.* **2012**, *41*, 6516–6527. [CrossRef] [PubMed]

54. Curran, R.; Lenehan, J.; McCann, M.; Kavanagh, K.; Devereux, M.; Egan, D.A.; Clifford, G.; Keane, K.; Creaven, B.S.; McKee, V. $[Ag_2(aca)_2]_n$ and $[Ag_4(aca)_4(NH_3)_2]$ (acaH = 9-anthracenecarboxylic acid): Synthesis, X-ray crystal structures, antimicrobial and anti-cancer activities. *Inorg. Chem. Commun.* **2007**, *10*, 1149–1153. [CrossRef]
55. Leal, A.L.A.B.; Machado, A.J.T.; Bezerra, C.F.; Inácio, C.E.S.; Rocha, J.E.; Sales, D.L.; de Freitas, T.S.; de Oliveira Almeida, W.; do Amaral, W.; da Silva, L.E. Chemical identification and antimicrobial potential of essential oil of *Piper rivinoides* kunth (BETIS-WHITE). *Food Chem. Toxicol.* **2019**, *131*, 110559. [CrossRef] [PubMed]
56. Abdellatif, K.R.; Lamie, P.F.; Omar, H.A. 3-Methyl-2-phenyl-1-substituted-indole derivatives as indomethacin analogs: Design, synthesis and biological evaluation as potential anti-inflammatory and analgesic agents. *J. Enzym. Inhib. Med. Chem.* **2016**, *31*, 318–324. [CrossRef] [PubMed]
57. Gao, E.; Zhou, Z.-Q.; Zou, J.; Yu, Y.; Feng, X.-L.; Chen, G.-D.; He, R.-R.; Yao, X.-S.; Gao, H. Bioactive asarone-derived phenylpropanoids from the rhizome of *Acorus tatarinowii* Schott. *J. Nat. Prod.* **2017**, *80*, 2923–2929. [CrossRef] [PubMed]
58. Chellian, R.; Pandy, V.; Mohamed, Z. Pharmacology and toxicology of α-and β-Asarone: A review of preclinical evidence. *Phytomedicine* **2017**, *32*, 41–58. [CrossRef] [PubMed]
59. Das, S.; Singh, V.K.; Dwivedy, A.K.; Chaudhari, A.K.; Dubey, N.K. Exploration of some potential bioactive essential oil components as green food preservative. *LWT* **2021**, *137*, 110498. [CrossRef]
60. Sarker, S.D.; Nahar, L. Progress in the chemistry of naturally occurring coumarins. *Prog. Chem. Org. Nat. Prod.* **2017**, *106*, 241–304. [PubMed]
61. Hua, X.; Yang, Q.; Zhang, W.; Dong, Z.; Yu, S.; Schwarz, S.; Liu, S. Antibacterial activity and mechanism of action of aspidinol against multi-drug-resistant methicillin-resistant *Staphylococcus aureus*. *Front. Pharmacol.* **2018**, *9*, 619. [CrossRef] [PubMed]
62. Freitas, T.S.d.; Xavier, J.d.C.; Pereira, R.L.; Rocha, J.E.; Muniz, D.F.; da Silva, P.T.; da Hora, J.P.; Dos Santos, H.S.; Bandeira, P.N.; Nogueira, C.E. Direct antibacterial and antibiotic resistance modulatory activity of chalcones synthesized from the natural product 2-hydroxy-3, 4, 6-trimethoxyacetophenone. *FEMS Microbiol. Lett.* **2020**, *367*, fnaa124. [CrossRef] [PubMed]
63. Ingallina, C.; Capitani, D.; Mannina, L.; Carradori, S.; Locatelli, M.; Di Sotto, A.; Di Giacomo, S.; Toniolo, C.; Pasqua, G.; Valletta, A. Phytochemical and biological characterization of Italian "sedano bianco di Sperlonga" Protected Geographical Indication celery ecotype: A multimethodological approach. *Food Chem.* **2020**, *309*, 125649. [CrossRef]
64. Dhivya, K.; Vengateswari, G.; Arunthirumeni, M.; Karthi, S.; Senthil-Nathan, S.; Shivakumar, M.S. Bioprospecting of Prosopis juliflora (Sw.) DC seed pod extract effect on antioxidant and immune system of *Spodoptera litura* (*Lepidoptera*: *Noctuidae*). *Physiol. Mol. Plant Pathol.* **2018**, *101*, 45–53. [CrossRef]
65. Alqahtani, F.Y.; Aleanizy, F.S.; Mahmoud, A.Z.; Farshori, N.N.; Alfaraj, R.; Al-Sheddi, E.S.; Alsarra, I.A. Chemical composition and antimicrobial, antioxidant, and anti-inflammatory activities of *Lepidium sativum* seed oil. *Saudi J. Biol. Sci.* **2019**, *26*, 1089–1092. [CrossRef] [PubMed]
66. Moreira, N.M.; Martelli, L.S.; Corrêa, A.G. Asymmetric organocatalyzed synthesis of coumarin derivatives. *Beilstein J. Org. Chem.* **2021**, *17*, 1952–1980. [CrossRef] [PubMed]
67. Chandrasekaran, M.; Kannathasan, K.; Venkatesalu, V. Antimicrobial activity of fatty acid methyl esters of some members of Chenopodiaceae. *Z. Für Nat. C* **2008**, *63*, 331–336. [CrossRef] [PubMed]
68. Kushwaha, P.; Yadav, S.S.; Singh, V.; Dwivedi, L. Phytochemical screening and GC-MS studies of the methanolic extract of Tridax procumbens. *Int. J. Pharm. Sci. Res.* **2019**, *10*, 2492–2496.
69. Abubakar, M.N.; Majinda, R.R. GC-MS analysis and preliminary antimicrobial activity of *Albizia adianthifolia* (Schumach) and *Pterocarpus angolensis* (DC). *Medicines* **2016**, *3*, 3. [CrossRef] [PubMed]
70. Rahman, M.; Ahmad, S.; Mohamed, M.; Ab Rahman, M. Antimicrobial compounds from leaf extracts of *Jatropha curcas*, *Psidium guajava*, and *Andrographis paniculata*. *Sci. World J.* **2014**, *2014*, 635240. [CrossRef]
71. Yu, F.-R.; Lian, X.-Z.; Guo, H.-Y.; McGuire, P.M.; Li, R.-D.; Wang, R.; Yu, F.-H. Isolation and characterization of methyl esters and derivatives from *Euphorbia kansui* (*Euphorbiaceae*) and their inhibitory effects on the human SGC-7901 cells. *J. Pharm. Pharm. Sci.* **2005**, *8*, 528–535. [PubMed]
72. El-Naggar, M.E.; Soliman, R.A.; Morsy, O.M.; Abdel-Aziz, M.S. Nanoemulsion of Capsicum fruit extract as an eco-friendly antimicrobial agent for production of medical bandages. *Biocatal. Agric. Biotechnol.* **2020**, *23*, 101516. [CrossRef]
73. Kurashov, E.A.; Fedorova, E.V.; Krylova, J.V.; Mitrukova, G.G. Assessment of the potential biological activity of low molecular weight metabolites of freshwater macrophytes with QSAR. *Scientifica* **2016**, *2016*, 1205680. [CrossRef]
74. Wanyama, A.W. Evaluation of Phytoconstituents, Antioxidants Potential, Cytotoxic, Antimicrobial Activities and Mineral Composition of *Vigna subterranea* (L) Verdic. Extracts. Ph.D. Thesis, Jomo Kenyatta University of Agriculture and Technology, Juja, Kenya, 2018.
75. Sui, G.; Li, T.; Zhang, B.; Wang, R.; Hao, H.; Zhou, W. Recent advances on synthesis and biological activities of aurones. *Bioorg. Med. Chem.* **2021**, *29*, 115895. [CrossRef]
76. Abbasi, M.; Omrani, M.; Raiatparvar Malieki, L.; Sonboli, A.; Nejad Ebrahimi, S. Phytochemical studies of *Tetrataenium nephrophyllum* and anti-acetylcholinesterase activities. *Trends Phytochem. Res.* **2021**, *5*, 210–217.
77. Yang, J.; Liu, C.; Cai, H.; Gu, D.; Ji, Z.; Guo, X.; Tian, J.; Meng, J.; Yang, Y. Identification and theoretical explanation of chemical composition against α-amylase in the n-hexane extract from *Sargassum fusiforme*. *Algal Res.* **2019**, *43*, 101642. [CrossRef]
78. Abuabaker, M.; Guo, H.; Shi, J.; Farah, A.; Zhang, J. Gas Chromatography-Mass spectrum and Infra-Red spectral analysis of Fixed Oil from Sudanese Adansonia digitata Seeds. *Chem. Methodol.* **2021**, *5*, 240–249.

79. Rychlicka, M.; Rot, A.; Gliszczyńska, A. Biological Properties, Health Benefits and Enzymatic Modifications of Dietary Methoxylated Derivatives of Cinnamic Acid. *Foods* **2021**, *10*, 1417. [CrossRef]
80. Al-Mudhafr, A.W.H. Detection of Active Compounds By Gc/Ms in Extract of Red Cabbage (*Brassica oleracea*) and Their Effect in the Preservation of Raw Milk. *Plant Arch.* **2020**, *20*, 4–8.
81. Mohamed, S.S.; Saber, A.A. Antifungal potential of the bioactive constituents in extracts of the mostly untapped brown seaweed *Hormophysa cuneiformis* from the Egyptian coastal waters. *Egypt. J. Bot.* **2019**, *59*, 695–708. [CrossRef]
82. Naik, B.; Maurya, V.K.; Kumar, V.; Kumar, V.; Upadhyay, S.; Gupta, S. Phytochemical analysis of *Diplazium esculentum* reveals the presence of medically important components. *Curr. Nutr. Food Sci.* **2021**, *17*, 210–215. [CrossRef]
83. Suresh, P.S.; Bhatt, V.; Singh, P.P.; Sharma, U. Steroidal sapogenins from genus Trillium: Chemistry, synthesis, and opportunities in neuro-active steroids designing. *Stud. Nat. Prod. Chem.* **2021**, *68*, 67–95.
84. Jiang, Q. Metabolism of natural forms of vitamin E and biological actions of vitamin E metabolites. *Free. Radic. Biol. Med.* **2022**, *179*, 375–387. [CrossRef]
85. Kumar, V.; Singh, S.; Singh, A.; Dixit, A.K.; Shrivastava, B.; Kondalkar, S.A.; Subhose, V. Determination of phytochemical, antioxidant, antimicrobial, and protein binding qualities of hydroethanolic extract of *Celastrus paniculatus*. *J. Appl. Biol. Biotechnol.* **2018**, *6*, 11–17.
86. Glevitzky, I.; Dumitrel, G.A.; Glevitzky, M.; Pasca, B.; Otrisal, P.; Bungau, S.; Cioca, G.; Pantis, C.; Popa, M.J.R.C. Statistical analysis of the relationship between antioxidant activity and the structure of flavonoid compounds. *Rev. Chim.* **2019**, *70*, 3103–3107. [CrossRef]
87. Lee, Y.H.; Choo, C.; Watawana, M.I.; Jayawardena, N.; Waisundara, V.Y. An appraisal of eighteen commonly consumed edible plants as functional food based on their antioxidant and starch hydrolase inhibitory activities. *J. Sci. Food Agric.* **2015**, *95*, 2956–2964. [CrossRef] [PubMed]
88. Kumar, N.; Goel, N. Phenolic acids: Natural versatile molecules with promising therapeutic applications. *Biotechnol. Rep.* **2019**, *24*, e00370. [CrossRef]
89. Bendary, E.; Francis, R.; Ali, H.; Sarwat, M.; El Hady, S. Antioxidant and structure–activity relationships (SARs) of some phenolic and anilines compounds. *Ann. Agric. Sci.* **2013**, *58*, 173–181. [CrossRef]
90. Shousha, W.G.; Aboulthana, W.M.; Salama, A.H.; Saleh, M.H.; Essawy, E.A. Evaluation of the biological activity of *Moringa oleifera* leaves extract after incorporating silver nanoparticles, in vitro study. *Bull. Natl. Res. Cent.* **2019**, *43*, 1–13. [CrossRef]
91. Dias, W.; do Vale Junior, E.; de Oliveira, M.d.D.A.; Barbosa, Y.; do Nascimento Silva, J.; da Costa Júnior, J.; de Almeida, P.; Martins, F. Cytogenotoxic effect, phytochemical screening and antioxidant potential of *Jatropha mollissima* (Pohl) Baill leaves. *S. Afr. J. Bot.* **2019**, *123*, 30–35. [CrossRef]
92. Mani, J.S.; Johnson, J.B.; Hosking, H.; Ashwath, N.; Walsh, K.B.; Neilsen, P.M.; Broszczak, D.A.; Naiker, M. Antioxidative and therapeutic potential of selected Australian plants: A review. *J. Ethnopharmacol.* **2021**, *268*, 113580. [CrossRef]
93. Susan, A.; Rajendran, K.; Sathyasivam, K.; Krishnan, U.M. An overview of plant-based interventions to ameliorate arsenic toxicity. *Biomed. Pharmacother.* **2019**, *109*, 838–852. [CrossRef]
94. Mehrandish, R.; Rahimian, A.; Shahriary, A. Heavy metals detoxification: A review of herbal compounds for chelation therapy in heavy metals toxicity. *J. Herbmed Pharmacol.* **2019**, *8*, 69–77. [CrossRef]
95. Karim, F.A.; Suleiman, M.; Rahmat, A.; Bakar, M.A. Phytochemicals, antioxidant and antiproliferative properties of five moss species from Sabah, Malaysia. *Int. J. Pharm. Pharm. Sci* **2014**, *6*, 292–297.
96. Mohandas, G.G.; Kumaraswamy, M. Antioxidant activities of terpenoids from *Thuidium tamariscellum* (*C. muell.*) Bosch. and Sande-Lac. a Moss. *Pharmacogn. J.* **2018**, *10*, 645–649. [CrossRef]
97. Khan, S.; Nazir, M.; Raiz, N.; Saleem, M.; Zengin, G.; Fazal, G.; Saleem, H.; Mukhtar, M.; Tousif, M.I.; Tareen, R.B. Phytochemical profiling, in vitro biological properties and in silico studies on *Caragana ambigua* stocks (Fabaceae): A comprehensive approach. *Ind. Crops Prod.* **2019**, *131*, 117–124. [CrossRef]
98. Khan, S.; Nazir, M.; Saleem, H.; Raiz, N.; Saleem, M.; Anjum, S.M.M.; Zengin, G.; Mukhtar, M.; Tousif, M.I.; Mahomoodally, F.M. Valorization of the antioxidant, enzyme inhibition and phytochemical propensities of *Berberis calliobotrys* Bien. ex Koehne: A multifunctional approach to probe for bioactive natural products. *Ind. Crops Prod.* **2019**, *141*, 111693. [CrossRef]
99. Colovic, M.B.; Krstic, D.Z.; Lazarevic-Pasti, T.D.; Bondzic, A.M.; Vasic, V.M. Acetylcholinesterase inhibitors: Pharmacology and toxicology. *Curr. Neuropharmacol.* **2013**, *11*, 315–335. [CrossRef]
100. Zuo, A.-R.; Dong, H.-H.; Yu, Y.-Y.; Shu, Q.-L.; Zheng, L.-X.; Yu, X.-Y.; Cao, S.-W. The antityrosinase and antioxidant activities of flavonoids dominated by the number and location of phenolic hydroxyl groups. *Chin. Med.* **2018**, *13*, 1–12. [CrossRef]
101. Uchida, R.; Ishikawa, S.; Tomoda, H. Inhibition of tyrosinase activity and melanine pigmentation by 2-hydroxytyrosol. *Acta Pharm. Sin. B* **2014**, *4*, 141–145. [CrossRef]
102. Corradi, I.; De Souza, E.; Sande, D.; Takahashi, J.A. Correlation between phenolic compounds contents, anti-tyrosinase and antioxidant activities of plant extracts. *Chem. Eng. Trans.* **2018**, *64*, 109–114.
103. Kumar, S.C.; Ramesh, N.; Sreevatsan, S.; Joseph, B.; Alle, P.; Belani, K.G.; Osterholm, M.T. Knowledge, attitudes, and poultry-handling practices of poultry workers in relation to avian influenza in India. *Indian J. Occup. Environ. Med.* **2013**, *17*, 16. [CrossRef]
104. Ling, J.; Kim, Y.-B.; Kim, J.-H.; Shin, J.-D.; Yoon, S.-A.; Shim, M.-W.; Kim, E.-K. Screening of depigmenting agents from Philippine plants. *한국생물공학회 학술대회* **2008**, 183. [CrossRef]

105. Ya, W.; Chun-Meng, Z.; Tao, G.; Yi-Lin, Z.; Ping, Z. Preliminary screening of 44 plant extracts for anti-tyrosinase and antioxidant activities. *Pak. J. Pharm. Sci.* **2015**, *28*, 1737–1744. [PubMed]
106. Chung, I.M.; Rajakumar, G.; Subramanian, U.; Venkidasamy, B.; Khanna, V.G.; Thiruvengadam, M. Insights on the current status and advancement of diabetes mellitus type 2 and to avert complications: An overview. *Biotechnol. Appl. Biochem.* **2020**, *67*, 920–928. [CrossRef] [PubMed]
107. Trikkalinou, A.; Papazafiropoulou, A.K.; Melidonis, A. Type 2 diabetes and quality of life. *World J. Diabetes* **2017**, *8*, 120. [CrossRef]
108. Salazar, M.O.; Osella, M.I.; Arcusin, D.E.; Lescano, L.E.; Furlan, R.L. New α-glucosidase inhibitors from a chemically engineered essential oil of *Origanum vulgare* L. *Ind. Crops Prod.* **2020**, *156*, 112855. [CrossRef]
109. Zhang, Z.; Dai, L.; Wang, H.; Chang, X.; Ren, S.; Lai, H.; Liu, L. Phytochemical profiles and antioxidant, anticholinergic, and antidiabetic activities of *Odontites serotina* (Lam.) dum. *Eur. J. Integr. Med.* **2021**, *44*, 101340. [CrossRef]
110. Sen, A.; Kurkcuoglu, M.; Senkardes, I.; Bitis, L.; Baser, K.H.C. Chemical composition, antidiabetic, anti-inflammatory and antioxidant activity of *Inula ensifolia* L. essential oil. *J. Essent. Oil Bear. Plants* **2019**, *22*, 1048–1057. [CrossRef]
111. Bothon, F.T.; Debiton, E.; Avlessi, F.; Forestier, C.; Teulade, J.-C.; Sohounhloue, D.K. In vitro biological effects of two anti-diabetic medicinal plants used in Benin as folk medicine. *BMC Complementary Altern. Med.* **2013**, *13*, 1–8. [CrossRef]
112. Pandey, B.P.; Pradhan, S.P. Chemical composition, in vitro antioxidant, and enzymes inhibitory potential of three medicinally important plants from Nepal (*Lepisorus mehrae, Pleurospermum benthamii,* and *Roscoea auriculata*). *Adv. Tradit. Med.* **2020**, *22*, 75–90. [CrossRef]
113. Ahmed, S.H.; Shaikh, T.G.; Waseem, S.; Qadir, N.A.; Yousaf, Z.; Ullah, I. Vaccine-induced thrombotic thrombocytopenia following coronavirus vaccine: A narrative review. *Ann. Med. Surg.* **2022**, *73*, 102988. [CrossRef] [PubMed]
114. Alkarithi, G.; Duval, C.; Shi, Y.; Macrae, F.L.; Ariëns, R.A. Thrombus structural composition in cardiovascular disease. *Arterioscler. Thromb. Vasc. Biol.* **2021**, *41*, 2370–2383. [CrossRef] [PubMed]
115. Helms, J.; Severac, F.; Merdji, H.; Schenck, M.; Clere-Jehl, R.; Baldacini, M.; Ohana, M.; Grunebaum, L.; Castelain, V.; Anglés-Cano, E. Higher anticoagulation targets and risk of thrombotic events in severe COVID-19 patients: Bi-center cohort study. *Ann. Intensive Care* **2021**, *11*, 1–8. [CrossRef] [PubMed]
116. Vani, M.; Rani, P.V.; Madhuri, O.; Sree, M.V.S.; Ramya, L.S.; Chandrika, M.G.; Padmalatha, K.; Supriya, J. Phytochemical and in vitro thrombolytic activity evaluation of *Cassia siamea* L., Leguminosae leaf extracts, and pyrogallol. *Int. J. Green Pharm.* **2019**, *13*, 213–217.
117. Maqsood, M.; Mushtaq, Z.; Rasheed, T.; Nisa, Z.U.; Sher, F. Thrombolytic and cytotoxic activity of different bioactive extracts of *E. coli*. *Case Stud. Chem. Environ. Eng.* **2021**, *3*, 100080. [CrossRef]
118. Manju, P.; Pushpa, D.A. A Study On Thrombolytic and Cytotoxic Activity of Methanolic Extract of *Zingiber officinale*. *Int. J. Life Sci. Pharma. Res.* **2020**, *10*, 1–5.
119. Bhatia, P.; Sharma, A.; George, A.J.; Anvitha, D.; Kumar, P.; Dwivedi, V.P.; Chandra, N.S. Antibacterial activity of medicinal plants against ESKAPE: An update. *Heliyon* **2021**, *7*, e06310. [CrossRef] [PubMed]
120. Kessler, A.; Kalske, A. Plant secondary metabolite diversity and species interactions. *Annu. Rev. Ecol. Evol. Syst.* **2018**, *49*, 115–138. [CrossRef]
121. Alibi, S.; Crespo, D.; Navas, J. Plant-derivatives small molecules with antibacterial activity. *Antibiotics* **2021**, *10*, 231. [CrossRef]
122. Xiao, Y.; Zhu, S.; Wu, G.; Xie, Y.; Ishaq, M.; Sun, Y.; Yan, S.-K.; Qian, X.-P.; Jin, H.-Z. Chemical Constituents of *Vernonia parishii*. *Chem. Nat. Compd.* **2020**, *56*, 134–136. [CrossRef]
123. Wangchuk, P.; Pyne, S.G.; Keller, P.A.; Taweechotipatr, M.; Kamchonwongpaisan, S. Phenylpropanoids and furanocoumarins as antibacterial and antimalarial constituents of the Bhutanese medicinal plant *Pleurospermum amabile*. *Nat. Prod. Commun.* **2014**, *9*, 1934578X1400900719. [CrossRef]
124. Khan, I.; Abbas, T.; Anjum, K.; Abbas, S.Q.; Shagufta, B.I.; Ali Shah, S.A.; Akhter, N.; Hassan, S.S. Antimicrobial potential of aqueous extract of *Camellia sinensis* against representative microbes. *Pak. J. Pharm. Sci.* **2019**, *32*, 631–636. [PubMed]
125. Aziz, M.; Ahmad, S.; Iqbal, M.N.; Khurshid, U.; Saleem, H.; Alamri, A.; Anwar, S.; Alamri, A.S.; Chohan, T.A. Phytochemical, pharmacological, and In-silico molecular docking studies of *Strobilanthes glutinosus* Nees: An unexplored source of bioactive compounds. *S. Afr. J. Bot.* **2022**, *147*, 618–627. [CrossRef]
126. Hassan, S.S.U.; Zhang, W.-D.; Jin, H.-Z.; Basha, S.H.; Priya, S.V.S.S. In-silico anti-inflammatory potential of guaiane dimers from *Xylopia vielana* targeting COX-2. *J. Biomol. Struct. Dyn.* **2022**, *40*, 484–498. [CrossRef] [PubMed]
127. Jones, W.P.; Kinghorn, A.D. Extraction of plant secondary metabolites. In *Natural Products Isolation*; Springer: Berlin/Heidelberg, Germany, 2006; pp. 323–351.
128. Chang, X.; Ye, Y.; Pan, J.; Lin, Z.; Qiu, J.; Peng, C.; Guo, X.; Lu, Y. Comparative Analysis of Phytochemical Profiles and Antioxidant Activities between Sweet and Sour Wampee (*Clausena lansium*) Fruits. *Foods* **2022**, *11*, 1230. [CrossRef] [PubMed]
129. Muflihah, Y.M.; Gollavelli, G.; Ling, Y.-C. Correlation Study of Antioxidant Activity with Phenolic and Flavonoid Compounds in 12 Indonesian Indigenous Herbs. *Antioxidants* **2021**, *10*, 1530. [CrossRef] [PubMed]
130. Tabassum, S.; Ahmad, S.; Rehman Khan, K.u.; Tabassum, F.; Khursheed, A.; Zaman, Q.u.; Bukhari, N.A.; Alfagham, A.; Hatamleh, A.A.; Chen, Y. Phytochemical Profiling, Antioxidant, Anti-Inflammatory, Thrombolytic, Hemolytic Activity In Vitro and In Silico Potential of Portulacaria afra. *Molecules* **2022**, *27*, 2377. [CrossRef]
131. Zhang, C.; Liu, D.; Wu, L.; Zhang, J.; Li, X.; Wu, W. Chemical characterization and antioxidant properties of ethanolic extract and its fractions from sweet potato (*Ipomoea batatas* L.) leaves. *Foods* **2019**, *9*, 15. [CrossRef]

132. Saleem, M.; Shazmeen, N.; Nazir, M.; Riaz, N.; Zengin, G.; Ataullah, H.M.; Nisar, F.; Mukhtar, M.; Tousif, M.I.J.C. Investigation on the Phytochemical Composition, Antioxidant and Enzyme Inhibition Potential of *Polygonum Plebeium* R. Br: A Comprehensive Approach to Disclose New Nutraceutical and Functional Food Ingredients. *Chem. Biodivers* **2021**, *18*, e2100706. [CrossRef]
133. Ghalloo, B.A.; Khan, K.-u.-R.; Ahmad, S.; Aati, H.Y.; Al-Qahtani, J.H.; Ali, B.; Mukhtar, I.; Hussain, M.; Shahzad, M.N.; Ahmed, I. Phytochemical Profiling, In Vitro Biological Activities, and In Silico Molecular Docking Studies of *Dracaena reflexa*. *Molecules* **2022**, *27*, 913. [CrossRef]
134. Shoibe, M.; Chy, M.; Uddin, N.; Alam, M.; Adnan, M.; Islam, M.; Nihar, S.W.; Rahman, N.; Suez, E. In vitro and in vivo biological activities of *Cissus adnata* (Roxb.). *Biomedicines* **2017**, *5*, 63. [CrossRef] [PubMed]
135. Fernandes, C.; Palmeira, A.; Ramos, I.I.; Carneiro, C.; Afonso, C.; Tiritan, M.E.; Cidade, H.; Pinto, P.C.; Saraiva, M.L.M.; Reis, S.J.P. Chiral derivatives of xanthones: Investigation of the effect of enantioselectivity on inhibition of cyclooxygenases (COX-1 and COX-2) and binding interaction with human serum albumin. *Pharmaceuticals* **2017**, *10*, 50. [CrossRef] [PubMed]

molecules

MDPI

Article

Anti-Inflammatory, Analgesic and Antioxidant Potential of New (2*S*,3*S*)-2-(4-isopropylbenzyl)-2-methyl-4-nitro-3-phenylbutanals and Their Corresponding Carboxylic Acids through In Vitro, In Silico and In Vivo Studies

Fawad Mahmood [1], Jamshaid Ali Khan [1], Mater H. Mahnashi [2], Muhammad Saeed Jan [3], Muhammad Aamir Javed [4], Umer Rashid [4], Abdul Sadiq [5,*], Syed Shams ul Hassan [6,7,*] and Simona Bungau [8,*]

1 Department of Pharmacy, University of Peshawar, Peshawar 25120, KP, Pakistan; fawadpharmacist@gmail.com (F.M.); jamshaidkhan@uop.edu.pk (J.A.K.)
2 Department of Pharmaceutical Chemistry, College of Pharmacy, Najran University, Najran 55461, Saudi Arabia; matermaha@gmail.com
3 Department of Pharmacy, University of Swabi, Swabi 23561, KP, Pakistan; saeedjanpharmacist@gmail.com
4 Department of Chemistry, COMSATS University Islamabad, Abbottabad Campus, Abbottabad 22060, KP, Pakistan; amirjaved55@gmail.com (M.A.J.); umerrashid@cuiatd.edu.pk (U.R.)
5 Department of Pharmacy, Faculty of Biological Sciences, University of Malakand, Chakdara 18000, KP, Pakistan
6 Shanghai Key Laboratory for Molecular Engineering of Chiral Drugs, School of Pharmacy, Shanghai Jiao Tong University, Shanghai 200240, China
7 Department of Natural Product Chemistry, School of Pharmacy, Shanghai Jiao Tong University, Shanghai 200240, China
8 Department of Pharmacy, Faculty of Medicine and Pharmacy, University of Oradea, 410028 Oradea, Romania
* Correspondence: sadiquom@yahoo.com (A.S.); shams1327@yahoo.com (S.S.u.H.); simonabungau@gmail.com (S.B.)

Citation: Mahmood, F.; Khan, J.A.; Mahnashi, M.H.; Jan, M.S.; Javed, M.A.; Rashid, U.; Sadiq, A.; Hassan, S.S.u.; Bungau, S. Anti-Inflammatory, Analgesic and Antioxidant Potential of New (2*S*,3*S*)-2-(4-isopropylbenzyl)-2-methyl-4-nitro-3-phenylbutanals and Their Corresponding Carboxylic Acids through In Vitro, In Silico and In Vivo Studies. *Molecules* **2022**, *27*, 4068. https://doi.org/10.3390/molecules27134068

Academic Editor: Pierangela Ciuffreda

Received: 2 May 2022
Accepted: 22 June 2022
Published: 24 June 2022

Abstract: In the current study, a series of new (2*S*,3*S*)-2-(4-isopropylbenzyl)-2-methyl-4-nitro-3-phenylbutanals (**FM1-6**) with their corresponding carboxylic acid analogues (**FM7-12**) has been synthesized. Initially, the aldehydic derivatives were isolated in the diastereomeric form, and the structures were confirmed with NMR, MS and elemental analysis. Based on the encouraging results in in vitro COX 1/2, 5-LOX and antioxidant assays, we oxidized the compounds and obtained the pure single (major) diastereomer for activities. Among all the compounds, **FM4**, **FM10** and **FM12** were the leading compounds based on their potent IC_{50} values. The IC_{50} values of compounds **FM4**, **FM10** and **FM12** were 0.74, 0.69 and 0.18 μM, respectively, in COX-2 assay. Similarly, the IC_{50} values of these three compounds were also dominant in COX-1 assay. In 5-LOX assay, the majority of our compounds were potent inhibitors of the enzyme. Based on the potency and safety profiles, **FM10** and **FM12** were subjected to the in vivo experiments. The compounds **FM10** and **FM12** were observed with encouraging results in in vivo analgesic and anti-inflammatory models. The molecular docking studies of the selected compounds show binding interactions in the minimized pocked of the target proteins. It is obvious from the overall results that **FM10** and **FM12** are potent analgesic and anti-inflammatory agents.

Keywords: Michael products; anti-inflammatory; antioxidant; analgesic; carrageenan; COX-2; 5-LOX

1. Introduction

Pain and inflammation are closely associated to each other and occur due to complex pathological conditions [1]. Inflammation is basically a response of the cell defense system against tissue injuries or any external stimuli [2]. The onset of inflammation is associated with pain [3]. In the early ages of human development, plants had been used in the management of inflammation and its associated pain [4]. With the development in science and new research, acetylsalicylic acid was first commercialized as an anti-inflammatory drug [5,6]. After the discovery of aspirin, various drugs have been discovered for the management of pain and inflammation, among which NSAIDs (Nonsteroidal Anti-inflammatory Drugs) are the most important and well-known group [7,8]. The pharmacological effects of NSAIDs are due to the inhibition of cyclooxygenase (COX) and lipoxygenase (LOX) enzymes, which are responsible for the metabolism of Arachidonic acid (AA) in the cell membrane and formation of inflammatory mediators such as prostaglandin by COX and leukotrienes by LOX [9]. COX-1 and COX-2 are the two isoforms of cyclooxygenase enzymes that act on the same substrates and catalyze the same reaction but are different in their inhibitor selectivity [10]. COX-1 is mostly involved for maintaining the integrity of the kidney and stomach, while COX-2 produces prostaglandins which mediate pain and inflammation [11,12]. The adverse renal and gastrointestinal effects of NSAIDs are due to COX-1 inhibition, while the inhibition of COX-2 is responsible for the accounts for the therapeutic effects of NSAIDs [13]. In order to prevent such adverse effects of COX-1 inhibition, the scientists turned to design selective COX-2 inhibitors to protect the gastrointestinal tract [14,15].

The oxidation process that takes place in human bodies destroys various cells and tissue and, last, leads to severe illness [16]. It has been observed that the oxidation process may lead to serious conditions such as cancer, various heart diseases and skin problems [17]. Currently, various approaches and techniques are used to eradicate the effect of free radicals [18]. Some of the major sources of antioxidants are natural sources, which may also be helpful in unseen disorders such as stress [19,20]. Day by day, new antioxidants from natural and synthetic sources are improving for the sake of human benefit [21,22]. Most natural products, especially fruits, have specific compounds showing strong antioxidants; however, currently, some of the synthetic compounds also developed have a strong antioxidant capacity [23,24]. Some researchers claims that nitrogenous compounds having a carboxylic acid group show strong antioxidants activities [25].

The Michael reaction of addition nucleophilic moieties to nitro-olefins is a powerful synthetic tool for making the carbon–carbon bond formation [26–28]. The reaction has been explored from long ago, and there is time-to-time modification for new outcomes [29]. The organocatalytic Michael addition has been studied from two decades [30]. However, to date, there have been new avenues for the researchers. Modifications or the exploration of new organocatalysts, extending substrate boundaries and sometimes exploring new chemical or biological applications, still is interesting for researchers [31–34]. The literature shows very limited biological studies on phenylbutanals or their derivatives. In the research early ages, it has been reported as bactericidal [35]. The synthetic derivatives of phenylbutanals have been previously reported with protease inhibitory potentials [36]. This study has been designed to synthesize new Michael products (2*S*,3*S*)-2-(4-isopropylbenzyl)-2-methyl-4-nitro-3-phenylbutanals and their corresponding carboxylic acids for analgesic and anti-inflammatory studies.

2. Results

2.1. Chemistry of the (2S,3S)-2-(4-isopropylbenzyl)-2-methyl-4-nitro-3-phenylbutanal and Their Carboxylic Acids

In the initial synthesis, we have synthesized and purified six nitro-butanal type derivatives having aldehyde functionalities (**FM1-6**). These compounds were purified in the diastereomeric form as the spots on the TLC were not separable. Both the minor and major diastereomers can be seen in the same NMRs. For convenience, we have integrated the whole ^{1}H NMR (with both diastereomers) of compounds **FM1-6**. We also performed the preliminary pharmacological activities on these diastereomeric compounds. In the second step reaction, we oxidized the aldehydic derivatives to their corresponding carboxylic acids (**FM7-12**), as shown in Scheme 1. The carboxylic acid derivatives (**FM7-12**) were clearly separable, and only major diastereomers of these compounds were further used in in vitro and in vivo pharmacological assays. The spectra of compounds are provided in the Supplementary Materials.

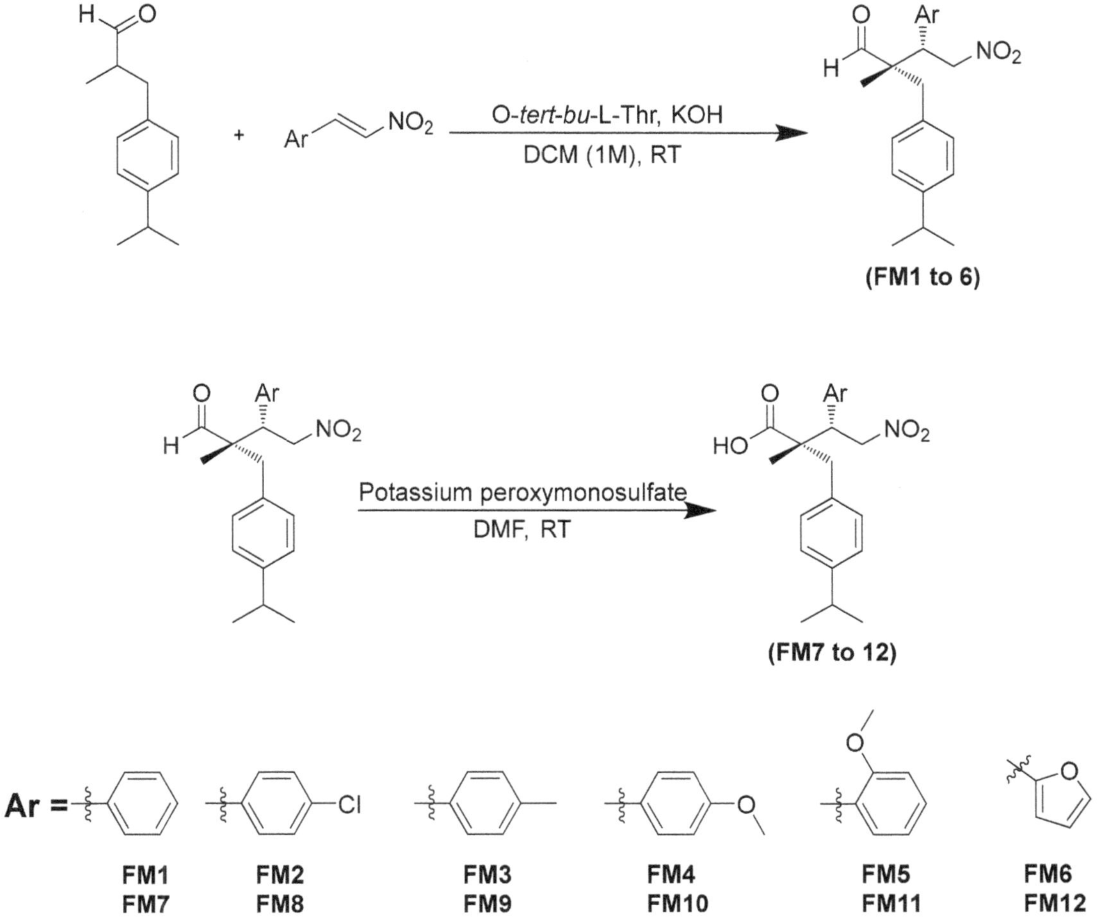

Scheme 1. Synthesis of (2*S*,3*S*)-2-(4-isopropylbenzyl)-2-methyl-4-nitro-3-phenylbutanals (**FM1-FM6**) and its corresponding carboxylic acids (**FM7-FM12**).

The individual details of the compounds (**FM1-12**) are given below.

2.1.1. (2*S*,3*S*)-2-(4-isopropylbenzyl)-2-methyl-4-nitro-3-phenylbutanal (FM1)

The compound **FM1** was isolated as a yellowish oil with 83% isolated yield in 24 h reaction time. The observed and calculated retardation factor value (R_f) was 0.32 in n-hexane and ethyl acetate (4:1). The observed melting point was 149–151 °C. ^{1}H NMR (In deuterated chloroform with 400 MHz): 9.65 (s, 1H), 7.37–7.31 (m, 3H), 7.25–7.22 (m, 1H), 7.15–7.10 (m, 3H), 6.91 (d, *J* = 6.62 Hz, 2H), 4.93–4.86 (m, 1H), 4.68 (dd, *J* = 3.85, 13.12 Hz, 1H), 3.85 (dd, *J* = 3.86, 13.17 Hz, 1H), 3.04 (d, *J* = 12.78 Hz, 1H), 2.92–2.81 (m, 1H), 2.41 (d, *J* = 12.73 Hz, 1H), 1.22 (d, *J* = 6.92 Hz, 6H) and 1.06 (s, 3H). ^{13}C NMR (In deuterated chloroform with 100 MHz): 206.09, 204.93, 147.79, 135.41, 132.40, 130.34, 130.31, 129.51, 129.34, 129.03, 128.95, 128.92, 128.47, 128.40, 126.71, 126.61, 52.48, 51.98, 49.65, 48.88, 42.13, 40.47, 33.78, 24.05, 17.95 and 16.24. LC-MS: m/z = 340.42 $[M + H]^+$; analysis calculated for $C_{21}H_{25}NO_3$. C, 74.31; H, 7.42; N, 4.13 and O, 14.14. Observed: C, 74.39; H, 7.40 and N, 4.10.

2.1.2. (2*S*,3*S*)-3-(4-chlorophenyl)-2-(4-isopropylbenzyl)-2-methyl-4-nitrobutanal (FM2)

The compound **FM2** was isolated as a clear, oily semisolid with 75% isolated yield in 30 h reaction time. The observed and calculated retardation factor value (R_f) was 0.35 in n-hexane and ethyl acetate (4:1). The observed melting point was 173–175 °C. ^{1}H NMR (In deuterated chloroform with 400 MHz): 9.65 (s, 1H), 7.29–7.07 (m, 6H), 6.94 (d, *J* = 8.25 Hz, 2H), 4.94 (dd, *J* = 11.12, 13.27 Hz, 1H), 4.68 (dd, *J* = 3.65, 13.32 Hz, 1H), 3.84 (dd, *J* = 3.66, 11.13 Hz, 1H), 3.11 (d, *J* = 12.33 Hz, 1H), 2.93–2.81 (m, 1H), 2.75–2.64 (m, 1H), 1.27 (d, *J* = 6.94 Hz, 6H) and 1.14 (s, 3H). ^{13}C NMR (In deuterated chloroform with 100 MHz): 206.24, 205.51, 145.65, 145.00, 138.34, 136.98, 132.88, 132.54, 132.04, 131.11, 130.74, 129.86, 129.75, 129.34, 129.17, 129.12, 127.14, 126.85, 126.64, 53.51, 52.14, 49.35, 43.32, 41.31, 37.21, 33.24, 33.00, 24.24, 24.10, 20.17, 18.98, 17.23 and 15.35. LC-MS: m/z = 374.87 $[M + H]^+$; analysis calculated for $C_{21}H_{24}ClNO_3$. C, 67.46; H, 6.47; Cl, 9.48; N, 3.75 and O, 12.84. Observed: C, 67.53; H, 6.45 and N, 3.73.

2.1.3. (2*S*,3*S*)-2-(4-isopropylbenzyl)-2-methyl-4-nitro-3-(p-tolyl)butanal (FM3)

The compound **FM3** was isolated as a white powder with 88% isolated yield in 20 h reaction time. The observed and calculated retardation factor value (R_f) was 0.38 in n-hexane and ethyl acetate (4:1). The observed melting point was 135–137 °C. ^{1}H NMR (in deuterated chloroform with 400 MHz): 9.67 (s, 1H), 7.20–7.12 (m, 6H), 6.93 (d, *J* = 8.11 Hz, 2H), 4.92 (dd, *J* = 11.61, 12.93 Hz, 1H), 4.68 (dd, *J* = 3.81, 13.02 Hz, 1H), 3.84 (dd, *J* = 3.95, 11.53 Hz, 1H), 3.07 (d, *J* = 13.80 Hz, 1H), 2.95–2.84 (m, 1H), 2.77–2.63 (m, 1H), 2.35 (s, 3H), 1.24 (d, *J* = 6.92 Hz, 6H) and 1.07 (s, 3H). ^{13}C NMR (In deuterated chloroform with 100 MHz): 206.17, 205.00, 147.69, 147.08, 138.14, 136.24, 132.98, 132.64, 132.24, 130.41, 130.34, 129.67, 129.62, 129.39, 129.21, 129.08, 126.68, 126.66, 126.59, 52.62, 52.15, 49.33, 48.56, 48.31, 42.02, 40.57, 36.35, 33.82, 33.80, 24.17, 24.07, 21.13, 17.77, 16.07 and 13.38. LC-MS: m/z = 354.45 $[M + H]^+$; analysis calculated for $C_{22}H_{27}NO_3$. C, 74.76; H, 7.70; N, 3.96 and O, 13.58. Observed: C, 74.86; H, 7.68 and N, 3.93.

2.1.4. (2*S*,3*S*)-2-(4-isopropylbenzyl)-3-(4-methoxyphenyl)-2-methyl-4-nitrobutanal (FM4)

The compound **FM4** was isolated as a white solid with 78% isolated yield in 24 h reaction time. The observed and calculated retardation factor value (R_f) was 0.30 in n-hexane and ethyl acetate (4:1). The observed melting point was 117–119 °C. ^{1}H NMR (In deuterated chloroform with 400 MHz): 9.64 (s, 1H), 7.34–7.15 (m, 6H), 6.91 (d, *J* = 8.21 Hz, 2H), 4.94 (dd, *J* = 11.04, 12.46 Hz, 1H), 4.72 (dd, *J* = 3.75, 12.52 Hz, 1H), 3.89 (dd, *J* = 3.75, 11.03 Hz, 1H), 3.85 (s, 3H), 3.10 (d, *J* = 11.14 Hz, 1H), 2.91–2.84 (m, 1H), 2.80–2.69 (m, 1H), 1.23 (d, *J* = 6.90 Hz, 6H) and 1.12 (s, 3H). ^{13}C NMR (In deuterated chloroform with 100 MHz): 207.14, 206.04, 148.24, 147.54, 139.34, 136.52, 132.99, 132.75, 132.54, 131.87, 130.94, 129.85, 129.74, 129.64, 129.51, 129.38, 127.26, 126.99, 126.45, 58.52, 56.54, 50.34, 49.44, 48.47, 40.54, 40.14, 36.05, 33.52, 32.52, 25.52, 24.99, 21.51, 16.51, 16.15 and 11.41. LC-MS: m/z = 370.45 $[M + H]^+$; analysis calculated for $C_{22}H_{27}NO_4$. C, 71.52; H, 7.37; N, 3.79 and O, 17.32. Observed: C, 71.63; H, 7.35 and N, 3.76.

2.1.5. (2*S*,3*S*)-2-(4-isopropylbenzyl)-3-(2-methoxyphenyl)-2-methyl-4-nitrobutanal (FM5)

The compound **FM5** was isolated as a half white powder with 72% isolated yield in 28 h reaction time. The observed and calculated retardation factor value (R_f) was 0.34 in n-hexane and ethyl acetate (4:1). The observed melting point was 129–131 °C. ^{1}H NMR (In deuterated chloroform with 400 MHz): 9.61 (s, 1H), 7.39–7.26 (m, 4H), 7.18–7.02 (m, 2H) 6.92 (d, *J* = 7.54 Hz, 2H), 4.91 (dd, *J* = 12.51, 13.40 Hz, 1H), 4.80 (dd, *J* = 3.94, 13.42 Hz, 1H), 3.96 (dd, *J* = 3.95, 12.53 Hz, 1H), 3.82 (s, 3H), 3.09 (d, *J* = 11.74 Hz, 1H), 2.90–2.81 (m, 1H), 2.75–2.63 (m, 1H), 1.22 (d, *J* = 6.91 Hz, 6H) and 1.09 (s, 3H). ^{13}C NMR (In deuterated chloroform with 100 MHz): 207.51, 206.31, 148.51, 147.30, 140.04, 138.50, 133.04, 132.92, 132.83, 132.64, 131.73, 130.54, 129.99, 129.90, 129.71, 129.58, 128.82, 127.52, 126.79, 57.30, 56.00, 52.74, 49.07, 48.37, 43.53, 41.43, 38.52, 32.89, 32.04, 24.16, 23.74, 20.43, 17.81, 16.63 and

11.74. LC-MS: m/z = 370.45 [M + H]$^+$; analysis calculated for $C_{22}H_{27}NO_4$. C, 71.52; H, 7.37; N, 3.79 and O, 17.32. Observed: C, 71.62; H, 7.35 and N, 3.77.

2.1.6. (2*S*,3*S*)-3-(furan-2-yl)-2-(4-isopropylbenzyl)-2-methyl-4-nitrobutanal (FM6)

The compound **FM6** was isolated a as yellowish semisolid with 95% isolated yield in 20 h reaction time. The observed and calculated retardation factor value (R_f) was 0.35 in n-hexane and ethyl acetate (4:1). The observed melting point was 161–163 °C. ^{1}H NMR (In deuterated chloroform with 400 MHz): 9.59 (s, 1H), 7.40 (dd, *J* = 4.59, 1.84 Hz, 1H), 7.14–7.10 (m, 2H), 7.01–6.93 (m, 2H), 6.34–6.32 (m, 1H), 6.24 (d, *J* = 2.65 Hz, 1H), 4.77 (dd, *J* = 11.31, 12.87 Hz, 1H), 4.57 (dd, *J* = 3.54, 12.88 Hz, 1H), 3.96 (dd, *J* = 3.53, 11.26 Hz, 1H), 2.99–2.94 (m, 1H), 2.89–2.78 (m, 1H), 2.52 (d, *J* = 13.92 Hz, 1H), 1.21 (d, J = 6.91 Hz, 6H) and 1.13 (s, 3H). ^{13}C NMR (In deuterated chloroform with 100 MHz): 205.14, 149.64, 147.87, 142.97, 130.40, 130.27, 126.65, 126.59, 110.54, 110.46, 75.32, 74.89, 52.09, 42.03, 41.52, 40.20, 33.73, 30.96, 23.94, 18.08 and 16.56. LC-MS: m/z = 330.39 [M + H]$^+$; analysis calculated for $C_{19}H_{23}NO_4$. C, 69.28; H, 7.04; N, 4.25 and O, 19.43. Observed: C, 69.40; H, 7.02 and N, 4.22.

2.1.7. (2*S*,3*S*)-2-(4-isopropylbenzyl)-2-methyl-4-nitro-3-phenylbutanoic Acid (FM7)

The compound **FM7** was isolated as a yellowish solid with 94% isolated yield. The observed and calculated retardation factor value (R_f) was 0.18 in n-hexane and ethyl acetate (4:1). The observed melting point was 243–245 °C. ^{1}H NMR (In deuterated chloroform with 400 MHz): 12.25 (s, 1H), 7.43–7.37 (m, 3H), 7.35–7.21 (m, 2H), 7.05 (d, *J* = 7.2 Hz, 2H), 6.95 (d, *J* = 6.7 Hz, 2H), 4.96 (dd, *J* = 4.6, 13.8 Hz, 1H), 4.74 (dd, *J* = 3.9, 13.8 Hz, 1H), 3.87 (dd, *J* = 4.6, 3.9 Hz, 1H), 3.07 (d, *J* = 11.2 Hz, 1H), 2.94 (sept, *J* = 6.9 Hz, 1H), 2.54 (d, *J* = 11.2 Hz, 1H), 1.22 (d, *J* = 6.92 Hz, 6H) and 1.06 (s, 3H). ^{13}C NMR (In deuterated chloroform with 100 MHz): 177.6, 145.4, 136.2, 132.1, 130.5, 130.0, 129.3, 129.2, 128.9, 128.4, 53.8, 48.4, 42.5, 35.1, 25.3, 15.1 and 14.9. LC-MS: m/z = 356.42 [M + H]$^+$; analysis calculated for $C_{21}H_{25}NO_4$. C, 70.96; H, 7.09; N, 3.94 and O, 18.01. Observed: C, 71.07; H, 7.07 and N, 3.92.

2.1.8. (2*S*,3*S*)-3-(4-chlorophenyl)-2-(4-isopropylbenzyl)-2-methyl-4-nitrobutanal (FM8)

The compound **FM8** was isolated as a half white solid with 95% isolated yield. The observed and calculated retardation factor value (R_f) was 0.22 in n-hexane and ethyl acetate (4:1). The observed melting point was 259–261 °C. ^{1}H NMR (In deuterated chloroform with 400 MHz): 12.21 (s, 1H), 7.30 (d, *J* = 8.1 Hz, 2H), 7.14 (d, *J* = 8.2, 2H), 7.04 (d, *J* = 6.7 Hz, 2H), 6.89 (d, *J* = 6.5 Hz, 2H), 4.94 (dd, *J* = 4.9, 13.4 Hz, 1H), 4.76 (dd, *J* = 5.9, 13.4 Hz, 1H), 3.83 (dd, *J* = 4.9, 5.9 Hz, 1H), 3.05 (d, *J* = 12.0 Hz, 1H), 2.95-2.87 (m, 1H), 2.57 (d, *J* = 11.9 Hz, 1H), 1.27 (d, *J* = 6.87 Hz, 6H) and 1.08 (s, 3H). ^{13}C NMR (In deuterated chloroform with 100 MHz): 177.4, 146.1, 137.8, 134.3, 131.4, 130.8, 129.7, 128.2, 128.1, 53.4, 47.9, 41.7, 34.6, 23.7, 16.4 and 13.8. LC-MS: m/z = 390.87 [M + H]$^+$; analysis calculated for $C_{21}H_{24}ClNO_4$. C, 64.69; H, 6.20; Cl, 9.09; N, 3.59 and O, 16.42. Observed: C, 64.78; H, 6.18 and N, 3.56.

2.1.9. (2*S*,3*S*)-2-(4-isopropylbenzyl)-2-methyl-4-nitro-3-(p-tolyl)butanal (FM9)

The compound **FM9** was isolated as a white solid with 90% isolated yield. The observed and calculated retardation factor value (R_f) was 0.25 in n-hexane and ethyl acetate (4:1). The observed melting point was 233–235 oC. ^{1}H NMR (In deuterated chloroform with 400 MHz): 12.23 (s, 1H), 7.24 (d, *J* = 6.5 Hz, 2H), 7.11 (d, J = 6.5, 2H), 7.06 (d, *J* = 5.4 Hz, 2H), 6.81 (d, *J* = 5.5 Hz, 2H), 4.92 (dd, *J* = 3.7, 12.1 Hz, 1H), 4.77 (dd, *J* = 4.6, 12.1 Hz, 1H), 3.85 (dd, *J* = 3.7, 4.6 Hz, 1H), 3.08 (d, *J* = 11.3 Hz, 1H), 2.93–2.85 (m, 1H), 2.55 (d, *J* = 11.0 Hz, 1H), 2.31 (s, 3H), 1.24 (d, *J* = 6.96 Hz, 6H) and 1.05 (s, 3H). ^{13}C NMR (In deuterated chloroform with 100 MHz): 175.2, 144.6, 136.0, 135.4, 131.4, 131.8, 130.7, 130.4, 129.1, 128.7, 52.0, 44.8, 42.2, 32.7, 26.3, 24.1, 15.3 and 14.9. LC-MS: m/z = 370.45 [M + H]$^+$; analysis calculated for $C_{22}H_{27}NO_4$. C, 71.52; H, 7.37; N, 3.79 and O, 17.32. Observed: C, 71.63; H, 7.35 and N, 3.76.

2.1.10. (2*S*,3*S*)-2-(4-isopropylbenzyl)-3-(4-methoxyphenyl)-2-methyl-4-nitrobutanal (FM10)

The compound **FM10** was isolated as a white solid with 93% isolated yield. The observed and calculated retardation factor value (R_f) was 0.20 in n-hexane and ethyl acetate (4:1). The observed melting point was 199–201 °C. ^{1}H NMR (In deuterated chloroform with 400 MHz): 12.18 (s, 1H), 7.28 (d, *J* = 8.3 Hz, 2H), 7.15 (d, *J* = 8.3, 2H), 7.09 (d, *J* = 5.7 Hz, 2H), 6.96 (d, *J* = 5.8 Hz, 2H), 4.90 (dd, *J* = 5.2, 13.9 Hz, 1H), 4.82 (dd, *J* = 6.1, 13.9 Hz, 1H), 3.87 (s, 3H), 3.81 (dd, *J* = 5.0, 5.9 Hz, 1H), 3.07 (d, *J* = 11.3 Hz, 1H), 2.90 (sept, *J* = 6.7 Hz, 1H), 2.56 (d, *J* = 12.4 Hz, 1H), 1.24 (d, *J* = 6.91 Hz, 6H) and 1.06 (s, 3H). ^{13}C NMR (In deuterated chloroform with 100 MHz): 175.3, 147.6, 140.1, 137.5, 132.6, 132.2, 129.2, 128.8, 128.6, 59.5, 51.0, 49.7, 40.0, 31.4, 21.0, 15.4 and 15.1. LC-MS: m/z = 386.45 [M + H]$^+$; analysis calculated for $C_{22}H_{27}NO_5$. C, 68.55; H, 7.06; N, 3.63 and O, 20.75. Observed: C, 68.68; H, 7.04 and N, 3.61.

2.1.11. (2*S*,3*S*)-2-(4-isopropylbenzyl)-3-(2-methoxyphenyl)-2-methyl-4-nitrobutanal (FM11)

The compound **FM11** was isolated as a white powder with 89% isolated yield. The observed and calculated retardation factor value (R_f) was 0.19 in n-hexane and ethyl acetate (4:1). The observed melting point was 211–213 °C. ^{1}H NMR (In deuterated chloroform with 400 MHz): 12.15 (s, 1H), 7.37–7.28 (m, 3H), 7.16–7.04 (m, 3H), 6.87 (d, *J* = 7.1 Hz, 2H), 4.89 (dd, *J* = 3.8, 12.1 Hz, 1H), 4.87 (dd, *J* = 5.8, 11.9 Hz, 1H), 3.84 (dd, *J* = 3.8, 5.6 Hz, 1H), 3.77 (s, 3H), 3.00 (d, *J* = 11.0 Hz, 1H), 2.87-2.80 (m, 1H), 2.61 (d, *J* = 11.0 Hz, 1H), 1.26 (d, *J* = 6.93 Hz, 6H) and 1.08 (s, 3H). ^{13}C NMR (In deuterated chloroform with 100 MHz): 178.2, 145.2, 143.5, 138.3, 133.4, 139.4, 129.0, 129.7, 128.6, 55.1, 50.6, 46.4, 42.4, 32.7, 24.2, 18.4 and 17.1. LC-MS: m/z = 386.45 [M + H]$^+$; analysis calculated for $C_{22}H_{27}NO_5$. C, 68.55; H, 7.06; N, 3.63 and O, 20.75. Observed: C, 68.69; H, 7.04 and N, 3.60.

2.1.12. (2*S*,3*S*)-3-(furan-2-yl)-2-(4-isopropylbenzyl)-2-methyl-4-nitrobutanal (FM12)

The compound **FM12** was isolated as a yellowish powder with 93% isolated yield. The observed and calculated retardation factor value (R_f) was 0.23 in n-hexane and ethyl acetate (4:1). The observed melting point was 251–253 °C. ^{1}H NMR (In deuterated chloroform with 400 MHz): 12.26 (s, 1H), 7.45 (d, *J* = 1.9 Hz, 1H), 7.07 (d, *J* = 5.5 Hz, 1H), 6.91 (d, *J* = 5.5 Hz, 1H), 6.31 (dd, *J* = 1.9, 3.2 Hz, 1H), 6.22 (d, *J* = 3.2 Hz, 1H), 4.78 (dd, *J* = 11.1, 4.87 Hz, 1H), 4.61 (dd, *J* = 4.2, 12.6 Hz, 1H), 3.94 (dd, *J* = 3.9, 11.0 Hz, 1H), 2.98 (d, *J* = 12.9 Hz, 1H), 2.85–2.78 (m, 1H), 2.55 (d, *J* = 13.5 Hz, 1H), 1.21 (d, *J* = 6.6 Hz, 6H) and 1.17 (s, 3H). ^{13}C NMR (In deuterated chloroform with 100 MHz): 205.14, 149.64, 147.87, 142.97, 130.40, 130.27, 126.65, 126.59, 110.54, 110.46, 75.32, 74.89, 52.09, 42.03, 41.52, 40.20, 33.73, 30.96, 23.94, 18.08 and 16.56. LC-MS: m/z = 346.38 [M + H]$^+$; analysis calculated for $C_{19}H_{23}NO_5$. C, 66.07; H, 6.71; N, 4.06 and O, 23.16. Observed: C, 66.21; H, 6.69 and N, 4.03.

2.2. Antioxidant Results

We tested the antioxidant activities of our compounds (**FM1-12**) using DPPH and ABTS standard methods, and the potencies are summarized in Table 1. We compared our activities with the standard gallic acid, which exhibited IC_{50} values of 09.02 and 03.23 μM against DPPH and ABTS free radicals, respectively. From our results, it can be easily depicted that the oxidized compounds (**FM7-12**) were comparatively potent antioxidants compared to their aldehydic derivatives (**FM1-6**). In aldehydic derivatives, **FM3** and **FM4** were found to be potent in both DPPH and ABTS assays. Similarly, in the oxidized form of compounds (**FM7-12**), compounds **FM10** and **FM12** were found with potent IC_{50} values. The observed IC_{50} values for compounds **FM10** and **FM12** were 08.36 and 15.30 μM in DPPH and 08.90 and 17.22 μM in ABTS assay, respectively. In comparison, the standard gallic acid exhibited IC_{50} values of 09.02 and 03.23 μM against DPPH and ABTS free radicals.

Table 1. ABTS and DPPH free radicals scavenging results of compounds **FM1-FM12**.

Samples	DPPH IC_{50} (μM)	ABTS IC_{50} (μM)
FM1	54.35	62.91
FM2	55.36	42.03
FM3	15.08	11.47
FM4	21.08	25.78
FM5	183.73	190.57
FM6	49.70	37.67
FM7	22.54	24.65
FM8	23.50	17.51
FM9	17.02	18.20
FM10	08.36	08.90
FM11	53.32	46.32
FM12	15.30	17.22
Gallic acid	09.02	03.23

2.3. Cyclo and Lipoxygenase Results

The results of in vitro cyclooxygenase and lipoxygenase enzymes' inhibitions obtained from our synthesized compounds (**FM1-FM12**) are summarized in Table 2. Our compounds were comparatively potent inhibitors of COX-2 enzymes compared to COX-1. In COX-2 results, we observed that the compound **FM4** and its corresponding carboxylic acid (**FM10**) were comparatively more potent, giving IC_{50} values of 0.74 and 0.69 μM, respectively. Both of these compounds have the para-methoxy substitution patterns, which probably have an effect in these specific enzymes' inhibitions. Similarly, the compound with carboxylic acid and furyl moieties (**FM12**) was the most potent, giving an IC_{50} value of 0.18 μM. On the other hand, the observed results of COX-1 were in a different pattern from that of the COX-2. In comparison, the standard celecoxib exhibited IC_{50} values of 0.042 and 10.87 μM against the COX-2 and -1 enzymes. The calculated selectivity index (SI) was highest for compounds **FM4** (42.8), **FM10** (62.7) and **FM12** (277.1). Though the potency was slightly lower than in standard drugs, however, the SI of our potent compound **FM12** (SI 277.1) was higher than that of standard celecoxib (SI 258.8). A comparatively high SI value shows that the compound would be a good choice specifically in cases of gastric ulcers. The lipoxygenase pathway was also assessed with the available enzyme, and the potencies of our compounds were compared with the zileuton standard drug. Overall, all of our tested compounds were potent inhibitors of 5-lipoxygenase, as can be depicted from the IC_{50} values in Table 2. In 5-LOX assay, five of our compounds were found to be most potent giving IC_{50} values less than one. Compounds **FM2**, **FM4**, **FM7**, **FM8** and **FM12** gave IC_{50} values of 0.64, 0.98, 0.73, 0.87 and 0.43 μM, respectively. The standard zileuton IC_{50} value was 0.30 μM against 5-LOX.

2.4. In Vivo Results

Based on the in vitro results, we selected three of our compounds **FM4**, **FM10** and **FM12** for the in vivo studies. In the acute toxicity studies of selected compounds, we only observed very mild seizures and disturbances in breath (temporary) at the highest dose (2000 mg/kg) of compound **FM4**. So, based on this very mild toxicity effect, we excluded the compound **FM4** from in vivo experiments. The other two compounds **FM10** and **12** were found safe even at the maximum dosage. In these two compounds, we observed no behavioral changes in experimental albino mice. A dose of 2000 mg/kg of the compounds was declared safe for animals use. The details of acute toxicity results are summarized in Table 3. According to the organization for economic cooperation and development (OECD) guidelines for the oral acute toxicity, an LD_{50} dose of the >300–2000 was categorized as category 4, and hence the drug was established to be safe.

Table 2. COX-2/1 and 5-LOX inhibitory potentials of the synthesized compounds (**FM1-12**).

Samples	IC_{50} (µM) ± SEM		SI	5LOX IC_{50} (µM) ± SEM
	COX-2	COX-1		
FM1	1.21 ± 0.03	14.76 ± 1.19	12.2	1.81 ± 0.11
FM2	2.51 ± 0.24	38.04 ± 1.65	15.1	0.64 ± 0.01
FM3	3.53 ± 0.18	12.79 ± 1.08	3.6	9.69 ± 0.99
FM4	0.74 ± 0.03	31.70 ± 1.37	42.8	0.98 ± 0.12
FM5	8.15 ± 0.98	58.37 ± 2.08	7.2	16.33 ± 1.02
FM6	4.72 ± 0.08	54.78 ± 1.95	11.6	6.17 ± 0.23
FM7	1.09 ± 0.01	25.31 ± 1.22	23.2	0.73 ± 0.05
FM8	1.99 ± 0.04	61.22 ± 1.84	30.8	0.87 ± 0.08
FM9	8.31 ± 1.01	50.07 ± 1.33	6.0	2.36 ± 0.33
FM10	0.69 ± 0.05	43.29 ± 1.16	62.7	1.77 ± 0.14
FM11	4.25 ± 0.21	35.02 ± 2.13	8.2	11.01 ± 1.14
FM12	0.18 ± 0.01	49.89 ± 1.91	277.1	0.43 ± 0.02
Celecoxib	0.042 ± 0.001	10.87 ± 1.15	258.8	------
Zileuton	------	------	------	0.50 ± 0.02

Table 3. Group of animals and drug quantities given for acute toxicity studies with various synthesized compounds.

Groups	Animals (Male/Female)	Compounds (FM4, FM10, FM12) mg/kg Body Weight
1	5	25
2	5	50
3	5	100
4	5	200
5	5	300
6	5	400
7	5	500
8	5	1000
9	5	2000

2.5. Carrageenan-Induced Inflammation Results

Based on the acute toxicity studies, we extended compound **FM10** and **FM12** for in vivo experiments. The carrageenan activity results on concentrations of 25, 50 and 75 mg/kg of the compounds and respective control groups are presented in Figure 1. Overall, our compounds have shown excellent anti-inflammatory activities in this assay. The observed and recorded activity of compound **FM10** was 54.54% at the first hour and remained in observations till the fourth hour. At the fourth hour, the activity was 64.92% at a concentration of 75 mg/kg. The activity profile of our compound was compared with the standard aspirin. The aspirin's activity was 52.77% at the first hour and 61.43% at the fourth hour of observations. Similarly, the compound **FM12** activity was 51.71 and 59.55% at the first and fourth hours, respectively.

2.6. Acetic Acid Induced Analgesic Results

A dose dependent analgesic activity profile was observed in the acetic acid induction writhing assay of analgesia. The analgesic potential was indomitable using the acetic acid induction writhing method, which displayed significant potential. Both tested samples were active on the doses of 25, 50 and 75 mg/kg b.wt. The tested compounds **FM10** and **FM12** at the highest doses (75 mg/kg) showed the highest activity when compared to the standard drug (acetyl salicylic acid) (Figure 2c). The standard drug (10 mg/kg) mean inhibition of writhes was 73.01%. **FM10** exhibited a mean inhibition of 85.52% at a high dose (75 mg/kg). Likewise, the compound **FM12** also showed a good inhibition (79.10%) at the same dose, which displayed the highest peripheral analgesic potential. The outcome also pointed out that compounds at a low dose, i.e., 25 as well as 50 mg/kg b.wt, also had moderate to good peripheral analgesic potential, which is displayed in Figure 2a,b.

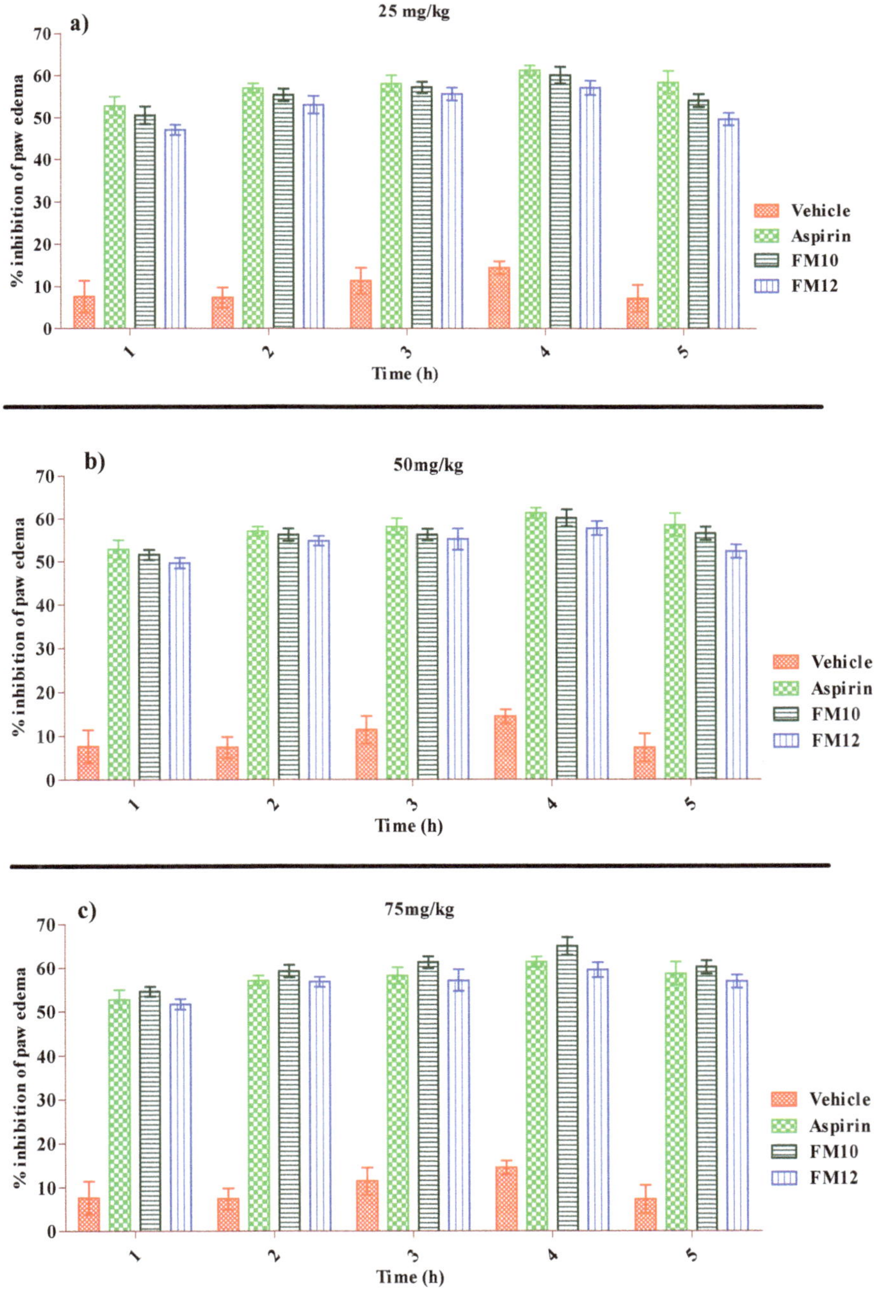

Figure 1. Results of carrageenan assays of compounds **FM10** and **12** at concentrations 25 (**a**), 50 (**b**) and 75 mg/kg (**c**).

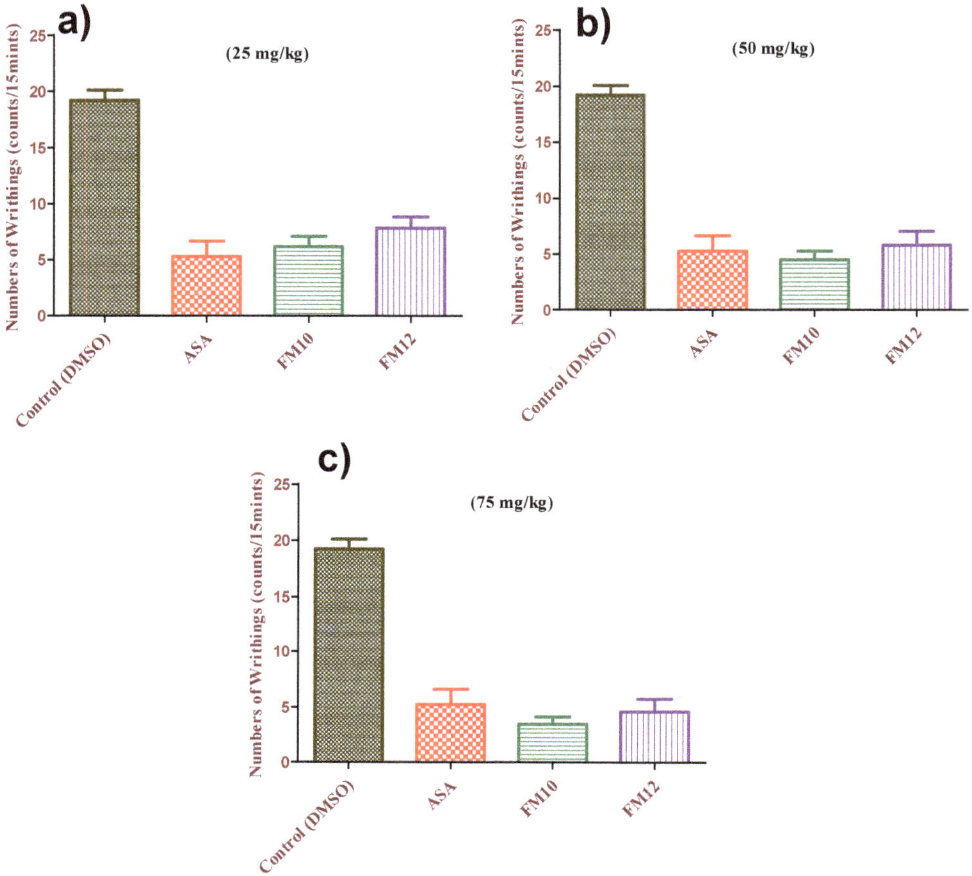

Figure 2. Acetic acid induced test result at different doses of compounds **FM10** and **12**. (**a**) 25 mg/kg, (**b**) 50 mg/kg and (**c**) 75 mg/kg.

2.7. Results of Formalin In Vivo Assay

The formalin (2%) intraplantar (*i.p*) induction to animals induces a classical biphasic licking response. The time for licking in early phase was 0 to 5 min, which was noted as 57.21 ± 0.42s, and for the late phase (15 to 30 min) it was recorded as 78.07 ± 0.43 s in the control tested group. The pre-treatment of tested compounds at different doses (i.e., 25, 50, 75 mg/kg i.p.) was checked. The compound **FM10** displayed outstanding activity, was significant next to the licking test in both stages, and had an obvious decrease of 87.59% and 76.41% inhibition in the early as well as late phase, as displayed in the Table 4. Likewise, the morphine (5 mg/kg, *i.p.*) injection exhibited clear action in the decrease of both phases of neurogenic pain (88.64% and 93.81%). So, our tested sample, especially compound **FM10,** was close to the standard drug at phase I. Likewise, compound **FM12** in phase I displayed 58.62, 72.78 and 83.54% inhibitions, whilst in phase II it showed 46.94, 60.86 and 72.02% inhibition at various doses such as 25, 50 and 75 mg/kg, correspondingly. Morphine plus naloxone displayed 10.29% potential in the early phase, and in the late phase it exhibited 14.79% activity. The indomethacin with naloxone displayed 10.29% activity in the early phase and 14.79% in the late phase.

Table 4. The effects of selected compounds on formalin test.

Samples Names	Dose in mg/kg Body Weight	Time Spent (Licking)			
		0–5 (min.)	Percent (%) Inhibition	15–30 (min.)	Percent (%) Inhibition
Negative control	-	57.21 ± 0.42	-	78.07 ± 0.43	-
FM10	25	18.10 ± 0.59	68.37 ***	37.78 ± 0.73	51.61 ***
	50	12.31 ± 0.47	78.49 ***	25.92 ± 0.98	66.80 ***
	75	7.10 ± 0.92	87.59 ***	18.42 ± 0.56	76.41 ***
FM12	25	23.68 ± 0.68	58.62 ***	41.43 ± 0.92	46.94 ***
	50	15.58 ± 0.48	72.78 ***	30.56 ± 0.65	60.86 ***
	75	09.42 ± 0.57	83.54 ***	21.85 ± 0.87	72.02 ***
Morphine	5	6.50 ± 0.78	88.64 ***	4.83 ± 0.62	93.81 ***
Morphine + Nalaxone	5 + 02	51.32 ± 0.33	10.29 **	66.52 ± 0.40	14.79 ***
Indomethacin + Nalaxone	10 + 02	34.00 ± 0.20	40.57 ***	20.00 ± 0.74	74.38 ***

Data are shown as the mean ± S.E.M; values are significantly variant compared to the control group, and all the data were analyzed via ANOVA followed by Dunnett's test; $n = 5$, **: $p < 0.01$, ***: $p < 0.001$.

2.8. Hotplate Analgesic Results

The results of the analgesic potential of the compounds on the hotplate method are summed up in Table 5. The **FM10** was yet again found to display a significant increase in latency time contrast to the standard control (morphine). Primarily, at 15 min, the reaction time means of all three doses of **FM10** were noted as 8.50 ± 0.64, 10.78 ± 0.32 and 13.52 ± 0.65 correspondingly at the doses of 25, 50 and 75 mg/kg b.wt. After sixty (60) minutes, the mean reaction time of the three (3) doses was noted as 6.66 ± 0.33, 8.74 ± 0.46 and 10.36 ± 0.54 min, correspondingly. The initial time reaction at 15 min for the morphine (standard drug) at 5 mg/kg was eminent as 12.88 ±0.26 min, and at 60 min it was noted as 11.22 ± 0.45 min. Likewise, at 15 min, the mean reaction times for compound **FM12** were noted as 7.50 ± 0.64, 9.42 ± 0.74 and 12.44 ± 0.62 min at 25, 50 and 75 mg/kg. Similarly, at 60 min, the reaction times for **FM12** were calculated as 6.45 ± 0.74, 7.39 ± 0.67 and 9.36 ±0.54 min on the similar tested doses.

Table 5. Results of analgesic activity following hot plate model.

Samples	Dose mg/kg	Reaction Time on Hot Plate in (min)			
		15	30	45	60
-ve control	-	3.91 ± 0.52	4.95 ± 0.39	3.35 ± 0.59	1.73 ± 0.44
FM10	25	8.50 ± 0.64 ***	8.83 ± 0.64 ***	7.52 ± 0.76 ***	6.66 ± 0.33 ***
	50	10.78 ± 0.32 ***	9.26 ± 0.43 ***	9.10 ± 0.57 ***	8.74 ± 0.46 ***
	75	13.52 ± 0.65 ***	12.23 ± 0.44 ***	11.54 ± 0.64 ***	10.36 ± 0.54 ***
FM12	25	7.50 ± 0.64 ***	8.36 ± 0.49 ***	7.27 ± 0.48 ***	6.45 ± 0.74 ***
	50	9.42 ± 0.74 ***	8.52 ± 0.45 ***	8.26 ± 0.47 ***	7.39 ± 0.67 ***
	75	12.44 ± 0.62 ***	11.51 ± 0.62 ***	9.54 ± 0.75 ***	9.36 ± 0.54 ***
Morphine	5	12.88 ± 0.26 ***	12.31 ± 0.62 ***	11.86 ± 0.87 ***	11.22 ± 0.45 ***

The values are existing as the mean ± SEM ($n = 5$). The asterisks display significance level in comparison with negative control: data were analyzed via Dunnett's test; *** $p < 0.001$.

2.9. Molecular Docking Studies

In compound **FM4,** NO_2 and OCH_3 moieties form conventional H-bonds with Arg120, Tyr355 and Arg513, while the aromatic ring of anisole forms a π–lone pair interaction with Tyr355 and π–alkyl interaction with Val523 and Ala527. The aromatic part of Cumene shows an amide π-stacked interaction with Gly526 and π–alkyl interaction with Val523 and Ala527, while aliphatic moiety shows a π–alkyl interaction with Phe381, Tyr385 and Trp387 (Figure 3A). In compound **FM12,** NO_2 and carboxylic acid part form four H-bonds with Arg120, Tyr355 and Val523. The aromatic cumene ring shows a π–sigma and π–alkyl interaction with Ser353 and Val523, respectively, while furan moiety displays π–alkyl interaction with Val349 and Leu531. The compound also exhibits a π–alkyl interaction with His90, Tyr355 and Phe518 (Figure 3B). In compound **FM-10,** carboxylic acid moiety forms two conventional hydrogen bonds with Arg120, while methoxy moiety attached with an aromatic ring also shows a conventional hydrogen bond interaction with Tyr385. One of the aromatic rings shows a π–sigma interaction with Ser353, and FM-12 also shows π–alkyl interactions with His90, Leu352, Ser353, Tyr355, Phe518, Val349, Val523 and Ala527 (Figure 3C).

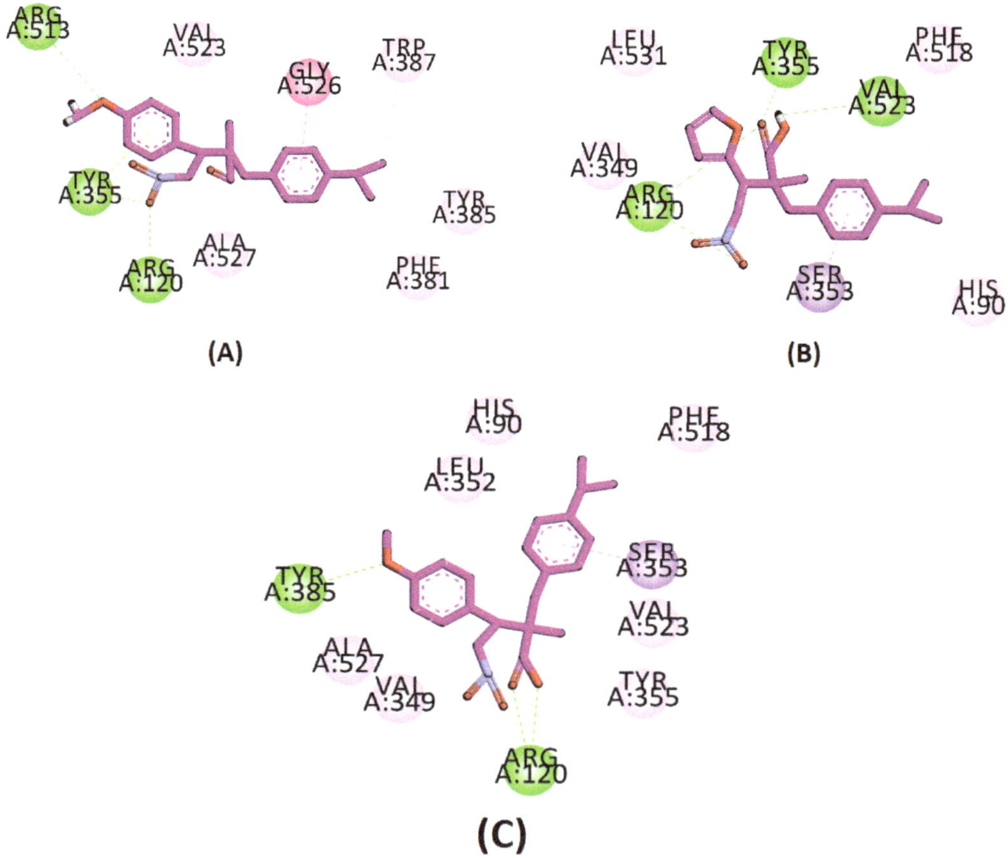

Figure 3. Two-dimensional interaction plots of (**A**) **FM4,** (**B**) **FM12** and (**C**) **FM-10** in active site of COX-2 (PDB ID = 1CX2).

Compound **FM2** shows a halogen interaction with Ile406 and Asn407 via chlorine moiety, while the aromatic ring of chlorobenzene shows a π–π T-shaped interaction with His372. NO2 moiety form a conventional H-bond with His367, while **FM2** also displays π–alkyl interactions with Phe359, Leu368, His372, Ala410, Trp599 and His600 (Figure 4A). In compound **FM12**, NO2 and carboxylic acid form H-bond interactions with Lys296 and His432; furan and a six membered aromatic ring show π–π stacked and π–π T-shaped interaction with His432 and Trp599, respectively; while **FM12** also shows π–π alkyl interactions with Leu414, His432, Trp599 and His600 (Figure 4B).

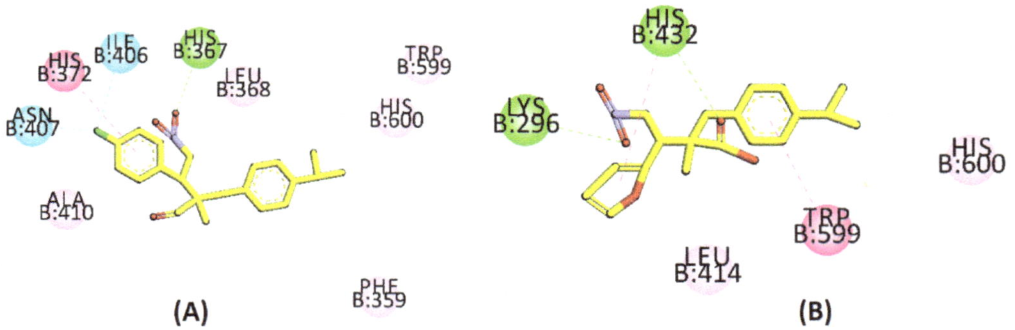

Figure 4. Two-dimensional interaction plots of (**A**) **FM2** and (**B**) **FM12** in active site of 5-LOX (PDB ID = 6N2W).

In compound **FM-6**, the furan moiety shows a π–sulfur interaction with Met61 and a π–alkyl interaction with Val218. NO2 moiety forms a metal acceptor bond with CU301, while **FM-6** also expresses a π–alkyl interaction with Pro201, His208 and Arg209 (Figure 5A). In compound **FM-7**, the NO_2 and carbonyl moiety of carboxylic acid form conventional hydrogen bonds with Lys296 and His432, respectively. The benzene ring shows a π–π stacked interaction with His432 and a π–alkyl interaction with Leu414 (Figure 5B). The chlorine moiety of **FM-8** forms a metal acceptor bond with Cu301, and the aromatic ring of chlorobenzene forms a π–π T-shaped and π–π stacked interaction with His60 and His208, respectively. The compound also shows a π–alkyl interaction with His42, Val218 and Phe227 (Figure 5C).

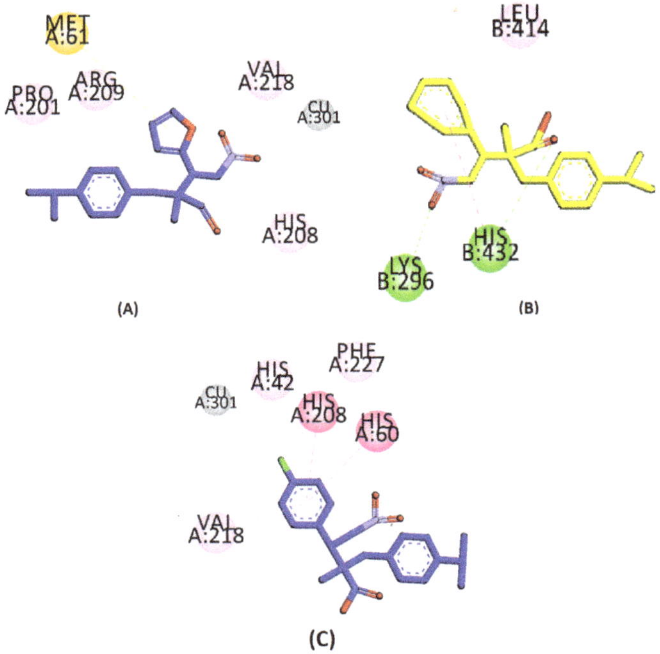

Figure 5. Two-dimensional interaction plots of (**A**) **FM6**, (**B**) **FM7** and (**C**) **FM8** in active site of 6N2W.

3. Discussion

The Michael addition is a powerful tool for synthesizing organic compounds having diverse chemical features [26,32]. The reaction combines a Michael donor and acceptor through C–C bond formation. A variety of Michael donors and acceptors has been studied to synthesized valuable molecules [37]. Enolizable aldehydes, ketones, ketoesters, cyanos and other nucleophilic substance are used as donor molecules. Similarly, nitroolefins, maleimides, vinyl sulfone and other α,β-unsaturated molecules with electron withdrawing groups are used as acceptors [38]. So, by changing any new Michael acceptor or donor, we can synthesize the new compounds. In this research, we reacted 3-(4-isopropylphenyl)-2-methylpropanal with different nitro-olefins to synthesize new compounds. Further, based on the literature, we noticed that the aldehydes are not stable drugs [39]. Therefore, we further oxidized our compounds by converting them into their corresponding carboxylic acids. The literature survey showed that the carboxylic acid-type drugs are potent inhibitors of COX and LOX pathways [9]. The first commercially available drug, aspirin, also has a carboxylic acid functional group.

The cyclooxygenase and lipoxygenase pathways are mainly involved in the inflammation and its associated pain [9,15]. The inhibitors of COX and LOX break up the prostaglandins and leukotrienes production [40]. The prostaglandins and leukotrienes are responsible for inflammation. Therefore, the dual inhibitions of COX and LOX pathways stop inflammation. Among the cyclooxygenases (i.e., COX-1 and COX-2), the selector inhibitors of COX-2 have the advantage of protecting stomach ulceration [41,42]. Therefore, COX-2 selectivity is very important for anti-inflammatory drugs. During our in vitro experiments, we observed that our compounds are selective inhibitors of COX-2. Specifically, by considering our two potent compounds **FM10** and **FM12**, we observed COX-2 selectivity indexes of 62.7 and 277, respectively. In this experiment, the COX-2 selectivity of commercially available standard drug celecoxib was 258.8. In the in vivo experiments, we observed that the carboxylic acid derivatives are comparatively more stable. The unwanted effect associated with aldehydic derivatives might be due to the unstable nature of aldehyde. The aldehyde serves as a pro-drug. Based on our experimental findings, we can claim that we have synthesized new (2*S*,3*S*)-2-(4-isopropylbenzyl)-2-methyl-4-nitro-3-phenylbutanals. Furthermore, we have modified all of our compounds into their respective carboxylic acids for enhance analgesic and anti-inflammatory potentials.

4. Materials and Methods

4.1. Equipments

The JEOL ECX 400 NMR spectrometer was used. The NMR operated at 400 MHz for proton NMR and 100 MHz for the carbon NMR. The LC-MS used was Agilent Technologies 1200 series (high performance liquid chromatography comprising of a G1315 diode array detector) and ion trap LC-MS G2445D SL. The elemental analyses were conducted with Elemental Vario EI III CHN analyzer. The melting points were determined with Gallenkamp 434.

4.2. Synthesis of (2S,3S)-3-aryl-2-(4-isopropylbenzyl)-2-methyl-4-nitrobutanals (FM 1–6)

In a small reaction vessel was added 3-(4-isopropylphenyl)-2-methylpropanal (2.0 mmol, 0.4 μL) in dichloromethane (1 M, 1 mL). To this solution was added further catalytic amounts of O-tertbutyl-L-threonine (0.1 mmol, 17.5 mg) and potassium hydroxide (0.1 mmol, 5.61 mg). The amino acid with KOH was stirred with the aldehyde for 2 to 3 min before adding the Michael acceptor to produce the nucleophilic enamine. Afterwards, the respective Michael acceptor (nitroolefinic compounds in 1.0 mmol) was further added with continued mixing at room temperature. The limiting reagent of the reaction (Michael acceptor) was checked by TLC analysis, and the reaction progress was attributed with the consumption of limiting reagent. At complete conversion (20–30 h), the reaction mixture was quenched with the aqueous portion (10 mL). The organic layer was diluted with dichloromethane (10 mL). The organic layer was separated by a separating funnel.

The procedure was repeated three times, and the dichloromethane layers (3 × 10 mL) were combined. Afterwards, anhydrous sodium sulfate was added to it to absorb any moisture. The sodium sulfate was then removed by filtration. The filtrate was washed with dichloromethane to obtain the crude product. The product was concentrated and purified by column chromatography. The structure of compound was confirmed with spectral analysis [26].

4.3. *Synthesis of (2S,3S)-3-aryl-2-(4-isopropylbenzyl)-2-methyl-4-nitrobutanoic Acids (FM 7–12)*

Each of the compounds synthesized in the previous step in one equivalent ratio was diluted in 10 mL DMF-anhydrous and to it was added potassium peroxy-mono-sulfate. The reaction was mixed at 25 °C. When the reaction was completed (3 h), 1 M of hydrochloric acid (HCl) was added to stop it. After workup of the reaction with hydrochloric acid, sodium sulfate anhydrous was added and was filtered. Then, the filtrate was washed out with organic solvent to obtain the crude product [9]. The final product was purified by column chromatography, and the structure was confirmed.

4.4. *DPPH Free Radical Scavenging Assay*

The protocol of Brand-Williams et al. was used for the DPPH assay with some modifications [43]. DPPH (4 mg) was dissolved in methanol (100 mL) to obtain a mixture of 0.01 mM 1,1-diphenyl,2-picrylhydrazyl (DPPH). The stock solution of the various synthesized compounds was prepared in methanol with 1 mg/mL concentration. This stock solution was used to prepare different concentrations of test samples ranging within 1000–62.5 µg/mL. The 0.1 mL of each concentration (1000–62.5 µg/mL) was combined with the DPPH (3 mL) solution in methanol. The solution was kept at 23 °C for 15 min incubation, followed by the absorbance measurement deliberated at 517 nm. Gallic acid was used as a standard drug in this assay. The percentage DPPH radical scavenging potential was measured via the formula [44]:

$$\% \text{ radical scavenging potential} = \frac{C_{Abs.} - S_{Abs.}}{C_{Abs.}} \times 100$$

where $C_{Abs.}$ is the absorbance of the control, and S_{Abs} is the absorbance of test samples/standard.

4.5. *ABTS Free Radical Scavenging Assay*

The total antioxidant activity of test compound (HBH) was estimated using the 2, 2-azinobis [3-ethylbenzthiazoline]-6-sulfonic acid (ABTS). The 100 mL of ABTS solution (7 mM) was added to 100 mL of potassium persulfate ($K_2S_2O_4$, 2.45 mM) solution, mixed, and kept in the dark for 12 h to generate free radicals. This activated, pre-generated ABTS solution was mixed with different concentrations of the various synthesized compounds (1000–62.5 µg/mL), followed by a suitable dilution with 50% methanol to produce an absorbance of 0.7 at 745 nm. Gallic acid at 2 mg/2 mL of water was used as a standard drug. Likewise, for the test sample, different concentrations (1000–62.5 µg/mL) of the standard drug were made for absorbance measurements at the same wavelength. The 300 µL of each test solution was added to 3 mL of ABTS solution to measure the absorbance at 745 nm through a UV-visible spectrophotometer. A similar volume of each standard solution was taken to determine the absorbance at the same wavelength. The ABTS percent scavenging potential was calculated via the above formula [45].

4.6. *Cyclooxygenase (COX-1/2) Assay*

The COX-1 and 2 enzymes' inhibitions assays on the synthesized compounds were carried out as per the standard reported method [46]. Initially, the respective enzyme solution was prepared in a concentration of 300 units/mL. The enzyme activation was started with keeping 10 µL of enzyme solution in the cold for up to 10 min. To this enzyme solution was added the substrate solution in HCl (0.1 M Trish buffer with pH of 8.0).

The co-factor 50 μL solution contained TMPD (*N*,*N*,*N*,*N*-tetramethyl-*p*-phenylenediamine dihydrochloride, 0.24 mM), hematin (1 mM) and glutathione (0.9 mM). Afterwards, the solutions from synthesized compounds (20 μL in concentration ranging from 31.25 to 1000 μg/mL) and the respective enzyme solution (60 μL) were kept at room temperature for five minutes. The reaction was initiated by adding arachidonic acid (20 μL, 30 mM). The overall solution was incubated for five minutes. Afterwards, the absorbance was recorded on a UV-visible instrument at 570 nm. From the absorbance value of every sample, the percentage inhibition was calculated as per the standard method [47].

4.7. 5-Lipoxygenase (5-LOX) Assay

The lipoxygenase inhibition assay on the synthesized compounds was carried out as per the standard reported method [48]. Solutions from synthesized compounds were prepared in concentrations ranging from 31.25 to 1000 μg/mL. The 5-LOX enzyme solution was also prepared at a strength of 10,000 units/mL. The linoleic acid (80 mM) was employed as a substrate in lipoxygenase assay. The buffer (phosphate) was also prepared for the assay having 50 mM strength and a pH of 6.3. The samples of synthesized compounds (250 μL), phosphate buffer (250 μL) and mixture of the enzyme were mixed and incubated for five minutes. Afterwards, the solution of the substrate (0.6 mM, 1000 μL) was mixed with lipoxygenase enzyme mixture with shaking. The absorbance was recorded on a UV-visible instrument at 234 nm. The zileuton was used a control drug in lipoxygenase assay. The percent inhibition was calculated as per the standard method.

4.8. Molecular Docking Studies

The molecular docking studies were performed using the MOE software [49–51]. Docking studies on the COX-2, 5-LOX and DPPH were carried out to assess binding orientation and ligand–enzyme interactions [9]. All the synthesized compounds were docked into active sites of DPPH, COX-2 and 5-LOX. Protein Data Bank accession codes 5I38, 1CX2 and 6N2W were used to explore crystal structures of DPPH, COX-2 and 5-LOX in complex with Kojic acid, SC-558 and NDGA, respectively. We evaluated docking reliability by re-docking native ligands prior to determining the docking poses of novel compounds. The computed RMSD values (<2.0 Å) were within acceptable ranges.

4.9. In Vivo Studies

4.9.1. Experimental Animals

Swiss albino mice of both sexes with an average weight of 30 to 35 g were obtained from the respective section of NIH (National Institute of Health) Islamabad, Pakistan. Written approval was obtained from the Departmental Ethical Committee (No. DREC/20). The animals were reserved in an animal house with the approval of the ethical committee. Throughout the experiments, standard ethical guidelines were followed [52].

4.9.2. Acute Toxicity

Before testing our selected compounds for in vivo experiments, we performed the toxicity test as per the protocol [53]. Four groups of animals were labelled, with eight animals in each group. The control group was given normal saline, while other groups were given different concentrations of the selected compounds. As per the standard protocols, the animals' behaviors were observed for allergic reactions and mortalities.

4.9.3. Carrageenan-Induced Inflammation

After the acute toxicity studies, the carrageenan-induced inflammation assay was performed on the compounds having a safety profile within limit. Forty (40) albino mice of both sexes were alienated into five different groups, with eight mice in each group. Group I was tagged as the negative control group and was administered dimethylsulfoxide (10 mL/kg, 10% *v*/*v*) and phosphate buffer (150 μL). Group II was tagged as the standard/positive control group and received a dose of aspirin (100 mg/kg in 0.9% normal saline). The remaining groups (III, IV and V) were tagged as experimental groups and received synthesized compounds (25, 50 and 100 mg/kg in DMSO) and Tween-80 in normal saline. After half an hour, carrageenan suspension (0.05 mL, 1% w/v in saline) was injected into the animals. After the injection of the irritant/carrageenan, the inflammation in the paws was measured by a plethysmometer in intervals (1 to 5 h). The inflammation in the paws of animals in different groups was compared with that of the vehicle, and the percent anti-inflammatory activities were recorded as per the standard method [54].

4.9.4. Acetic Acid Induced Writhing Test

The acetic induced analgesic assay on the compounds **FM10** and **FM12** was performed to determine the role of the peripheral pathway. The albino mice of both sexes were divided into two groups. Compounds **FM10** and **FM12** were administered to both the groups in doses of 25 and 50 mg/kg. After 1 h, acetic acid was injected intraperitoneally in the strength of 10 mL/kg. The negative control was Tween 80 1% solution in the strength of 10 mL/kg. The positive control was diclofenac sodium in the strength of 50 mg/kg intraperitoneally. The activities in animals were determined from the number of stretchings and writings [53].

4.9.5. Formalin-Induced Paw-Licking Test

In this assay, the mice were tagged, and compounds **FM10** and **FM12** were given in concentrations of 25 and 50 mg/kg. Formalin (20 μL, 2.5%) was injected into the animals after 30 min of the compounds. The early phase was initially five minutes, while the late phase was 15-30 min. In both the phases, the mice were under observation for licking. As per the protocols, naloxone (2 mg/kg), indomethacin (10 mg/kg) and morphine (5 mg/kg) were used [28].

4.9.6. Hot Plate Test

The selected compounds (**FM10** and **FM12**) were also tested for anti-nociceptive potentials using a hot plate apparatus. Briefly, test compounds at concentrations of 25 and 50 mg/kg were administered 30 min before observation to the animals and were placed on the surface of hot plate analgesia meter, which was maintained at a temperature of 55 ± 0.2 °C. The response latency, which is a measure of the time taken by animals after the placement of animals on a plate and the licking of paws or jumping, were observed. Morphine (5 mg/kg) was used as a positive. Observations were made after 30, 60 and 90 min of drugs administration [28].

5. Conclusions

From our current results, it can be concluded that (2*S*,3*S*)-2-(4-isopropylbenzyl)-2-methyl-4-nitro-3-phenylbutanals (**FM1-6**) and their corresponding carboxylic acids (**FM7-12**) are potential compounds to treat analgesia and inflammation. All of our synthesized compounds (**FM1-12**) are new and were synthesized for the first time. All the compounds are equally potent against the tested in vitro COX-1, COX-2 and 5-LOX targets. The observed IC_{50} values of our most potent compound **FM12** were 0.18, 49.89 and 0.43 μM against COX-1, COX-2 and 5-LOX enzymes. In comparison, the standard celecoxib exhibited IC_{50} values of 0.042 and 10.87 μM (against COX-1 and 2 enzymes), while zileuton gave 0.50 μM against the 5-LOX enzyme. The free radicals within the body can complicate inflammation and the associated pain. Therefore, as a supplementary target, the compounds have also been tested for the in vitro antioxidant assays. We observed that (2*S*,3*S*)-2-(4-isopropylbenzyl)-2-methyl-4-nitro-3-phenylbutanoic acids (**FM7-12**) were comparatively safer in experimental animals. So, based on these observations, we extended potential compounds **FM10** and **FM12** to in vivo studies of analgesia and inflammation. The selected compounds showed a very excellent activity profile in the tested in vivo experiments. We also performed the molecular docking studies of the selected compounds with the target proteins of the respective enzymes. The binding energies showed that our designed compounds are suitable for the COX and LOX targets and can inhibit both of them to treat analgesia and inflammation.

Supplementary Materials: The following supporting information can be downloaded at: https://www.mdpi.com/article/10.3390/molecules27134068/s1. The supporting information contains a section on chemicals and drugs. Moreover, the following representative spectra of the compounds are provided. Figure S1: ^{1}H NMR spectrum of compound **FM1**. Figure S2: ^{13}C NMR spectrum of compound **FM1**. Figure S3: ^{1}H NMR spectrum of compound **FM3**. Figure S4: ^{13}C NMR spectrum of compound **FM3**. Figure S5: ^{1}H NMR spectrum of compound **FM6**. Figure S6: ^{13}C NMR spectrum of compound **FM6**. Figure S7: ^{1}H NMR spectrum of compound **FM12**.

Author Contributions: F.M., M.S.J., M.H.M. and M.A.J. synthesized the compounds and contributed to pharmacological assays under the supervision of J.A.K. and U.R. performed the molecular docking studies and also contributed to the chemistry part. S.S.u.H. helped in in-silico and writing of paper. S.B. contributed in facilitation our experimental and final draft corrections. A.S. supervised the overall project and drafted and refined the manuscript for publication. All authors have read and agreed to the published version of the manuscript.

Funding: This research was funded by Najran University KSA grant number [NU-IF/INT/01/006].

Institutional Review Board Statement: Not applicable.

Informed Consent Statement: Not applicable.

Data Availability Statement: Not applicable.

Acknowledgments: Authors would like to acknowledge the support of the Deputy for Research and Innovation—Ministry of Education, Kingdom of Saudi Arabia for this research through a grant (NU-IF/INT/01/006) under the institutional Funding Committee at Najran University, Kingdom of Saudi Arabia. We are also thankful to the Higher Education Commission (HEC) Pakistan for their financial support via Project No. 10562/KPK/R&D/HEC/2017. The authors wish to thank to University of Oradea, Oradea, Romania for financial support in publishing this paper.

Conflicts of Interest: The authors declare no conflict of interest.

References

1. Walsh, D.A.; Mapp, P.I.; Kelly, S. Calcitonin gene-related peptide in the joint: Contributions to pain and inflammation. *Br. J. Clin. Pharmacol.* **2015**, *80*, 965. [CrossRef] [PubMed]
2. Ul Hassan, S.S.; Ishaq, M.; Zhang, W.; Jin, H.-Z. An overview of the mechanisms of marine fungi-derived antiinflammatory and anti-tumor agents and their novel role in drug targeting. *Curr. Pharm. Des.* **2021**, *27*, 2605–2614. [CrossRef] [PubMed]
3. Greig, S.L.; Garnock-Jones, K.P. Loxoprofen: A review in pain and inflammation. *Clin. Drug Investig.* **2016**, *36*, 771. [CrossRef] [PubMed]
4. Muhammad, I.; ul Hassan, S.S.; Cheung, S.; Li, X.; Wang, R.; Zhang, W.D.; Yan, S.K.; Zhang, Y.; Jin, H.Z. Phytochemical study of Ligularia subspicata and valuation of its anti-inflammatory activity. *Fitoterapia* **2021**, *148*, 104800. [CrossRef]
5. Gadek, T.R.; Nicholas, J.B. Small molecule antagonists of proteins. *Biochem. Pharmacol.* **2003**, *65*, 1. [CrossRef]
6. Sneader, W. The discovery of aspirin: A reappraisal. *BMJ* **2000**, *321*, 1591. [CrossRef]
7. Hart, F.D.; Huskisson, E.C. Non-steroidal anti-inflammatory drugs. *Drugs* **1984**, *27*, 232. [CrossRef]
8. Gøtzsche, P.C. Non-steroidal anti-inflammatory drugs. *BMJ* **2000**, *320*, 1058. [CrossRef]
9. Sadiq, A.; Mahnashi, M.H.; Alyami, B.A.; Alqahtani, Y.S.; Alqarni, A.O.; Rashid, U. Tailoring the substitution pattern of Pyrrolidine-2, 5-dione for discovery of new structural template for dual COX/LOX inhibition. *Bioorg. Chem.* **2021**, *112*, 104969. [CrossRef]
10. Hawkey, C.J. COX-1 and COX-2 inhibitors. *Best Pract. Res. Clin. Gastroenterol.* **2001**, *15*, 801. [CrossRef]
11. Shams, S.; Zhang, W.; Jin, H.; Basha, S.H.; Priya, S.V.S.S. In-silico anti-inflammatory potential of guaiane dimers from *Xylopia vielana* targeting COX-2. *J. Biomol. Struct. Dyn.* **2020**, *40*, 484–498. [CrossRef]
12. FitzGerald, G.A. COX-2 and beyond: Approaches to prostaglandin inhibition in human disease. *Nat. Rev. Drug Discov.* **2003**, *2*, 879. [CrossRef]
13. Suleyman, H.; Demircan, B.; Karagoz, Y. Anti-inflammatory and side effects of cyclo-oxygenase inhibitors. *Pharmacol. Rep.* **2007**, *59*, 247.
14. Shi, S.; Klotz, U. Clinical use and pharmacological properties of selective COX-2 inhibitors. *Eur. J. Clin. Pharmacol.* **2008**, *64*, 233. [CrossRef]

15. Jan, M.S.; Ahmad, S.; Hussain, F.; Ahmad, A.; Mahmood, F.; Rashid, U.; Ullah, F.; Ayaz, M.; Sadiq, A. Design, synthesis, in-vitro, in-vivo and in-silico studies of pyrrolidine-2, 5-dione derivatives as multitarget anti-inflammatory agents. *Eur. J. Med. Chem.* **2020**, *186*, 111863. [CrossRef]
16. Shah, S.M.; Sadiq, A.; Shah, S.M.; Ullah, F. Antioxidant, total phenolic contents and antinociceptive potential of Teucrium stocksianum methanolic extract in different animal models. *BMC Complement. Altern. Med.* **2014**, *14*, 217. [CrossRef]
17. Mahnashi, M.H.; Alyami, B.A.; Alqahtani, Y.S.; Alqarni, A.O.; Jan, M.S.; Hussain, F.; Zafar, R.; Rashid, U.; Abbas, M.; Tariq, M.; et al. Antioxidant Molecules Isolated from Edible Prostrate Knotweed: Rational Derivatization to Produce More Potent Molecules. *Oxid. Med. Cell. Longev.* **2022**, *27*, 2022. [CrossRef]
18. Sadiq, A.; Mahmood, F.; Ullah, F.; Ayaz, M.; Ahmad, S.; Haq, F.U.; Khan, G.; Jan, M.S. Synthesis, anticholinesterase and antioxidant potentials of ketoesters derivatives of succinimides: A possible role in the management of Alzheimer's. *Chem. Cent. J.* **2015**, *9*, 1. [CrossRef]
19. Hassan, S.S.; Muhammad, I.; Abbas, S.Q.; Hassan, M.; Majid, M.; Jin, H.Z.; Bungau, S. Stress driven discovery of natural products from actinobacteria with anti-oxidant and cytotoxic activities including docking and admet properties. *Int. J. Mol. Sci.* **2021**, *22*, 11432. [CrossRef]
20. ul Hassan, S.S.; Jin, H.Z.; Abu-Izneid, T.; Rauf, A.; Ishaq, M.; Suleria, H.A. Stress-driven discovery in the natural products: A gateway towards new drugs. *Biomed. Pharmacother.* **2019**, *109*, 459. [CrossRef]
21. Shah, S.; Shah, S.M.; Ahmad, Z.; Yaseen, M.; Shah, R.; Sadiq, A.; Khan, S.; Khan, B. Phytochemicals, in vitro antioxidant, total phenolic contents and phytotoxic activity of Cornus macrophylla Wall bark collected from the North-West of Pakistan. *Pak. J. Pharm. Sci.* **2015**, *28*, 23.
22. Bibi, A.; Shah, T.; Sadiq, A.; Khalid, N.; Ullah, F.; Iqbal, A. L-isoleucine-catalyzed michael synthesis of N-alkylsuccinimide derivatives and their antioxidant activity assessment. *Russ. J. Org. Chem.* **2019**, *55*, 1749. [CrossRef]
23. Jabeen, M.; Ahmad, S.; Shahid, K.; Sadiq, A.; Rashid, U. Ursolic acid hydrazide based organometallic complexes: Synthesis, characterization, antibacterial, antioxidant, and docking studies. *Front. Chem.* **2018**, *6*, 55. [CrossRef]
24. Kumar, D.; Kumar, R.; Ramajayam, R.; Lee, K.W.; Shin, D.S. Synthesis, antioxidant and molecular docking studies of (-)-catechin derivatives. *J. Korean Chem. Soc.* **2021**, *65*, 106. [CrossRef]
25. Chen, J.; Yang, J.; Ma, L.; Li, J.; Shahzad, N.; Kim, C.K. Structure-antioxidant activity relationship of methoxy, phenolic hydroxyl, and carboxylic acid groups of phenolic acids. *Sci. Rep.* **2020**, *10*, 2611. [CrossRef]
26. Nugent, T.C.; Bibi, A.; Sadiq, A.; Shoaib, M.; Umar, M.N.; Tehrani, F.N. Chiral picolylamines for Michael and aldol reactions: Probing substrate boundaries. *Org. Biomol. Chem.* **2012**, *10*, 9287. [CrossRef]
27. Berner, O.M.; Tedeschi, L.; Enders, D. Asymmetric Michael additions to nitroalkenes. *Eur. J. Org. Chem.* **2002**, *2002*, 1877. [CrossRef]
28. Ahmad, S.; Mahnashi, M.H.; Alyami, B.A.; Alqahtani, Y.S.; Ullah, F.; Ayaz, M.; Tariq, M.; Sadiq, A.; Rashid, U. Synthesis of michael adducts as key building blocks for potential analgesic drugs: In vitro, in vivo and in silico explorations. *Drug Des. Dev. Ther.* **2021**, *15*, 1299. [CrossRef]
29. Pasuparthy, S.D.; Maiti, B. Enantioselective Organocatalytic Michael Addition Reactions Catalyzed by Proline/Prolinol/Supported Proline based Organocatalysts: An Overview. *ChemistrySelect* **2022**, *7*, e202104261. [CrossRef]
30. List, B.; Pojarliev, P.; Martin, H.J. Efficient proline-catalyzed Michael additions of unmodified ketones to nitro olefins. *Org. Lett.* **2001**, *3*, 2423. [CrossRef]
31. Das, T.; Mohapatra, S.; Mishra, N.P.; Nayak, S.; Raiguru, B.P. Recent Advances in Organocatalytic Asymmetric Michael Addition Reactions to α, β-Unsaturated Nitroolefins. *ChemistrySelect* **2021**, *6*, 3745. [CrossRef]
32. Sadiq, A.; Nugent, T.C. Catalytic access to succinimide products containing stereogenic quaternary carbons. *ChemistrySelect* **2020**, *5*, 11934. [CrossRef]
33. Nugent, T.C.; Sadiq, A.; Bibi, A.; Heine, T.; Zeonjuk, L.L.; Vankova, N.; Bassil, B.S. Noncovalent bifunctional organocatalysts: Powerful tools for contiguous quaternary-tertiary stereogenic carbon formation, scope, and origin of enantioselectivity. *Chem. Eur. J.* **2012**, *18*, 4088. [CrossRef] [PubMed]
34. Mahmood, F.; Jan, M.S.; Ahmad, S.; Rashid, U.; Ayaz, M.; Ullah, F.; Hussain, F.; Ahmad, A.; Khan, A.U.; Aasim, M.; et al. Ethyl 3-oxo-2-(2, 5-dioxopyrrolidin-3-yl) butanoate derivatives: Anthelmintic and cytotoxic potentials, antimicrobial, and docking studies. *Front. Chem.* **2017**, *5*, 119. [CrossRef]
35. McCulloch, A.C.; Stock, B.H. The bactericidal activity of some aromatic alcohols. *Australas. J. Pharm.* **1966**, *47*, 514.
36. Mühlman, A.; Lindberg, J.; Classon, B.; Unge, T.; Hallberg, A.; Samuelsson, B. Synthesis of Novel, Potent, Diol-Based HIV-1 Protease Inhibitors via Intermolecular Pinacol Homocoupling of (2 S)-2-Benzyloxymethyl-4-phenylbutanal. *J. Med. Chem.* **2001**, *44*, 3407. [CrossRef]
37. Seitzberg, J.G.; Knapp, A.E.; Lund, B.W.; Mandrup Bertozzi, S.; Currier, E.A.; Ma, J.N.; Sherbukhin, V.; Burstein, E.S.; Olsson, R. Discovery of potent and selective small-molecule PAR-2 agonists. *J. Med. Chem.* **2008**, *51*, 5490. [CrossRef]
38. Zhang, Y.; Wang, W. Recent advances in organocatalytic asymmetric Michael reactions. *Catal. Sci. Technol.* **2012**, *2*, 42. [CrossRef]
39. Fritz, K.S.; Petersen, D.R. An overview of the chemistry and biology of reactive aldehydes. *Free Radic. Biol. Med.* **2013**, *59*, 85. [CrossRef]
40. Leval, X.D.; Julémont, F.; Delarge, J.; Pirotte, B.; Dogné, J.M. New trends in dual 5-LOX/COX inhibition. *Curr. Med. Chem.* **2002**, *9*, 941. [CrossRef]

41. Mendes, R.T.; Stanczyk, C.P.; Sordi, R.; Otuki, M.F.; Santos, F.A.; Fernandes, D. Selective inhibition of cyclooxygenase-2: Risks and benefits. *Rev. Bras. Reumatol.* **2012**, *52*, 774. [CrossRef]
42. Hawkey, C.J. COX-2 inhibitors. *Lancet* **1999**, *353*, 307. [CrossRef]
43. Huneif, M.A.; Alshehri, D.B.; Alshaibari, K.S.; Dammaj, M.Z.; Mahnashi, M.H.; Majid, S.U.; Javed, M.A.; Ahmad, S.; Rashid, U.; Sadiq, A. Design, synthesis and bioevaluation of new vanillin hybrid as multitarget inhibitor of α-glucosidase, α-amylase, PTP-1B and DPP4 for the treatment of type-II diabetes. *Biomed. Pharmacother.* **2022**, *150*, 113038. [CrossRef]
44. Sadiq, A.; Rashid, U.; Ahmad, S.; Zahoor, M.; AlAjmi, M.F.; Ullah, R.; Noman, O.M.; Ullah, F.; Ayaz, M.; Khan, I.; et al. Treating hyperglycemia from Eryngium caeruleum M. Bieb: In-vitro α-glucosidase, antioxidant, in-vivo antidiabetic and molecular docking-based approaches. *Front. Chem.* **2020**, *8*, 1064. [CrossRef]
45. Sadiq, A.; Zeb, A.; Ullah, F.; Ahmad, S.; Ayaz, M.; Rashid, U.; Muhammad, N. Chemical characterization, analgesic, antioxidant, and anticholinesterase potentials of essential oils from Isodon rugosus Wall. ex. Benth. *Front. Pharmacol.* **2018**, *9*, 623. [CrossRef]
46. Munir, A.; Khushal, A.; Saeed, K.; Sadiq, A.; Ullah, R.; Ali, G.; Ashraf, Z.; Mughal, E.U.; Jan, M.S.; Rashid, U.; et al. Synthesis, in-vitro, in-vivo anti-inflammatory activities and molecular docking studies of acyl and salicylic acid hydrazide derivatives. *Bioorg. Chem.* **2020**, *104*, 104168. [CrossRef]
47. Mahnashi, M.H.; Alyami, B.A.; Alqahtani, Y.S.; Jan, M.S.; Rashid, U.; Sadiq, A.; Alqarni, A.O. Phytochemical profiling of bioactive compounds, anti-inflammatory and analgesic potentials of Habenaria digitata Lindl.: Molecular docking based synergistic effect of the identified compounds. *J. Ethnopharmacol.* **2021**, *273*, 113976. [CrossRef]
48. Javed, M.A.; Ashraf, N.; Saeed Jan, M.; Mahnashi, M.H.; Alqahtani, Y.S.; Alyami, B.A.; Alqarni, A.O.; Asiri, Y.I.; Ikram, M.; Sadiq, A.; et al. Structural Modification, In Vitro, In Vivo, Ex Vivo, and In Silico Exploration of Pyrimidine and Pyrrolidine Cores for Targeting Enzymes Associated with Neuroinflammation and Cholinergic Deficit in Alzheimer's Disease. *ACS Chem. Neurosci.* **2021**, *12*, 4123. [CrossRef]
49. Ahmad, S.; Iftikhar, F.; Ullah, F.; Sadiq, A.; Rashid, U. Rational design and synthesis of dihydropyrimidine based dual binding site acetylcholinesterase inhibitors. *Bioorg. Chem.* **2016**, *69*, 91. [CrossRef]
50. Sadiq, A.; Mahnashi, M.H.; Rashid, U.; Jan, M.S.; Alshahrani, M.A.; Huneif, M.A. 3-(((1*S*,3*S*)-3-((*R*)-hydroxy(4-(trifluoromethyl)phenyl) methyl)-4-oxocyclohexyl)methyl)pentane-2,4-dione: Design and synthesis of new stereopure multi-target antidiabetic agent. *Molecules* **2022**, *27*, 3265. [CrossRef]
51. Sarfraz, M.; Sultana, N.; Rashid, U.; Akram, M.S.; Sadiq, A.; Tariq, M.I. Synthesis, biological evaluation and docking studies of 2, 3-dihydroquinazolin-4 (1H)-one derivatives as inhibitors of cholinesterases. *Bioorg. Chem.* **2017**, *70*, 237. [CrossRef] [PubMed]
52. Aslam, H.; Khan, A.U.; Naureen, H.; Ali, F.; Ullah, F.; Sadiq, A. Potential application of Conyza canadensis (L) Cronquist in the management of diabetes: In vitro and in vivo evaluation. *Trop. J. Pharm. Res.* **2018**, *17*, 1287. [CrossRef]
53. Shah, S.M.; Ullah, F.; Shah, S.M.; Zahoor, M.; Sadiq, A. Analysis of chemical constituents and antinociceptive potential of essential oil of Teucrium Stocksianum bioss collected from the North West of Pakistan. *BMC Complement. Altern. Med.* **2012**, *12*, 351. [CrossRef] [PubMed]
54. Alam, F.; Din, K.M.; Rasheed, R.; Sadiq, A.; Jan, M.S.; Minhas, A.M.; Khan, A. Phytochemical investigation, anti-inflammatory, antipyretic and antinociceptive activities of Zanthoxylum armatum DC extracts-in vivo and in vitro experiments. *Heliyon* **2020**, *6*, e05571. [CrossRef]

molecules

MDPI

Article

Pharmacological Inhibition of Endogenous Hydrogen Sulfide Attenuates Breast Cancer Progression

Nazeer Hussain Khan [1,2], Di Wang [1], Wenkang Wang [3], Muhammad Shahid [4], Saadullah Khattak [1,2], Ebenezeri Erasto Ngowi [1,5], Muhammad Sarfraz [1,6], Xin-Ying Ji [1,7,*], Chun-Yang Zhang [8,9,*] and Dong-Dong Wu [1,10,*]

1 Henan International Joint Laboratory for Nuclear Protein Regulation, School of Basic Medical Sciences, Henan University, Kaifeng 475004, China; kakakhan3514@gmail.com (N.H.K.); wangdi20220316@gmail.com (D.W.); saadullah271@gmail.com (S.K.); ebenezerngowi92@gmail.com (E.E.N.); chiefpharm@gmail.com (M.S.)
2 School of Life Sciences, Henan University, Kaifeng 475004, China
3 Department of Breast Surgery, The First Affiliated Hospital, Zhengzhou University, Zhengzhou 450052, China; victor1125@foxmail.com
4 Department of Biological Sciences and Biotechnology, Faculty of Science and Technology, Universiti Kebangsaan Malaysia, Bangi 43600, Malaysia; mshahdaslam@gmail.com
5 Department of Biological Sciences, Faculty of Science, Dar es Salaam University College of Education, Dar es Salaam 11101, Tanzania
6 Faculty of Pharmacy, The University of Lahore, Lahore 54590, Pakistan
7 Kaifeng Key Laboratory of Infection and Biological Safety, College of Medicine, Henan University, Kaifeng 475004, China
8 Department of Thoracic Surgery, The First Affiliated Hospital of Zhengzhou University, Zhengzhou 450052, China
9 Department of General Thoracic Surgery, Hami Central Hospital, Hami 839000, China
10 School of Stomatology, Henan University, Kaifeng 475004, China
* Correspondence: 10190096@vip.henu.edu.cn (X.-Y.J.); zcy198200@163.com (C.-Y.Z.); ddwubiomed2010@163.com (D.-D.W.)

Citation: Khan, N.H.; Wang, D.; Wang, W.; Shahid, M.; Khattak, S.; Ngowi, E.E.; Sarfraz, M.; Ji, X.-Y.; Zhang, C.-Y.; Wu, D.-D. Pharmacological Inhibition of Endogenous Hydrogen Sulfide Attenuates Breast Cancer Progression. *Molecules* **2022**, *27*, 4049. https://doi.org/10.3390/molecules27134049

Academic Editors: Piyali Dasgupta and Syed Shams ul Hassan

Received: 28 February 2022
Accepted: 18 May 2022
Published: 23 June 2022

Abstract: Hydrogen sulfide (H_2S), a gaseous signaling molecule, is associated with the development of various malignancies via modulating various cellular signaling cascades. Published research has established the fact that inhibition of endogenous H_2S production or exposure of H_2S donors is an effective approach against cancer progression. However, the effect of pharmacological inhibition of endogenous H_2S-producing enzymes (cystathionine-γ-lyase (CSE), cystathionine-β-synthase (CBS), and 3-mercaptopyruvate sulfurtransferase (3-MPST)) on the growth of breast cancer (BC) remains unknown. In the present study, DL-propargylglycine (PAG, inhibitor of CSE), aminooxyacetic acid (AOAA, inhibitor of CBS), and L-aspartic acid (L-Asp, inhibitor of 3-MPST) were used to determine the role of endogenous H_2S in the growth of BC by in vitro and in vivo experiments. An in silico study was also performed to confirm the results. Corresponding to each enzyme in separate groups, we treated BC cells (MCF-7 and MDA-MB-231) with 10 mM of PAG, AOAA, and L-Asp for 24 h. Findings reveal that the combined dose (PAG + AOAA + L-Asp) group showed exclusive inhibitory effects on BC cells' viability, proliferation, migration, and invasion compared to the control group. Further, treated cells exhibited increased apoptosis and a reduced level of phospho (p)-extracellular signal-regulated protein kinases such as p-AKT, p-PI3K, and p-mTOR. Moreover, the combined group exhibited potent inhibitory effects on the growth of BC xenograft tumors in nude mice, without obvious toxicity. The molecular docking results were consistent with the wet lab experiments and enhanced the reliability of the drugs. In conclusion, our results demonstrate that the inhibition of endogenous H_2S production can significantly inhibit the growth of human breast cancer cells via the AKT/PI3K/mTOR pathway and suggest that endogenous H_2S may act as a promising therapeutic target in human BC cells. Our study also empowers the rationale to design novel H_2S-based anti-tumor drugs to cure BC.

Keywords: endogenous hydrogen sulfide; breast cancer; apoptosis; signaling pathway; tumor growth

1. Introduction

Breast cancer (BC) is a prevalent and growing concern of malignancy in women worldwide [1]. According to recent reports, approximately 2.3 million individuals are diagnosed with BC each year, with an annual mortality rate of around 450,000 [2,3]. Leading associated risk factors of BC include age, the aggravate prevalence of genetic mutations in predisposition genes (*BRCA1* and *BRCA2*), lifestyle base-modified (non-genetic) risk factors, nulliparity, early menarche, first pregnancy in women older than 30 years of age, older age onset of menopause, usage of oral contraceptives, and personal or familial history of BC and other clinical complaints [4–6].

Hydrogen sulfide (H_2S), together with nitric oxide (NO) and carbon monoxide (CO), is involved in modulating multiple physiological and pathological processes. It has been widely regarded as an endogenous gasotransmitter molecule [7]. Under normal physiological conditions, three enzymes that are widely expressed in mammalian tissues and cells—namely cystathionine-β-synthase (CBS), cystathionine-γ-lyase (CSE), and 3-mercaptopyruvate sulfurtransferase (3-MST)—produce H_2S [8]. H_2S exhibits a pleiotropic and often dose-dependent effect after being released in the form of acid-labile sulfur and bound sulfane sulfur [9]. It has been well established that H_2S is involved in mediating essential cellular mechanisms and plays a crucial role in the regulation of many physiological conditions, including energy production, neuroprotection, vasorelaxation, glucose homeostasis, and angiogenesis [10–12]. In particular to the role of H_2S in cancer, it has been reported that endogenous H_2S is involved in cancer development and progression [13,14].

In terms of cytoprotective biological response, H_2S is considered to be a bidirectional target in cancer research. On the one hand, studies have shown that exogenous exposure of H_2S donors prevents tumor development [15,16]. On the other hand, a number of studies have shown that reducing H_2S levels by downregulating the production of endogenous H_2S in cancer cells leads to a decline in cancer progression [17,18]. With these potent therapeutic characteristics in cancer research, H_2S has gained the attention of many researchers, and drugs that can repress or trigger the synthesis or promote the release of H_2S have attained enormous value in preclinical and clinical settings.

To facilitate the future clinical translation research of H_2S and the proposition of promising anticancer approaches for therapeutic manipulation of H_2S, a variety of novel inhibitors have been synthesized to suppress the activity of H_2S-producing enzymes. Currently, three pharmacologically well-characterized compounds —namely DL-propargylglycine (PAG), L-aspartic acid (L-Asp), and aminooxyacetic acid (AOAA)—are considered to be competitive, potent inhibitors of CSE, 3-MPST, and CBS, respectively. PAG ($C_5H_7NO_2$) is a selective inhibitor compound which basically blocks the active site of the CSE enzyme, thereby preventing it from binding to its original substrate. The solubility of this drug is significantly high in both water and PBS [19,20]. L-Aspartic acid (L-Asp) ($C4H_{14}NO_4$) works in similar manner to PAG. The solubility of the drug is relatively low—nearly half that of PAG in water [21]. Similarly, in many studies, AOAA has also been reported to reduce H_2S synthesis by inhibiting the function of CBS in both acidic and in prodrug forms. AOAA has been reported to reduce intracellular adenosine triphosphate levels and decrease glycolysis rate, which in doing so, regulates cellular activities. Studies have claimed the selectivity of AOAA; however, it also inhibits CSE. The effect of AOAA is concentration dependent. In addition, AOAA is highly soluble in both water and PBS [20,22,23]. Given the encouraging results in inhibiting H_2S synthesis, it is worth investigating the electrostatics (i.e., protonation or deprotonation states) of these drugs both in solution and physiological condition.

These inhibitors have been well studied in various experimental and clinical studies as potential future anti-cancer therapies [23–25].

However, there is no study on the pharmacological inhibition of H_2S in BC cells using these inhibitors. Therefore, by using PAG, L-Asp, and AOAA as potent inhibitors of H_2S-producing enzymes, the present research focused on investigating the effect of pharmacological inhibition of endogenous H_2S on human BC cell proliferation, invasion, and migration by using various in vitro and in vivo approaches.

2. Material Method

2.1. Cell Culture

Human BC cell lines MCF-7 and MDA-MB-231 were purchased from Fenghbio Biosciences (Changsha, Hunan, China). The cells were grown in DMEM media supplemented with 10% fetal bovine serum, 100 ug/mL streptomycin, and 100 U/mL penicillin and were maintained at 37 °C in a humidified atmosphere of 5% CO_2 and 95% air.

2.2. Drugs Formulations/Treatment

To obtain a concentration of drugs (DL-propargylglycine (PAG, inhibitor of CSE), aminooxyacetic acid (AOAA, inhibitor of CBS), and L-aspartic acid (L-Asp, inhibitor of 3-MST)) that inhibited cellular H_2S synthesis but was noncytotoxic to cell survival, cells were treated with a gradient of concentrations (1, 2, 5, 10, and 20 mM) for 24 h. Followed by the 24 h treatment of individual drugs, optical density (OD) of the treated cells using a CCK-8 kit was measured. Drug concentrations and cell toxicity response were measured for all treatments. Additionally, the combined dose effects of three inhibitors were investigated at a concentration of 10 mM and a higher concentration of individual drugs (30 Mm) by using a CCK-8 test. Phosphate buffer saline (PBS) was used to treat the control group cells.

2.3. Cell Viability Assay

The CCK-8 assay was used to detect cell viability according to the manufacturer's instructions. In brief, 10,000 MCF-7 and MDA-MB-231 cells were seeded in 96-well plates and cultured in normal medium for 24 h after the treatment. At the check point, 10% CCK-8 solution was added to each well of the 96-well plate before OD readings, as done in previous studies [26].

2.4. Detection of H_2S Level

To determine the inhibitory efficacy of drugs after the 24 h treatment, the enzyme-linked immunosorbent assay kit (LanpaiBio, Shanghai, China) was used to detect H_2S levels in MCF-7 and MDA-MB-231 cells [27]. The cell culture supernatants were collected to test the levels of H_2S. Next, the standard controls were prepared accordingly, and samples were treated with reagents and incubated for 0.5 h at 37 °C. Then, color-developing agents were added to each well and incubated for 15 min at 37 °C. The OD of each well was measured with a microplate reader at 450 nm. Using the equation derived from the standard controls and the absorbance of the samples obtained from the microplate reader, the concentration of H_2S was determined. The experiments were repeated three times.

2.5. Cell Proliferation Assay

For assessment of cell proliferation, the Light EdU Apollo 567 in Vitro Imaging Kit (RiboBio, Guangzhou, Guangdong, China) was used to determine the proliferation. In brief, 10^4 cells/well were seeded in a 96-well plate, incubated, and treated with corresponding drugs for 24 h. After the treatment, the growth media was replaced with pure media containing EdU A and incubated for 2 h. Then, cells were fixed with 4% paraformaldehyde (PFA) and were washed with PBS. By adopting the kit protocol, the Apollo step was conducted under dark conditions. After washing with PBS and methanol, cells were treated with DNA staining (Hoechst reagent) under dark conditions at room temperature. Imaging was performed using a fluorescent microscope, and the proliferation rate was calculated using the following formula: Percentage cell proliferation = (EdU − positive cells/total cells × 100).

2.6. Wound Healing Assay

The scratch wound-healing motility assay was performed to evaluate the migration ability of treated BC cells. In a 6-well plate, after the cells reached 80–90% confluence, a sterile pipette tip was scraped across the monolayer cells. The cells were then returned to the incubator after the drug treatment until the indicated time. An Olympus CKK41 microscope was used to photograph the reprehensive sites, which were analyzed by ImageJ software to measure the migration. The migration rate (MR) was estimated using the following formula: MR (%) = [(A − B)/A] × 100, where A and B are the widths at 0 and 24 h, respectively.

2.7. Colony Formation Assay

To assess the effect of drug treatment on colony formation, we seeded the cell lines (5×10^2) into 6-well plates and cultured them for 14 days at 37 °C. At the end of experiment, methanol was used to fix the colonies, which were then stained at room temperature with 0.5% crystal violet. The colony number was then calculated by scanning the plate.

2.8. Migration and Invasion Assay

To determine the migration and invasion ability of treated cells, we firstly spread the media and cell suspension (200 µL) in the upper chamber of a 24-well specialized plate. The lower chamber contained 20% fetal bovine serum (600 µL), used as a chemoattractant. After incubation for 24 h, cells were fixed with methanol for 20 min and then stained with crystal violet dye solution for 40 min. Finally, images for transwell invasion were taken using a Zeiss Axioskop 2 plus microscope (Carl Zeiss, Thornwood, NY, USA).

2.9. Cell Death Assay

For the detection of apoptosis/necrosis of 24 h treated BC cells, TDT-mediated dUTP-biotin nick end labeling (TUNEL) assay was performed. In brief, 1×10^5 cells/well seeded in a 96-well plate were treated and incubated for 24 h. Cells were washed with PBS once and fixed with 4% PFA for 30 min at room temperature. After fixation, the cells were washed with PBS and treated with 1% triton for 5 min. Then, Tunel solution 50 µL was added per well and incubated for 80 min at 37 °C. After careful washing, cells were subjected to DAPI for 5 min and photographs were taken with a fluorescence microscope. Using ImageJ software, cell death rate was counted by using the ratio of TUNEL-positive cells in total cells.

2.10. Western Blot Analysis

The protein level was determined using a western blot. The principal antibodies (anti-B-cell lymphoma-extra-large (Bcl-xl), anti-B-cell lymphoma-2 (Bcl-2), anti-Bcl-xl/Bcl-2-associated death promoter (Bad), anti-Bcl-2-associated X protein (Bax), anti-cleaved poly adenosine diphosphate-ribose polymerase (PARP), anti-cleaved caspase-3, N and E cadherin, and Vimentin antibodies) were procured from ProteinTech (Chicago, IL, USA). Antibodies against PI3K/AKT/mTOR and their corresponding anti-phospho were obtained from cell Signaling technology (CST, Danvers, MA, USA), as the secondary antibody conjugated with horse. The internal control was chosen to be β-actin. An enhanced chemilumescence setup (Thermo Fisher Scientific, Rockford, IL, USA), was used to acquire the photographs. ImageJ software was used to calculate the band intensities.

2.11. Animal Study

The Henan University School of Medicine's Committee on Medical Ethics and Welfare for the experimental Animals (HUSOM-2019-168) gave its approval for the animal research. With slight adjustments, the animal research was carried out as described earlier [28]. Vital River Laboratory Animal Technology Co., Ltd. (Beijing, China) provided the BALB/C nude mice (male, 4 weeks old). Nude mice's right flanks were injected subcutaneously with MCF-7 and MDA-MA231 cells (5×10^6 cells in 200 µL PBS). The mice were divided into five groups at random (n = 6 each group). For 28 days, 10 mM PAG, L-Asp, AOAA, and PAG + AOAA + L-Asp (combined group) were injected subcutaneously near the tumor. PBS was injected subcutaneously into the control animals. Every day, the body weights and tumor volumes were measured. Tumor volumes were calculated using the following formula: volume = length × width^2/2 [29]. At the end of the trial, the mice were anaesthetized with 3% isoflurane and executed by cervical dislocation. The tumor growth inhibition rate

(IR) was calculated as IR (%) = [(A − B)/A] × 100, where A and B represent the control and treatment groups' respective average tumor weights [16].

2.12. Tumor Tissue Staining

Tumor specimens were fixed in paraffin embedded in 10% neutral-buffered formalin, and then tissues were segmented to thicknesses of 5 mm by adopting the hematoxylin and esosin (HE) staining procedures [26]. Images were taken with a Zeiss Axioskop 2plus microscope.

2.13. Immunohistochemistry (IHC)

The tumor microvessel density (MVD) was determined by using the Cluster of Differentiation 31 (CD31), a significant biomarker for vascular endothelial cells [30]. Tissues were stained with the CD31 antibody (CST), and tumor vessels were detected and measured with a Zeiss Axiokop 2plus microscope. In addition, tumor tissues were stained with anti-Ki67 antibody (CST), and Ki67 antibodies and positive cells were imaged using a Zeiss Axioskop 2 plus microscope. The proliferation index (PI) was calculated by dividing the number of Ki67-positive cells by the total number of cells [31]. Similarly, to find the apoptotic index (PI), tumor tissues were stained with the cleaved caspase-3 antibody. Cell death rate was calculated by the ratio of cleaved caspse-3-positive cells to the total cells [32].

2.14. Statistics Analysis

The mean ± standard error of the mean (SEM) was used to express all experimental data. The two-tailed Student's t test was used to determine the difference between the two groups. Using SPSS 17.0 software, the difference between different groups was examined using one-way analysis of variance, followed by Tukey's test. Statistical significance was defined as $p < 0.05$.

3. In silico Validation of Drugs/Inhibitors

3.1. Ligand and Receptor Protein Preparation

The structures of L-aspartic acid (L-Asp), DL-Propargylglycine (PAG), and aminooxyacetic acid (AOAA) in the Spatial Data File (SDF) were retrieved from the PubChem database (https://pubchem.ncbi.nlm.nih.gov/ (accessed on 20 December 2021)) with PubChem IDs 5960, 95575, and 286, respectively. Prior to PDBQT file generation, all the SDF files were prepared with the addition of Gasteiger charges and polar hydrogens by employing a set of AutoDock tools (version 1.5.6) and were saved in PDBQT format using the Open Babble tool integrated into PyRx software (version 0.8). The X-ray crystal structures of the target receptor proteins were retrieved from the Protein Data Bank (PDB) (https://www.rcsb.org/ (accessed on 21 December 2021)) with PDB IDs 3COG, 3OLH, and 4COO for Cystathionine gamma-lyase (CSE), 3-mercaptopyruvate sulfurtransferase (3-MPST), and Cystathionine beta-synthase (CBS), respectively. Before molecular docking, all the protein PDBs were preprocessed by removing the non-amino acid residues and water molecules, adding polar hydrogen atoms, and optimising by adding Kollman charges, utilizing the AutoDock tool [33,34].

3.2. Druggable Pockets and Molecular Docking Analysis

The druggable pockets of each selected protein receptor were identified through the P2RANK server (https://prankweb.cz/ (accessed on 26 December 2021)), which uses a template-independent machine learning-based method. The top-ranked predicted pockets were chosen and considered for the molecular docking simulation study. The detailed results of P2RANK are tabulated in Supplementary Table S1. The AutoDock Vina tool was utilized to execute molecular docking simulations due to its rapid, stochastic optimizations by operating multiple CPU cores, which allowed it to overcome the confines of the maximum number of rotatable bonds, atoms, and grid map size [35–37]. Docking grids of all macromolecule receptors were adjusted into squares of 24 Å with x, y, and z coordinates of −21.7641, 33.6885, and −18.7002 for Cystathionine gamma-lyase; −35.004,

−20.3346, and −41.463 for 3-mercaptopyruvate sulfurtransferase; and −9.2621, 22.9917, and −18.5675 for Cystathionine beta-synthase, respectively, to define the ligand binding sites. The interaction of the ligand–protein complexes was visualized using the BIOVIA Discovery Studio Visualizer version 20.1.0 tool.

4. Results

4.1. Inhibition of Endogenous H_2S Attenuates the Viability and Proliferation of Human BC Cells

As presented in Figure 1, treatment of 10 mM of PAG, AOAA, and L-Asp reduced the cell viability in a dose-dependent manner, with a sharp decrease in H_2S content in corresponding treated human cell (MCF-7 and MDA-MB 231) groups (Figure 1a,b). It was shown that there is a concentration-dependent decrease in the cell viability of cells (Supplementary Figure S1). Furthermore, the combined dose treatment (10 mM L-Asp + 10 mM PAG + 10 mM AOAA) had a stronger effect compared to treatment with 30 mM PAG and 30 mM L-Asp (Supplementary Figure S2). These preliminary findings on the drugs and cell viability suggest that the 10 mM treatment is the optimum dose for measuring the pharmacological effect on cells. Furthermore, this decrease in H_2S content in treated cells led to a significant decline in the cell proliferation and colony formation ability of BC cells (Figure 1c–f). According to these results, the combined dose treatment and AOAA exhibit a strong inhibitory effect compared to L-Asp and PAG. Taken all together, these findings suggest that upon treatment with these drugs, either separately or in combination form, they can suppress endogenous H_2S production, which may reduce viability, proliferation, and colony formation, indicating that H_2S is involved in the growth and progression of human BC.

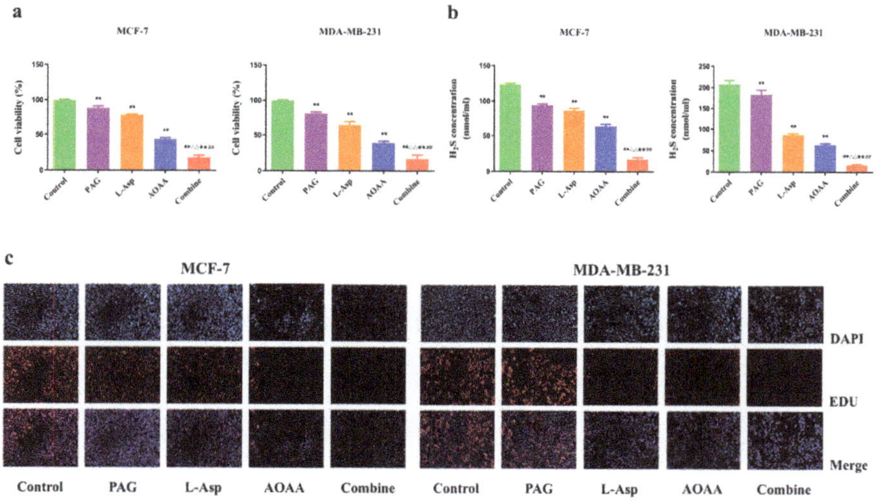

Figure 1. *Cont.*

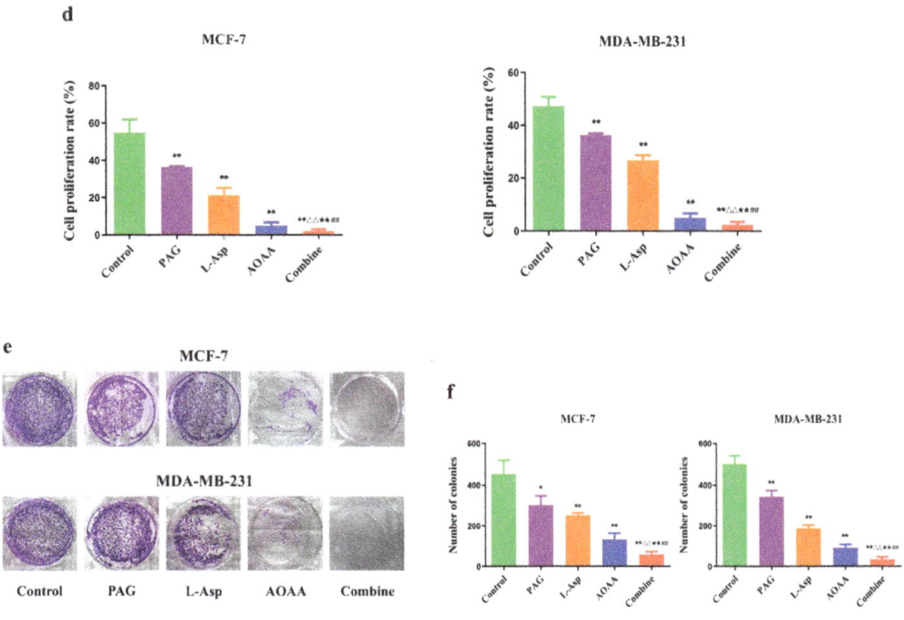

Figure 1. Effects of PAG, AOAA, and L-Asp on viability, H_2S content, proliferation, and colony formation in human BC cells. (**a**) The CC-K8 assay was used to determine the percentage of viable cells. The cell viability of each group without PAG, AOAA, or L-Asp treatment was normalized as 100% and considered to be the control group. (**b**) The levels of H_2S in BC cells were detected. (**c**) The proliferation rate of each group was analyzed. (**d**) DNA replication activities of BC cells in each group were examined by EdU assay (original magnification ×200). (**e**) The clonogenic capacity was determined in BC cells. (**f**) The number of colonies was calculated. The experiments were performed in triplicate. Data are presented as mean ± SEM. * $p < 0.05$, ** $p < 0.01$ compared with the control group; △△ $p < 0.01$ compared with PAG group; ★★ $p < 0.01$ compared with AOAA group; ## $p < 0.01$ compared with L-Asp group.

4.2. Inhibition of Endogenous H_2S Reduces the Migration and Invasion Rate of Human BC Cells

Followed by the 24 h treatment, it was observed that the migration and invasion of MCF-7 and MDA-MB-231 cells were reduced in each of the PAG-, L-Asp-, and AOAA-treated groups when compared to the control group. Furthermore, when we used the drugs in combination, a greater inhibitory effect on MCF-7 and MDA-MB-231 cell migration and invasion was recorded compared to the separately treated groups (Figure 2a,b). These results demonstrate that pharmacological treatment with these drugs could significantly inhibit the migration and invasion of human BC cells.

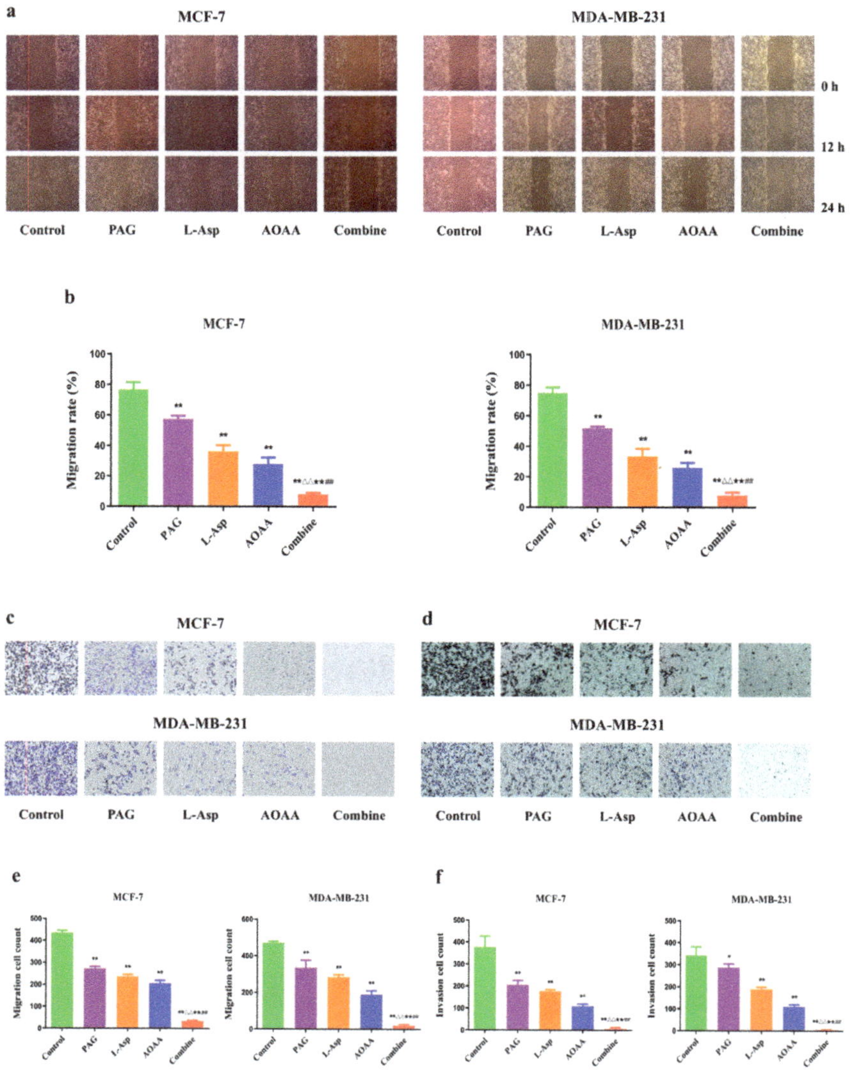

Figure 2. Effects of PAG, AOAA, and L-Asp on the migration and invasion of human BC cells. (**a**) Cell migration was measured by wound-healing assay (original magnification ×100). (**b**) The number of the migrated cells was calculated. (**c**,**d**) Transwell assay was performed to assess the migration and invasion of BC cells (original magnification ×200). (**e**,**f**) The number of the migrated and invasive cells was calculated. The experiments were performed in triplicate. Data are presented as mean ± SEM. ** $p < 0.01$ compared with the control group; △△ $p < 0.01$ compared with PAG group; ★★ $p < 0.01$ compared with AOAA group; ## $p < 0.01$ compared with L-Asp group.

4.3. Suppression of Endogenous H_2S Induces Apoptosis in Human BC Cells

The results of the present study demonstrate that, compared to the control group, treatment with PAG, L-Asp, or AOAA induce apoptosis in BC cells, as shown in Figure 3a,b. Findings from the TUNEL assay reveal that the combination group has a strong apoptotic index, followed by AOAA, L-Asp, and PAG. Regarding the mitochondrial apoptotic process, which is commonly mediated by cleaved caspase-3 [38], PARP is considered to be a crucial target for exploring this cell death mechanism [39]. In our study, both cleaved caspase-3 and PARP had higher expression in the drug-treated cells compared to the control group. Furthermore, they were higher in the PAG, L-Asp and AOAA separately treated groups than in the control group. Moreover, the amount of these two proteins (cleaved caspase-3 and cleaved PARP) in the combination group was noted to be higher than in the separately treated groups. Collectively, these results clearly indicate the apoptosis induction effect of these drug treatments.

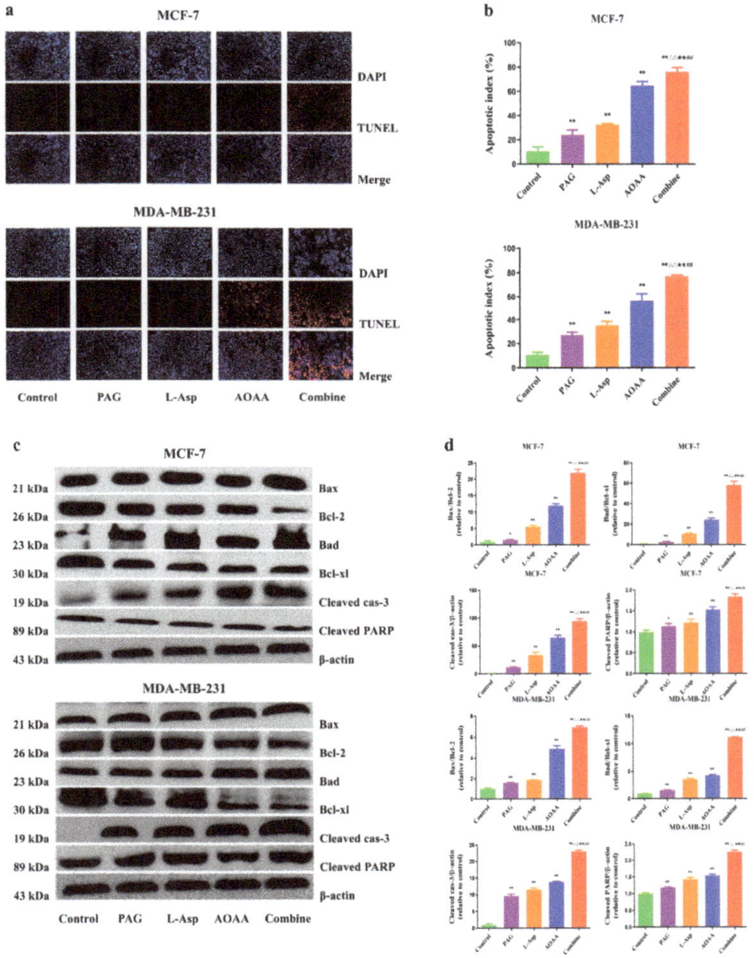

Figure 3. Effects of PAG, AOAA, and L-Asp on the cell death of human BC cells (**a**) The apoptotic level was measured by TUNEL staining (original magnification ×200). (**b**) Apoptotic index was calculated. (**c**) The expression levels of Bax, Bcl-2, Bad, Bcl-xl, cleaved caspase-3, and cleaved PARP were detected by Western blot. β-actin was used as a loading control. (**d**) Densitometric quantification was performed, normalized to the level of β-actin. The experiments were performed in triplicate. Data are presented as mean ± SEM. ** $p < 0.01$ compared with the control group; △△ $p < 0.01$ compared with PAG group; ★★ $p < 0.01$ compared with AOAA group; ## $p < 0.01$ compared with L-Asp group.

4.4. Suppression of Endogenous H_2S Inhibits Epithelial–Mesenchymal Transition in Human BC Cells

Epithelial mesenchymal transition (EMT) has been shown to be crucial in tumorigenesis, where the EMT program enhances metastasis. An aberrant expression of N-cadherin and E-cadherin in many types of tumors is regarded as a hallmark of EMT and considered to be a therapeutic target for inhibiting cancer cell migration [40,41]. Figure 4 shows the EMT rates in BC cells. When PAG, L-Asp, and AOAA were applied for 24 h, there was a significant increase in the expression of E-cadherin and a decrease in N-cadherin and Vimentin compared to the control group. Furthermore, analysis indicates that the effect was much stronger in the combined group, followed by the AOAA, L-Asp, and PAG groups (Figure 4). These findings reveal that suppression of endogenous H_2S could be a way forward to decrease cancer metastasis in human BC cells.

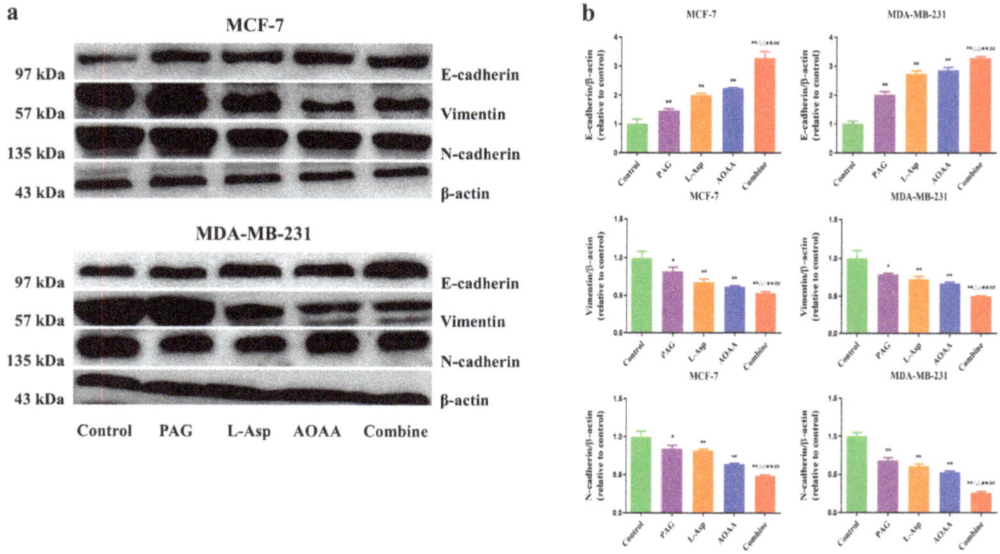

Figure 4. Effects of PAG, AOAA, and L-Asp on the metastasis of ability human BC cells. (**a**) The expression levels of N-Cadherin, N-Cadherin, and Vimentin were detected by Western blot. β-actin was used as a loading control. (**b**) Densitometric quantification was performed, normalized to the level of β-actin. The experiments were performed in triplicate. Data are presented as mean ± SEM. * $p < 0.05$, ** $p < 0.01$ compared with the control group; $^{\triangle}$ $p < 0.05$, $^{\triangle\triangle}$ $p < 0.01$ compared with PAG group; $^{\bigstar}$ $p < 0.05$, $^{\bigstar\bigstar}$ $p < 0.01$ compared with AOAA group; $^{\#\#}$ $p < 0.01$ compared with L-Asp group.

4.5. Suppression of Endogenous H_2S Disrupts the PI3K/AKT/mTOR Pathway in Human BC Cells

The PI3K/AKT/mTOR signaling pathway is a key signal transduction pathway involved in the many hallmarks of cancer, including survival, metabolism, motility, and genome stability [42,43]. This pathway contributes to several cancer-promoting aspects of the tumor environment, such as angiogenesis and inflammatory cell recruitment [44]. As presented in Figure 5, phosphorylation of PI3K, AKT, and mTOR decreased in the separately treated PAG, L-Asp, and AOAA groups compared to control, which may lead to both decreased cellular proliferation and increased cell death and suppress tumor growth [45]. Overall, these results demonstrate that pharmacological inhibition of endogenous hydrogen sulfide production suppresses BC progression via weakening of AKT, PI3K, and mTOR phosphorylation.

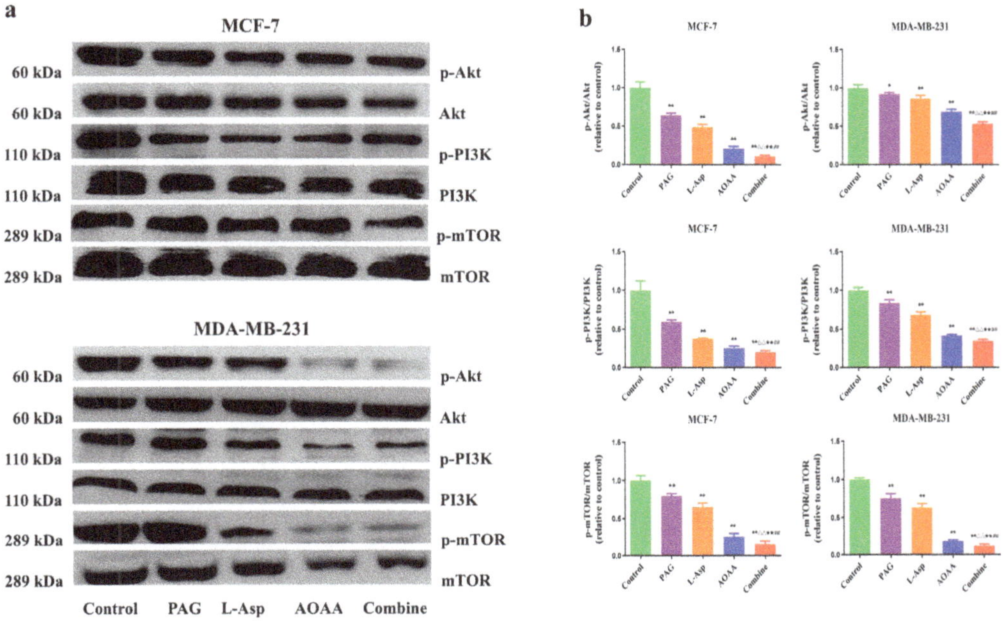

Figure 5. Effects of PAG, AOAA, and L-Asp on the PI3K/AKT/mTOR signaling pathway in human BC cells. (**a**) The expression levels of p-PI3K, p-AKT, and p-mTOR were detected by Western blot. (**b**) Densitometric quantification was performed, normalized to the level of respective non-phosphorylated candidate protein. The experiments were performed in triplicate. Data are presented as mean ± SEM. * $p < 0.05$, ** $p < 0.01$ compared with the control group; △ $p < 0.05$, △△ $p < 0.01$ compared with PAG group; ★ $p < 0.05$, ★★ $p < 0.01$ compared with AOAA group; ## $p < 0.01$ compared with L-Asp group.

4.6. Suppression of Endogenous H_2S Inhibits the Angiogenesis and Growth of Human BC Xenograft Tumors

Nude mice tumor models have been successfully established using MCF-7 and MDA-MB-231 cells [46,47]. The effect of endogenous H_2S inhibition on BC xenograft tumor growth was subsequently investigated. The tumor sizes and weights in the PAG, AOAA, and L-Asp groups were significantly lower than in the control group (Figure 6a,b). In addition, the tumor volume and weight were lower in the combination group than in the PAG, AOAA, and L-Asp groups, but the tumor inhibitory rate was higher (Figure 6c,d). It has been observed that HE, CD31, Ki67, and cleaved caspase-3 staining analyses on human BC xenograft tumors also support the effect of drugs, with respect to in vivo results. MVD and PI values were lower in the PAG, AOAA, and L-Asp groups while AI was higher compared to the non-treated control group. Furthermore, in the combination group, the MVD and PI were lower but the AI was higher compared to the PAG, AOAA, and L-Asp groups (Figure 6e,f).

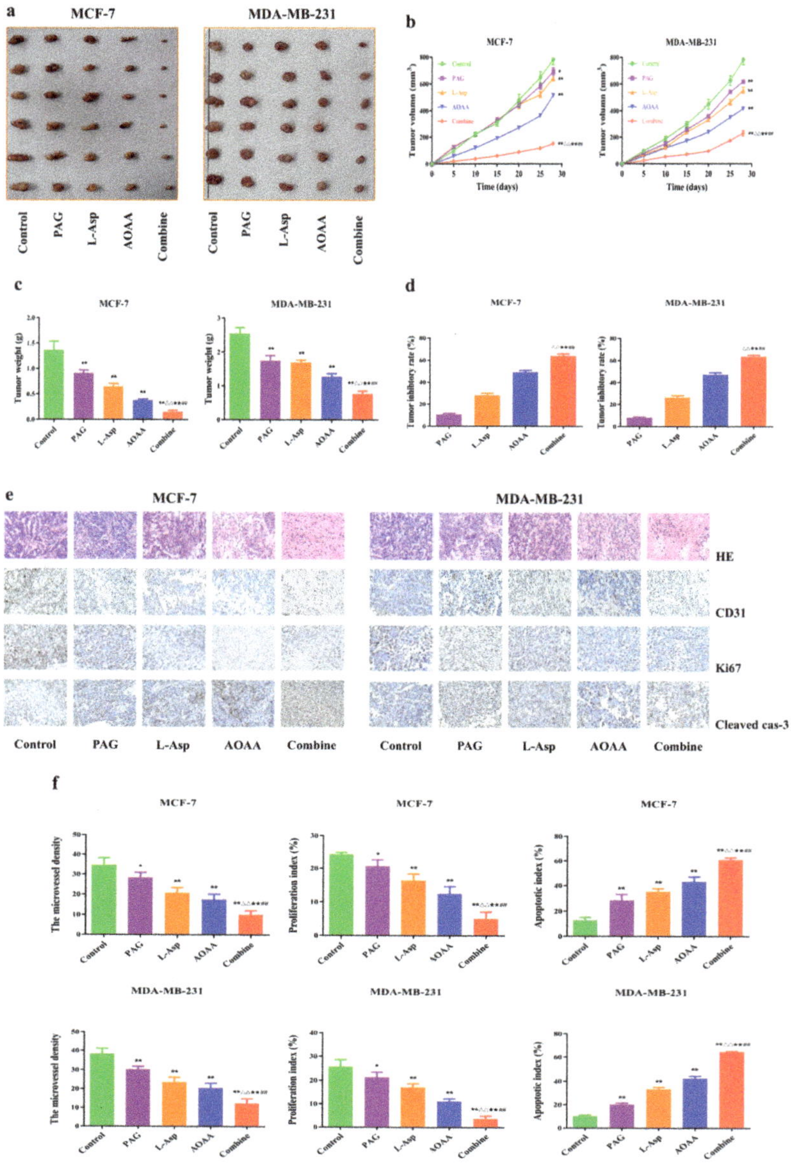

Figure 6. Effects of PAG, AOAA, and L-Asp on the growth of human BC xenograft tumors in nude mice. (**a**) The representative tumor samples from each group are shown. (**b**) The tumor volumes of human BC xenograft tumors were measured (n = 6). (**c**) The tumors were weighed (n = 6). (**d**) The inhibition rate of tumor growth was calculated (n = 6). (**e**) Representative photographs of HE, CD31, Ki67, and cleaved caspase-3 staining in human BC xenograft tumors (original magnification ×400). (**f**) The PI, MVD, and apoptotic index were calculated (n = 3). Data are presented as mean ± SEM. * $p < 0.05$, ** $p < 0.01$ compared with the control group; △ $p < 0.05$, △△ $p < 0.01$ compared with PAG group; ★★ $p < 0.01$ compared with AOAA group; ## $p < 0.01$ compared with L-Asp group.

4.7. Molecular Docking Analysis

The implementation of computational studies plays a vital role in the early stages of drug discovery and development [48]. In the present study, a molecular docking simulation was carried out to identify the ligand–protein interaction. All the ligands were docked with each of the selected receptor proteins (CSE, CBS, and MPST) to ascertain the optimal conformational binding area by employing the AutoDock Vina tool. This tool provides the minimum binding energy value in kcal/mol, where the lower the binding energy value, the higher the bonding affinity. Also, the greater the number of bonds (i.e., hydrogen and carbon), the stronger the binding interaction between ligand and protein (Table 1). The results of the present study revealed that all the ligands were found in their respective druggable pockets and supported our experimental work findings. Visualization of drug–protein interactions after molecular docking are shown in Figure 7.

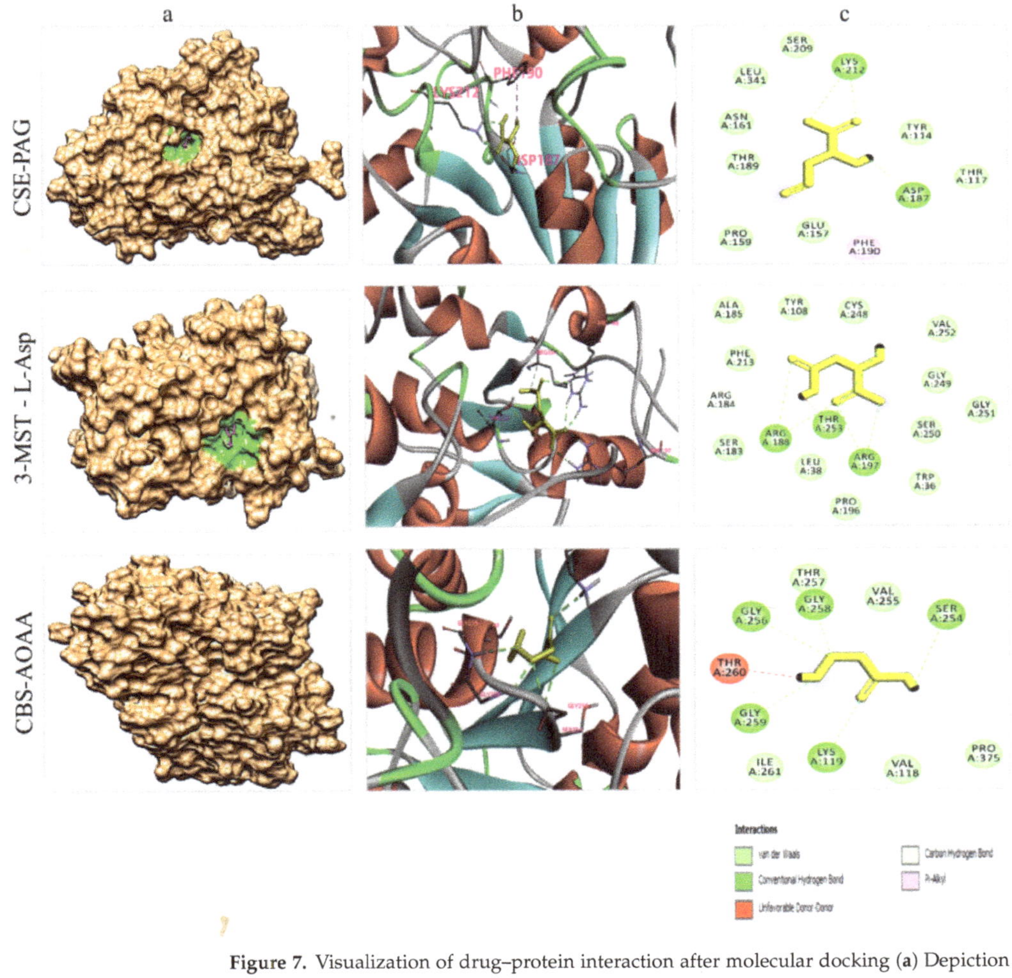

Figure 7. Visualization of drug–protein interaction after molecular docking (**a**) Depiction of each protein docking interaction pocket (magenta: drug molecule, green: drug pocket area in the corresponding enzymes). Note: In the CBS-AOAA complex, the pocket cannot be seen because the drug interaction area is in the cleft of the protein. (**b**) The 3D interaction sites of drugs with corresponding enzymes. (**c**) The 2D interaction of the interacting amino acid residues of the proteins with each drug molecule.

Table 1. The binding energy value (in kcal/mol) of drugs (PAG, L-Asp, and AOAA) to their respective enzymes (CSE, 3-MPST, and AOAA) in the drug docking exercise.

Drug	PubChem ID	Cystathionine Gamma-Lyase	3-Mercaptopyruvate Sulfurtransferase	Cystathionine Beta-Synthase
L-aspartic acid	5960	−4.9	−5.3	−4.5
DL-Propargylglycine	95575	−5.4	−5.1	−4.8
Aminooxyacetic acid	286	−4	−4.6	−4.6

5. Discussion

H_2S is known to be a third gaseous molecule that is predominantly involved in various pathophysiological conditions [49–51]. Published research has given the utmost importance to the physiological workings of H_2S in cancer cells. Given the several novel findings on H_2S donors' and inhibitors' roles in cancer [16,52], the rapid development of various H_2S-based therapeutics reflects the excitement that this unique mediator has sparked, as well as encouraging findings generated in preclinical and early clinical trials. Endogenous H_2S can promote cancer growth by inducing angiogenesis, regulating mitochondrial bioenergetics, accelerating cell cycle progression, and inhibiting apoptosis. In the present study, we inhibited endogenous H_2S synthesis by inhibiting H_2S-generating enzymes in human breast cancer cells.

Breast cancer is one the most frequent malignancies diagnosed in women worldwide [1,2]. There is no promising therapy available that can be used for the medication of individuals presenting advanced stages of breast cancer [53]. Therefore, it is essential to discover putative therapeutic drugs that can be used for the effective management and prevention of breast cancer. In the present study, we investigated the anti-tumor effect of the pharmacological inhibition of H_2S synthesis by employing the respective inhibitors for H_2S-generating enzymes in breast cancer in vitro and in vivo, as well as their in silico docking intensities.

It has been reported that exogenous treatment with PAG and AOAA (1–10 mM) for up to 3 days results in a reduction of cancer cell viability in a dose-dependent manner [52,54]. Adapting to the rationale of published studies, we treated the human BCs (MDA-MB-231 and MCF-7) with 1–10 mM of CSE, MPST, and MPST inhibitors, both individually and in combined solution form. Compared to the untreated cells, our findings demonstrated a remarkable reduction in cancer cell viability and growth rate in treated cells. In the follow-up experiments, we opted for a single concentration (10 mM) and checked the following results. In treated cells, we observed a conspicuous effect on the growth rate, invasion, and migration of BC cells compared to the control group. However, we found that this effect varied in the four groups of treated cells. PAG is the least effective, while the combined group has the strongest and most potent anti-tumor effect against the invasion and migration of both selected types of BC cells. The results showed that pharmacological inhibition of endogenous H_2S synthesis could reduce the viability, growth, migration, and invasion of both types of human BC cells, empowering the synthesis of H_2S-based chemopreventive drugs.

Moreover, the molecular docking simulation results also validated the findings of the undertaken study. Results of the binding pockets of each drug molecule predicted by P2RANK web servers were similar to those previously reported by studies on these enzymes. In a previous study carried out by Sun et al. (2009), in addition to its structural basis, they reported the inhibition mechanism of the CSE enzyme [55]. They documented that CSE is a tetramer when it is in PLP (pyridoxal-5-phosphate)-bound states. Each monomer of CSE contains two domains: (a) a larger PLP-binding domain ranging from 9 to 263 residues, and (b) a smaller domain covering 264–401 residues. The PAG bound to the CSE enzyme occupies the space of the substrate's side chain and covalently binds to the Tyr114 residue, which is thought to block substrate accessibility, thereby inhibiting CSE activity. Additionally, during this inhibition, Asp187 and Lys212 facilitate this process to accomplish the task [55]. Likewise, Meier et al. (2001) conducted a study to elucidate the structure of the CBS enzyme and its mechanism of inhibition [56]. They reported the crucial roles of Lys119, Gly256, Thr258, Gly259, and Thr260 residues in the active site of the enzyme. They revealed that these amino acid residues have significant importance in the working or inhibition process of the CBS enzyme [56]. Similarly, Yadav et al. (2013) extensively studied the structure of the MPST enzyme and highlighted the roles of Tyr108, Arg188, Arg197, and Thy253 amino acid residues in the functional behavior of this enzyme [57]. Overall, our drug–protein docking analysis explained the selectively and competitiveness of these drugs in inhibiting enzyme physiology, enhancing the reliability of these drugs to be used as future anti-cancer medicines. It was observed that PAG had the highest binding affinity (−5.4 kcal/mol) towards the CSE protein, followed by L-Asp and AOAA. Similarly, L-Asp was found to have the best binding interaction, with a binding energy of −5.3 kcal/mol, compared to other inhibitors with MPST protein. Likewise, AOAA was found to be good inhibitor of CBS protein and had a binding energy of −4.6 kcal/mol (Table 1). Figure 7 depicts the types of bonds and interacting amino acid residues in the ligand–protein interactions. The overall results of all the selected inhibitors' binding affinities with the three selected proteins are moderately convincing and support our experimental work findings.

Furthermore, apoptosis plays a crucial role in the progression and dynamic balance of complex organization in multicellular organizations [58]. At the cellular level, apoptosis occurs via both intrinsic and extrinsic pathways, wherein the former is mediated by mitochondria and the latter is stimulated by death receptors. This results in numerous morphological modifications, including genetic material condensation, nuclear fragmentation, and cell shrinkage [59]. Proapoptotic proteins (such as Bad and Bax) and members of the Bcl-2 family (such as Bcl-xl and Bcl-2) are the primary mediators of intrinsic cell death. However, caspases and PARP activation by certain stimuli also contribute to the occurrence of apoptotic cascades in the cell [39,58]. Going through the published literature, it has been shown that AOAA and PAG treatment causes cell death by promoting apoptotic cascades in the A549 and 95D human non-small cell lung cancer cells [13,60].

In the present study, following treatment, we observed the ratios of Bad/Bax-xl and Bax/Bcl-2 and the increased expression at the protein level of cleaved PARP and caspase-3 in the PAG, L-Asp, and AOAA groups. The combined treatment cells exhibited a stronger effect on the ratios of Bad/Bax-xl and Bax/Bcl-2 and higher protein expression of cleaved PARP and caspase-3 than the PAG, L-Asp, and AOAA groups. These results suggest that inhibition of endogenous H_2S synthesis induces apoptosis in human BC cells.

Similarly, following the 24 h treatment of PAG, L-Asp, and AOAA, significant downregulation of N-cadherin and Vimentin and increased expression of E-cadherin were observed compared to the control group. We found a more potent effect in the combined group, followed by the AOAA-treated group. Considering the inevitable roles of E and N cadherins in cancer metastasis [40,41], these findings reveal that suppression of endogenous H_2S could be a way forward for decreasing cancer metastasis in human BC cells.

The PI3K/AKT/mTOR pathway is essential in the regulation of cell growth, viability, mortality, and protein synthesis [61]. Research reveals that activation of this pathway contributes to tumor predisposition and increased risk of death [62]. It has also been found that inhibition of PI3K/AKT/mTOR leads to a decrease in proliferation index and shows potent efficacy in the treatment of BC [63]. Our findings show that PAG, L-Asp, and AOAA exposure disrupt PI3K/AKT/mTOR and reduce its level compared to the control group. Following the apoptosis result trends, we observed that the combined group has a potent inhibitory effect on the downregulation of the PI3K/AKT/mTOR pathway. Taken together, these results show that inhibition of endogenous H_2S decreases the expression of p-PI3K, p-mTOR, and p-AKT, indicating that they inhibit the proliferation, invasion, and migration of human BC cancer cells via blocking the PI3K/AKT/mTOR pathway. In recent research studies, MDA-MB-231 and MCF-7 cells have been successfully used to establish subcutaneous xenograft models [46,47]. Previously, the effect of endogenous H_2S inhibition by employing AOAA and PAG on xenograft tumor growth has been reported in human chronic myeloid leukemia and astrocytoma. They observed dramatically decreased tumor growth in the treated animals compared to control [64,65]. Our study showed that PAG, AOAA, and L-Asp significantly inhibited BC xenograft tumor growth, and the combined group showed stronger preventive effects on tumor growth.

Studies have established the fact that CD31 is a key biomarker for vascular endothelial cells, and that tumor MVD may be reflected in CD31 stains [30]. Similarly, Ki67 and cleaved caspase-3 are well known for detecting proliferation index and the apoptotic index, respectively [31,32]. Supporting the trends, our IHC results of tumor tissue show that PAG, L-Asp, and AOAA reduced CD31 and Ki67 levels but increased cleavage caspase-3 levels. In addition, the combined group showed lower expression of CD31 and elevated levels of cleavaged-caspase-3 than PAG, L-Asp, and AOAA groups.

6. Conclusions

In conclusion, the results of the present study suggest that reducing endogenous H_2S levels by pharmacologically inhibiting H_2S-producing enzymes with their respective synthesized inhibitors attenuates the growth of human BC cells in vitro and in vivo. H_2S inhibition leads to enhanced cell death and disrupts the PI3K/AKT/mTOR pathway, which eventually reduces the progression of BC cells. The docking potentials of these drugs to the active sites of their respective enzymes are consistent with the wet lab experiments, enhancing the reliability of these drugs. These findings, especially those for the combined dose with treatments of PAG, L-Asp, and AOAA, give a new insight towards suppressing H_2S levels in cells to control the progression of tumors. Given the pervasive nature of H_2S and the rapid development of various H_2S-based therapeutics approaches, the present study supports the findings from preclinical and early clinical trials and strongly endorses the rationale of empowering H_2S-based chemo-preventive drug synthesis against different malignancies. It is worth exploring the development of innovative H_2S-based therapeutic and diagnostic approaches for pre-clinical findings, which hold the promise of increased efficacy, lower toxicity, or both, to start trials in clinical settings.

Supplementary Materials: The following supporting information can be downloaded at: https://www.mdpi.com/article/10.3390/molecules27134049/s1, Figure S1: Dose dependent effects of PAG, AOAA, and L-Asp on viability of human breast cancer cells (MCF-7, MDA-MB-231); Figure S2: An intra- dose effect comparison of cell viability between combined low concentration (10 mL/mol) of PAG, AOAA, and L-Asp and higher concentration of individual drugs (30 Mm) by using a CCK-8 test on human breast cancer cells (MCF-7, MDA-MB-231); Table S1.

Author Contributions: Conceptualization, N.H.K., D.-D.W. and X.-Y.J.; methodology, N.H.K.; software, M.S. (Muhammad Shahid); validation, N.H.K. and M.S. (Muhammad Sarfraz); formal analysis, D.W.; investigation, N.H.K.; resources, C.-Y.Z.; data curation, W.W.; writing—original draft preparation, S.K. and E.E.N.; writing—review and editing, N.H.K.; visualization, D.-D.W. and X.-Y.J.; supervision, D.-D.W.; project administration, X.-Y.J.; funding acquisition, X.-Y.J. and C.-Y.Z. All authors have read and agreed to the published version of the manuscript.

Funding: This work was supported by grants from the National Natural Science Foundation of China (Nos. 81802718, U1504817); the Training Program for Young Backbone Teachers of Institutions of Higher Learning in Henan Province, China (No. 2020GGJS038); the Natural Science Foundation of Edu-cation Department of Henan Province, China (No. 21A310003); and the Foundation of Science and Technology Department of Henan Province, China (Nos. 222102310490, 222102310495).

Institutional Review Board Statement: Not applicable.

Informed Consent Statement: The animal experiment was approved by the Committee of Medical Ethics and Welfare for Experimental Animals of Henan University School of Medicine (HUSOM-2019–165).

Data Availability Statement: All data generated or analyzed in this study were included in this article.

Conflicts of Interest: The authors declare that they have no conflict of interest related to this work.

Abbreviations

H_2S—Hydrogen sulfide; L-Cys—L-cysteine; CSE—Cystathionine-γ-lyase; CBS—Cystathionine-β-synthase; 3-MST—3-Mercaptopyruvate sulfurtransferase; CAT—Cysteine aminotransferase; BC —Breast cancer; PAG—DL-propargylglycine; AOAA—Aminooxyacetic acid; L-Asp—L-aspartic acid; PBS—Phosphate buffer saline; MTT—3-(4,5)-Dimethylthiahiazo(-z-y1)-3,5-di-phenytetrazoliumromide; EdU—5-Ethynyl-2′-deoxyuridine; MR—Migration rate; TUNEL—TdT-mediated dUTP-biotin nick end labeling; Bcl-xl—B-cell lymphoma-extra-large; PI3K—hosphatidylinositol-3-Kinase; Bcl-2—B-cell lymphoma-2; Bad—Bcl-xl/Bcl-2-associated death promoter; Bax—Bcl-2-associated X protein; PARP—Poly adenosine diphosphate-ribose polymerase; CST—Cell Signaling Technology; mTOR—mammalian target of rapamycin; IR—Inhibition rate; HE—Hematoxylin and eosin; IHC—Immunohistochemistry; CD31—Cluster of differentiation 31; SEM—Standard error of the mean.

References

1. Fleege, N.M.; Cobain, E.F. Breast Cancer Management in 2021: A Primer for the OB GYN. *Best Pract. Res. Clin. Obstet. Gynaecol.* **2022**, *in press*.
2. Ferlay, J.; Colombet, M.; Soerjomataram, I.; Parkin, D.M.; Piñeros, M.; Znaor, A.; Bray, F. Cancer statistics for the year 2020: An overview. *Int. J. Cancer* **2021**, *149*, 778–789. [CrossRef] [PubMed]
3. Sung, H.; Ferlay, J.; Siegel, R.L.; Laversanne, M.; Soerjomataram, I.; Jemal, A.; Bray, F. Global Cancer Statistics 2020: GLOBOCAN Estimates of Incidence and Mortality Worldwide for 36 Cancers in 185 Countries. *CA Cancer J. Clin.* **2021**, *71*, 209–249. [CrossRef] [PubMed]
4. Metcalfe, K.; Lubinski, J.; Lynch, H.T.; Ghadirian, P.; Foulkes, W.; Kim-Sing, C.; Neuhausen, S.; Tung, N.; Rosen, B.; Gronwald, J.; et al. Family History of Cancer and Cancer Risks in Women with BRCA1 or BRCA2 Mutations. *JNCI J. Natl. Cancer Inst.* **2010**, *102*, 1874–1878. [CrossRef] [PubMed]
5. Britt, K.L.; Cuzick, J.; Phillips, K.-A. Key steps for effective breast cancer prevention. *Nat. Cancer* **2020**, *20*, 417–436. [CrossRef]
6. Khan, N.H.; Duan, S.-F.; Wu, D.-D.; Ji, X.-Y. Better Reporting and Awareness Campaigns Needed for Breast Cancer in Pakistani Women. *Cancer Manag. Res.* **2021**, *13*, 2125–2129. [CrossRef]
7. Filipovic, M.R.; Zivanovic, J.; Alvarez, B.; Banerjee, R. Chemical Biology of H_2S Signaling through Persulfidation. *Chem. Rev.* **2017**, *118*, 1253–1337. [CrossRef]
8. Hine, C.; Harputlugil, E.; Zhang, Y.; Ruckenstuhl, C.; Lee, B.C.; Brace, L.; Longchamp, A.; Treviño-Villarreal, J.H.; Mejia, P.; Ozaki, C.K.; et al. Endogenous Hydrogen Sulfide Production Is Essential for Dietary Restriction Benefits. *Cell* **2014**, *160*, 132–144. [CrossRef]
9. Paul, B.D.; Snyder, S.H. H_2S signalling through protein sulfhydration and beyond. *Nat. Rev. Mol. Cell Biol.* **2012**, *13*, 499–507. [CrossRef]
10. Zhang, H.; Huang, Y.; Chen, S.; Tang, C.; Wang, G.; Du, J.; Jin, H. Hydrogen sulfide regulates insulin secretion and insulin resistance in diabetes mellitus, a new promising target for diabetes mellitus treatment? A review. *J. Adv. Res.* **2020**, *27*, 19–30. [CrossRef]
11. Giovinazzo, D.; Bursac, B.; Sbodio, J.I.; Nalluru, S.; Vignane, T.; Snowman, A.M.; Albacarys, L.M.; Sedlak, T.W.; Torregrossa, R.; Whiteman, M.; et al. Hydrogen sulfide is neuroprotective in Alzheimer's disease by sulfhydrating GSK3β and inhibiting Tau hyperphosphorylation. *Proc. Natl. Acad. Sci. USA* **2021**, *118*, e2017225118. [CrossRef] [PubMed]
12. Yang, G.; Wu, L.; Jiang, B.; Yang, W.; Qi, J.; Cao, K.; Meng, Q.; Mustafa, A.K.; Mu, W.; Zhang, S.; et al. H2S as a physiologic vasorelaxant: Hypertension in mice with deletion of cystathionine γ-lyase. *Science* **2008**, *322*, 587–590. [CrossRef] [PubMed]
13. Szabo, C. Hydrogen Sulfide, an Endogenous Stimulator of Mitochondrial Function in Cancer Cells. *Cells* **2021**, *10*, 220. [CrossRef] [PubMed]
14. Shackelford, R.E.; Mohammad, I.Z.; Meram, A.T.; Kim, D.; Alotaibi, F.; Patel, S.; Ghali, G.E.; Kevil, C.G. Molecular Functions of Hydrogen Sulfide in Cancer. *Pathophysiology* **2021**, *28*, 28. [CrossRef] [PubMed]
15. Cai, F.-F.; Xu, H.-R.; Yu, S.-H.; Li, P.; Lu, Y.-Y.; Chen, J.; Bi, Z.-Q.; Sun, H.-S.; Cheng, J.; Zhuang, H.-Q.; et al. ADT-OH inhibits malignant melanoma metastasis in mice via suppressing CSE/CBS and FAK/Paxillin signaling pathway. *Acta Pharmacol. Sin.* **2021**, *42*, 1–14. [CrossRef] [PubMed]
16. Dong, Q.; Yang, B.; Han, J.-G.; Zhang, M.-M.; Liu, W.; Zhang, X.; Yu, H.-L.; Liu, Z.-G.; Zhang, S.-H.; Li, T.; et al. A novel hydrogen sulfide-releasing donor, HA-ADT, suppresses the growth of human breast cancer cells through inhibiting the PI3K/AKT/mTOR and Ras/Raf/MEK/ERK signaling pathways. *Cancer Lett.* **2019**, *455*, 60–72. [CrossRef] [PubMed]
17. Sonke, E.; Verrydt, M.; Postenka, C.O.; Pardhan, S.; Willie, C.J.; Mazzola, C.R.; Hammers, M.D.; Pluth, M.D.; Lobb, I.; Power, N.E.; et al. Inhibition of endogenous hydrogen sulfide production in clear-cell renal cell carcinoma cell lines and xenografts restricts their growth, survival and angiogenic potential. *Nitric Oxide* **2015**, *49*, 26–39. [CrossRef] [PubMed]
18. Chen, S.; Yue, T.; Huang, Z.; Zhu, J.; Bu, D.; Wang, X.; Pan, Y.; Liu, Y.; Wang, P. Inhibition of hydrogen sulfide synthesis reverses acquired resistance to 5-FU through miR-215-5p-EREG/TYMS axis in colon cancer cells. *Cancer Lett.* **2019**, *466*, 49–60. [CrossRef]

19. Whiteman, M.; Winyard, P.G. Hydrogen sulfide and inflammation: The good, the bad, the ugly and the promising. *Expert Rev. Clin. Pharmacol.* **2011**, *4*, 13–32. [CrossRef]
20. Asimakopoulou, A.; Panopoulos, P.; Chasapis, C.T.; Coletta, C.; Zhou, Z.; Cirino, G.; Giannis, A.; Szabo, C.; Spyroulias, G.A.; Papapetropoulos, A. Selectivity of commonly used pharmacological inhibitors for cystathionine β synthase (CBS) and cystathionine γ lyase (CSE). *Br. J. Pharmacol.* **2013**, *169*, 922–932. [CrossRef]
21. Zhang, F.; Chen, S.; Wen, J.Y.; Chen, Z.W. 3-Mercaptopyruvate sulfurtransferase/hydrogen sulfide protects cerebral endo-thelial cells against oxygen-glucose deprivation/reoxygenation-induced injury via mitoprotection and inhibition of the RhoA/ROCK pathway. *Am. J. Physiol. Cell Physiol.* **2020**, *319*, C720–C733. [CrossRef] [PubMed]
22. Hellmich, M.; Chao, C.; Módis, K.; Ding, Y.; Zatarain, J.; Thanki, K.; Maskey, M.; Druzhyna, N.; Untereiner, A.; Ahmad, A.; et al. Efficacy of Novel Aminooxyacetic Acid Prodrugs in Colon Cancer Models: Towards Clinical Translation of the Cystathionine β-Synthase Inhibition Concept. *Biomolecules* **2021**, *11*, 1073. [CrossRef] [PubMed]
23. Roy, A.; Khan, A.; Islam, M.; Prieto, M.; Majid, D. Interdependency of cystathione gam-ma-lyase and cystathione beta-synthase in hydrogen sulfide-induced blood pressure regulation in rats. *Am. J. Hypertens.* **2011**, *25*, 74–81. [CrossRef] [PubMed]
24. Bai, X.; Ihara, E.; Hirano, K.; Tanaka, Y.; Nakano, K.; Kita, S.; Ogawa, Y. Endogenous hydrogen sulfide contributes to tone generation in porcine lower esophageal sphincter via Na^+/Ca_2^+ exchanger. *Cell. Mol. Gastroenterol. Hepatol.* **2018**, *5*, 209–221. [CrossRef] [PubMed]
25. Yue, T.; Zuo, S.; Bu, D.; Zhu, J.; Chen, S.; Ma, Y.; Ma, J.; Guo, S.; Wen, L.; Zhang, X.; et al. Aminooxyacetic acid (AOAA) sensitizes colon cancer cells to oxaliplatin via exaggerating apoptosis induced by ROS. *J. Cancer* **2020**, *11*, 1828–1838. [CrossRef]
26. Wu, D.; Li, M.; Tian, W.; Wang, S.; Cui, L.; Li, H.; Wang, H.; Ji, A.; Li, Y. Hydrogen sulfide acts as a double-edged sword in human hepatocellular carcinoma cells through EGFR/ERK/MMP-2 and PTEN/AKT signaling pathways. *Sci. Rep.* **2017**, *7*, 5134. [CrossRef]
27. Wu, D.; Zhong, P.; Wang, Y.; Zhang, Q.; Li, J.; Liu, Z.; Ji, A.; Li, Y. Hydrogen sulfide attenuates high-fat diet-induced non-alcoholic fatty liver disease by inhibiting apoptosis and promoting autophagy via reactive oxygen species/phosphatidylinositol 3-kinase/AKT/mammalian target of rapamycin signaling pathway. *Front. Pharmacol.* **2020**, *11*, 585860. [CrossRef]
28. Wu, D.; Li, M.; Gao, Y.; Tian, W.; Li, J.; Zhang, Q.; Liu, Z.; Zheng, M.; Wang, H.; Wang, J.; et al. Peptide V3 Inhibits the Growth of Human Hepatocellular Carcinoma by Inhibiting the Ras/Raf/MEK/ERK Signaling Pathway. *J. Cancer* **2019**, *10*, 1693–1706. [CrossRef]
29. Dong, P.; Fu, H.; Chen, L.; Zhang, S.; Zhang, X.; Li, H.; Wu, D.; Ji, X. PCNP promotes ovarian cancer progression by accelerating β-catenin nuclear accumulation and triggering EMT transition. *J. Cell. Mol. Med.* **2020**, *24*, 8221–8235. [CrossRef]
30. Mbagwu, S.I.; Filgueira, L. Differential Expression of CD31 and Von Willebrand Factor on Endothelial Cells in Different Regions of the Human Brain: Potential Implications for Cerebral Malaria Pathogenesis. *Brain Sci.* **2020**, *10*, 31. [CrossRef]
31. Lanng, M.B.; Møller, C.B.; Andersen, A.-S.H.; Pálsdóttir, A.; Røge, R.; Østergaard, L.R.; Jørgensen, A.S. Quality assessment of Ki67 staining using cell line proliferation index and stain intensity features. *Cytom. Part A* **2018**, *95*, 381–388. [CrossRef] [PubMed]
32. Kobayashi, T.; Masumoto, J.; Tada, T.; Nomiyama, T.; Hongo, K.; Nakayama, J. Prognostic significance of the immunohisto-chemical staining of cleaved caspase-3, an activated form of caspase-3, in gliomas. *Clin. Cancer Res.* **2007**, *13*, 3868–3874. [CrossRef] [PubMed]
33. Asghar, A.; Tan, Y.-C.; Shahid, M.; Yow, Y.-Y.; Lahiri, C. Metabolite Profiling of Malaysian Gracilaria edulis Reveals Eplerenone as Novel Antibacterial Compound for Drug Repurposing Against MDR Bacteria. *Front. Microbiol.* **2021**, *12*, 1378. [CrossRef] [PubMed]
34. Hassan, S.S.U.; Zhang, W.D.; Jin, H.Z.; Basha, S.H.; Priya, S.S. In-silico anti-inflammatory potential of guaiane dimers from Xylo-pia vielana targeting COX-2. *J. Biomol. Struct. Dyn.* **2022**, *40*, 484–498. [CrossRef]
35. Azfaralariff, A.; Farahfaiqah, F.; Shahid, M.; Sanusi, S.A.; Law, D.; Isa, A.R.M.; Muhamad, M.; Tsui, T.T.; Fazry, S. Marantodes pumilum: Systematic computational approach to identify their therapeutic potential and effectiveness. *J. Ethnopharmacol.* **2021**, *283*, 114751. [CrossRef]
36. Trott, O.; Olson, A.J. AutoDock Vina: Improving the speed and accuracy of docking with a new scoring function, efficient optimization, and multithreading. *J. Comput. Chem.* **2010**, *31*, 455–461. [CrossRef]
37. Shams ul Hassan, S.; Abbas, S.Q.; Hassan, M.; Jin, H.Z. Computational Exploration of Anti-Cancer Potential of GUAIANE Dimers from Xylopia vielana by Targeting B-Raf Kinase Using Chemo-Informatics, Molecular Docking, and MD Simulation Studies. *Anti-Cancer Agents Med. Chem. (Former Curr. Med. Chem. Anti-Cancer Agents)* **2022**, *22*, 731–746. [CrossRef]
38. Man, S.M.; Kanneganti, T.-D. Converging roles of caspases in inflammasome activation, cell death and innate immunity. *Nat. Rev. Immunol.* **2015**, *16*, 7–21. [CrossRef]
39. Wu, D.; Liu, Z.; Wang, Y.; Zhang, Q.; Li, J.; Zhong, P.; Xie, Z.; Ji, A.; Li, Y. Epigallocatechin-3-Gallate Alleviates High-Fat Diet-Induced Nonalcoholic Fatty Liver Disease via Inhibition of Apoptosis and Promotion of Autophagy through the ROS/MAPK Signaling Pathway. *Oxidative Med. Cell. Longev.* **2021**, *2021*, 5599997. [CrossRef]
40. Mrozik, K.M.; Blaschuk, O.W.; Cheong, C.M.; Zannettino, A.C.W.; VanDyke, K. N-cadherin in cancer metastasis, its emerging role in haematological malignancies and potential as a therapeutic target in cancer. *BMC Cancer* **2018**, *18*, 939. [CrossRef]
41. Wang, D.; Wang, Y.; Wu, X.; Kong, X.; Li, J.; Dong, C. RNF20 Is Critical for Snail-Mediated E-Cadherin Repression in Human Breast Cancer. *Front. Oncol.* **2020**, *10*, 2762. [CrossRef] [PubMed]
42. Janku, F.; Yap, T.A.; Meric-Bernstam, F. Targeting the PI3K pathway in cancer: Are we making headway? *Nat. Rev. Clin. Oncol.* **2018**, *15*, 273–291. [CrossRef] [PubMed]
43. Fruman, D.A.; Rommel, C. PI3K and cancer: Lessons, challenges and opportunities. *Nat. Rev. Drug Discov.* **2014**, *13*, 140–156. [CrossRef]
44. Tan, X.; Zhang, Z.; Yao, H.; Shen, L. Tim-4 promotes the growth of colorectal cancer by activating angiogenesis and recruiting tumor-associated macrophages via the PI3K/AKT/mTOR signaling pathway. *Cancer Lett.* **2018**, *436*, 119–128. [CrossRef] [PubMed]

45. Miricescu, D.; Totan, A.; Stanescu-Spinu, I.-I.; Badoiu, S.C.; Stefani, C.; Greabu, M. PI3K/AKT/mTOR Signaling Pathway in Breast Cancer: From Molecular Landscape to Clinical Aspects. *Int. J. Mol. Sci.* **2020**, *22*, 173. [CrossRef]
46. Ren, G.; Shi, Z.; Teng, C.; Yao, Y. Antiproliferative Activity of Combined Biochanin A and Ginsenoside Rh2 on MDA-MB-231 and MCF-7 Human Breast Cancer Cells. *Molecules* **2018**, *23*, 2908. [CrossRef]
47. Jia, H.; Wang, X.; Liu, W.; Qin, X.; Hu, B.; Ma, Q.; Lv, C.; Lu, J. *Cimicifuga dahurica* extract inhibits the proliferation, migration and invasion of breast cancer cells MDA-MB-231 and MCF-7 in vitro and in vivo. *J. Ethnopharmacol.* **2021**, *277*, 114057. [CrossRef]
48. Shahid, M.; Azfaralariff, A.; Law, D.; Najm, A.A.; Sanusi, S.A.; Lim, S.J.; Cheah, Y.H.; Fazry, S. Comprehensive computational target fishing approach to identify Xanthorrhizol putative targets. *Sci. Rep.* **2021**, *11*, 1594. [CrossRef]
49. Yuan, A.; Hao, C.; Wu, X.; Sun, M.; Qu, A.; Xu, L.; Xu, C. Chiral CuxOS@ ZIF-8 Nanostructures for Ultrasensitive Quantification of Hydrogen Sulfide In Vivo. *Adv. Mater.* **2020**, *32*, 1906580. [CrossRef]
50. Ngowi, E.E.; Afzal, A.; Sarfraz, M.; Khattak, S.; Zaman, S.U.; Khan, N.H.; Li, T.; Jiang, Q.-Y.; Zhang, X.; Duan, S.-F.; et al. Role of hydrogen sulfide donors in cancer development and progression. *Int. J. Biol. Sci.* **2021**, *17*, 73–88. [CrossRef]
51. Ngowi, E.E.; Sarfraz, M.; Afzal, A.; Khan, N.H.; Khattak, S.; Zhang, X.; Li, T.; Duan, S.-F.; Ji, X.-Y.; Wu, D.-D. Roles of Hydrogen Sulfide Donors in Common Kidney Diseases. *Front. Pharmacol.* **2020**, *11*, 564281. [CrossRef] [PubMed]
52. Ascenção, K.; Dilek, N.; Augsburger, F.; Panagaki, T.; Zuhra, K.; Szabo, C. Pharmacological induction of mesenchymal-epithelial transition via inhibition of H2S biosynthesis and consequent suppression of ACLY activity in colon cancer cells. *Pharmacol. Res.* **2021**, *165*, 105393. [CrossRef] [PubMed]
53. Aggarwal, S.; Verma, S.S.; Aggarwal, S.; Gupta, S.C. Drug repurposing for breast cancer therapy: Old weapon for new battle. In *Seminars in Cancer Biology*; Elsevier: Amsterdam, The Netherlands, 2021.
54. Untereiner, A.; Pavlidou, A.; Druzhyna, N.; Papapetropoulos, A.; Hellmich, M.R.; Szabo, C. Drug resistance induces the upregulation of H2S-producing enzymes in HCT116 colon cancer cells. *Biochem. Pharmacol.* **2018**, *149*, 174–185. [CrossRef] [PubMed]
55. Sun, Q.; Collins, R.; Huang, S.; Holmberg-Schiavone, L.; Anand, G.S.; Tan, C.H.; Van-den-Berg, S.; Deng, L.W.; Moore, P.K.; Karlberg, T.; et al. Structural basis for the inhibition mechanism of human cystathi-onine γ-lyase, an enzyme responsible for the production of H2S. *J. Biol. Chem.* **2009**, *284*, 3076–3085. [CrossRef]
56. Meier, M.; Janosik, M.; Kery, V.; Kraus, J.P.; Burkhard, P. Structure of human cystathionine β-synthase: A unique pyridoxal 5′-phosphate-dependent heme protein. *EMBO J.* **2001**, *20*, 3910–3916. [CrossRef]
57. Yadav, P.K.; Yamada, K.; Chiku, T.; Koutmos, M.; Banerjee, R. Structure and kinetic analysis of H_2S produc-tion by human mercaptopyruvate sulfurtransferase. *J. Biol. Chem.* **2013**, *288*, 20002–20013. [CrossRef]
58. Tang, C.; Zhao, C.-C.; Yi, H.; Geng, Z.-J.; Wu, X.-Y.; Zhang, Y.; Liu, Y.; Fan, G. Traditional Tibetan Medicine in Cancer Therapy by Targeting Apoptosis Pathways. *Front. Pharmacol.* **2020**, *11*, 976. [CrossRef]
59. Koff, J.L.; Ramachandiran, S.; Bernal-Mizrachi, L. A Time to Kill: Targeting Apoptosis in Cancer. *Int. J. Mol. Sci.* **2015**, *16*, 2942–2955. [CrossRef]
60. Wang, M.; Yan, J.; Cao, X.; Hua, P.; Li, Z. Hydrogen sulfide modulates epithelial-mesenchymal transition and angiogenesis in non-small cell lung cancer via HIF-1α activation. *Biochem. Pharmacol.* **2019**, *172*, 113775. [CrossRef]
61. Rodon, J.; Dienstmann, R.; Serra, V.; Tabernero, J. Development of PI3K inhibitors: Lessons learned from early clinical trials. *Nat. Rev. Clin. Oncol.* **2013**, *10*, 143–153. [CrossRef]
62. Sun, C.-H.; Chang, Y.-H.; Pan, C.-C. Activation of the PI3K/Akt/mTOR pathway correlates with tumour progression and reduced survival in patients with urothelial carcinoma of the urinary bladder. *Histopathology* **2011**, *58*, 1054–1063. [CrossRef] [PubMed]
63. Wang, Y.; Nie, H.; Zhao, X.; Qin, Y.; Gong, X. Bicyclol induces cell cycle arrest and autophagy in HepG2 human hepatocellular carcinoma cells through the PI3K/AKT and Ras/Raf/MEK/ERK pathways. *BMC Cancer* **2016**, *16*, 742. [CrossRef] [PubMed]
64. Wang, D.; Yang, H.; Zhang, Y.; Hu, R.; Hu, D.; Wang, Q.; Liu, Y.; Liu, M.; Meng, Z.; Zhou, W.; et al. Inhibition of cystathionine β-synthase promotes apoptosis and reduces cell proliferation in chronic myeloid leukemia. *Signal Transduct. Target. Ther.* **2021**, *6*, 1–11. [CrossRef]
65. Cano-Galiano, A.; Oudin, A.; Fack, F.; Allega, M.F.; Sumpton, D.; Martinez-Garcia, E.; Niclou, S.P. Cystathionine-γ-lyase drives antioxidant defense in cysteine-restricted IDH1 mutant astrocytomas. *Neuro-Oncol. Adv.* **2021**, *3*, vdab057. [CrossRef] [PubMed]

molecules

Article

Anabasis articulata (Forssk.) Moq: A Good Source of Phytochemicals with Antibacterial, Antioxidant, and Antidiabetic Potential

Fakhria A. Al-Joufi [1], Marwa Jan [2], Muhammad Zahoor [2,*], Nausheen Nazir [2], Sumaira Naz [2], Muhammad Talha [2], Abdul Sadiq [3], Asif Nawaz [3] and Farhat Ali Khan [4]

1 Department of Pharmacology, College of Pharmacy, Jouf University, Sakaka 72341, Aljouf, Saudi Arabia; faaljoufi@ju.edu.sa

2 Department of Biochemistry, University of Malakand, Chakdara Dir Lower 18800, KPK, Pakistan; jmarwa084@gmail.com (M.J.); nausheen.nazir@uom.edu.pk (N.N.); sumaira.biochem@gmail.com (S.N.); livingontheedge36837@gmail.com (M.T.)

3 Department of Pharmacy, University of Malakand, Chakdara Dir Lower 18800, KPK, Pakistan; sadiquom@yahoo.com (A.S.); asifnawaz2445@gmail.com (A.N.)

4 Department of Pharmacy, Shaheed Benazir Bhutto University, Sheringal Dir Upper 18050, KPK, Pakistan; farhatkhan2k9@yahoo.com

* Correspondence: mohammadzahoorus@yahoo.com

Abstract: *Anabasis articulata* is medicinally used to treat various diseases. In this study, *A. articulata* was initially subjected to extraction, and the resultant extracts were then evaluated for their antimicrobial, antioxidant, and antidiabetic potentials. After obtaining the methanolic extract, it was subjected to a silica gel column for separation, and fractions were collected at equal intervals. Out of the obtained fractions (most rich in bioactive compounds confirmed through HPLC), designated as A, B, C, and D as well hexane fraction, were subjected to GC-MS analysis, and a number of valuable bioactive compounds were identified from the chromatograms. The preliminary phytochemical tests were positive for the extracts where fraction A exhibited the highest total phenolic and flavonoid contents. The hexane fraction as antimicrobial agent was the most potent, followed by the crude extract, fraction A, and fraction D. DPPH and ABTS assays were used to estimate the free radical scavenging potential of the extracts. Fraction C was found to contain potent inhibitors of both the tested radicals, followed by fraction D. The potential antidiabetic extracts were determined using α-glucosidase and amylase as probe enzymes. The former was inhibited by crude extract, hexane, and A, B, C and D fractions to the extent of 85.32 ± 0.20, 61.14 ± 0.49, 62.15 ± 0.84, 78.51 ± 0.45, 72.57 ± 0.92 and 70.61 ± 0.91%, respectively, at the highest tested concentration of 1000 μg/mL with their IC_{50} values 32, 180, 200, 60, 120 and 140 μg/mL correspondingly, whereas α-amylase was inhibited to the extent of 83.98 ± 0.21, 58.14 ± 0.75, 59.34 ± 0.89, 81.32 ± 0.09, 74.52 ± 0.13 and 72.51 ± 0.02% (IC_{50} values; 34, 220, 240, 58, 180, and 200 μg/mL, respectively). The observed biological potentials might be due to high phenolic and flavonoid content as detected in the extracts. The *A. articulata* might thus be considered an efficient therapeutic candidate and could further be investigated for other biological potentials along with the isolation of pure responsible ingredients.

Keywords: diabetes; antibacterial activity; total phenolic contents; total flavonoid contents; DPPH; ABTS; antidiabetic enzymes; HPLC-UV; GC-MS

Citation: Al-Joufi, F.A.; Jan, M.; Zahoor, M.; Nazir, N.; Naz, S.; Talha, M.; Sadiq, A.; Nawaz, A.; Khan, F.A. *Anabasis articulata* (Forssk.) Moq: A Good Source of Phytochemicals with Antibacterial, Antioxidant, and Antidiabetic Potential. *Molecules* **2022**, 27, 3526. https://doi.org/10.3390/molecules27113526

Academic Editors: Syed Shams ul Hassan, Mohamed M. Abdel-Daim, Tapan Behl and Simona Bungau

Received: 10 May 2022
Accepted: 26 May 2022
Published: 30 May 2022

1. Introduction

Humans have employed medicinal plants not only for therapeutic purposes but also for other applications. Since the beginning of human life on earth, they have constantly been utilized by humans for medicinal purposes, which is thus considered the beginning of the exploration of plants for medicinal purposes. Medicinal plants are valuable sources of biodiversity for humanity in providing multiple bioactive secondary metabolites such as sterols, saponins, tri-terpenes, alkaloids, polyphenols, flavonoids, tannins, and essential oils [1]. The mentioned phytochemicals cause different physiological and therapeutic effects when utilized by a human. These effects are broadly summed up as antioxidant, antimicrobial, anti-constitutive, anti-plasmodial, antidiabetic, spasmolytic, and neuroprotective potentials [2–4].

Free radicals are chemical species with unpaired electrons that are capable of attacking other chemical substances, especially those containing double bonds. The oxygen and nitrogen-based free radicals are constantly produced in the human body, which can attack biologically important substances, such as DNA and protein, causing a number of health complications from aging to life-threatening cancer and diabetes mellites. Oxidative stress is a general term used to describe such health complications, whereas antioxidants are chemical substances capable of scavenging the responsible free radicals. Most antioxidants contain a benzene ring, which can delocalize the free electrons associated with free radicals. Most of the plant's secondary metabolites fall into these categories, especially flavonoids and phenolics [5].

Diabetes mellitus is one of the top 10 causative factors of human deaths globally. In individuals with diabetes, there is either less/no production of insulin (type-1) or resistance to the reception of insulin by its receptors (type-2) [6,7]. Type 2 is more prevalent and generally appears as a result of a combination of resistance to insulin action along with an inadequate insulin secretory response [6–8]. Although a number of therapies are used to control this dreadful disease, a 100% efficient therapy is still not available. Scientists around the globe are constantly exploring plants for their antidiabetic potentials, and few of them have produced far-reaching results. Extensive research in this regard is still needed as, according to the world health organization, 1.6 million deaths occurred due to diabetes mellitus in 2016. For the treatment and management of this disorder, either insulin is taken or other strategies collectively known as non-insulin treatment are followed. In the non-insulin treatment category, the most popular approach used is to inhibit the carbohydrate metabolic enzymes (α-amylase and α-glucosidase), thus resulting in the minimum release of glucose molecules into the bloodstream [9,10]. Several synthetic inhibitors are commercially available that are taken orally by patients, and although effective, they are associated with more side effects as compared to natural products [11]. Importantly, the trend of pharmacological screening for hypoglycemic and antidiabetic potential has increased manyfold in the last few decades [11]. For ages, medicinal plants have been used as anti-hyperglycemic agents in folk medicine [12–14].

Anabasis articulata (*A. articulata*; Figure 1) belongs to the genus *Anabasis*, and the family Amaranthaceae is a medicinally valuable plant that is subjected to scientific exploration very little and needs to be explored in line with modern approaches. It is a xerophyte primarily found in deserts. In many parts of the world, it is used in folk medicine to treat skin conditions such as eczema and other ailments, including diabetes, headache, and fever [8].

As mentioned above, *A. articulata* has been reported to have medicinal properties by very few researchers, and thus, the present study is an attempt to explore *A. articulata* for its antimicrobial, antioxidant, and antidiabetic potential in connection to its phytochemical composition, which was investigated using preliminary phytochemical tests, HPLC and GC-MS analysis.

Figure 1. The *A. articulata* plant.

2. Materials and Methods

2.1. Plant Material Collection

The leaves and stem (1 kg; equal proportions) of *A. articulata* were collected from Mohmand Agency (34.5356° N, 71.2874° E, a deserted tribal area), Pakistan, in the year 2020–2021. The collected plant samples were identified by the Taxonomist, working as an expert in Herbarium, University of Malakand, Pakistan. A voucher specimen (BGH-UOM-190) was kept for record in the herbarium. After cleaning, the samples were shade dried, grounded, and then subjected to the extraction process.

2.2. Chemicals and Reagents

All chemicals used in the study were of analytical grade except those used as solvents in HPLC analysis. The reagents; 2,2-Diphenyle-1 picrylhydrazyl (DPPH), sodium carbonate, Folin-Ciocalteu (F-C) reagent were obtained from Sigma-Aldrich CHEMIE GmbH, St. Louis, MO, USA, whereas quercetin and 2,2′-Azino-Bis-3 ethyl benzothiazoline-6-sulfonic Acid (ABTS) aluminum chloride, methanol, sodium hydroxide, ethanol, sodium nitrite, and ascorbic acid were purchased from Sigma-Aldrich, Taufkirchen, Germany.

2.3. Extraction and Fractionation

Extraction and fractionation were carried out according to already reported protocols [6,13]. About 200 g of the powdered sample was soaked in methanol, filtered through the Whatman filter paper, and concentrated using a rotary evaporator (Rotavapor R-200, Buchi, Flawil, Switzerland). About 20 g of crude extract was obtained and used in subsequent experiments. The extract was eluted through a silica gel column for fractionation as per the following details: The extract was mixed with silica gel slurry and then allowed to dry in the air. The sample loaded silica was then carefully loaded to a large silica gel column with an internal diameter of 10 cm and packed height of 50 cm using a gradient of increasing polarity from *n*-hexane to ethyl acetate as the mobile phase. The oil fraction was extracted through *n*-hexane solvent. The effluents of columns were separated into four purified fractions designated as A, B, C, and D. Fraction A was separated by silica gel column chromatography in a solvent system of ethyl acetate and *n*-hexane (5:95), fraction B was separated by silica gel column chromatography in a solvent system of ethyl acetate and *n*-hexane (10:90), fraction C was separated in a solvent system of ethyl acetate and *n*-hexane (20:80), while fraction D was separated by silica gel column chromatography in a solvent system of ethyl acetate and *n*-hexane (30:70). The resultant extracts were stored at a temperature of 4 °C in a refrigerator till their screening for in vitro biological activities.

2.4. Preliminary Phytochemical Analysis

Reported protocols were followed for the identification of major phytochemicals in the extracts [15].

2.5. Estimation of Total Phenolic Content (TPC) and Total Flavonoid Content (TFC)

Shirazi et al.'s method was used for the estimation of TPC [16] in extract, whereas Kim et al.'s method [17] was followed for TFC estimation. From the stock sample (5 mg/5 mL), 1 mL was added to 9 mL of distilled water, to which 1 mL F-C reagent was added and incubated for 6 min, after which 10 mL of 7% sodium carbonate solution and 25 mL of distilled water were added. The absorbance was measured at 760 nm after 90 min of incubation time. The standard gallic acid solution's curve was used for the estimation of the TPC expressed as mg GAE (Gallic acid equivalent)/g of the dry sample.

In distilled water (500 µL), 100 µL of each sample, 150 µL of aluminum chloride, 100 µL of 5% sodium nitrate, and 200 µL of 1M sodium hydroxide were added for assessment of TFC. The absorbance of the mixture was recorded at 510 nm after being incubated for 5 min. The TFC was calculated as mg QE (quercetin equivalents)/g of the dry sample.

2.6. HPLC-UV Characterization

Methanolic extract, hexane, and purified fraction were each added to a distilled water:methanol (1:1) mixture, heated at 50 °C for 1 h, dually filtered, and poured into HPLC vials [18]. The extract's phytochemicals were separated using HPLC Agilent 1260 with an eclipsed C18 column (Santa Clara, CA, USA). The spectrums were recorded at 320 nm. Retention times of the available standards were employed to identify the unknown compounds present in the analyzed samples. Quantification of antioxidants was measured by formula (Equation (1)):

$$Cx = \frac{Ax \times Cs(\mu g/mL) \times V(\text{mL})}{As \times Sample\ (\text{wt. in g})} \tag{1}$$

where Cx = concentration of unknown sample, As = peak area of standard, Ax = peak area of unknown sample, Cs = concentration of standard (0.09 µg/mL).

2.7. GC-MS Analysis

The extracts were further analyzed for volatile components using GC-MS (Agilent Technological USA) analysis [19]. The mass spectra and retention time of the compounds present in samples were compared with those of Willy and NIST libraries [20].

2.8. Antibacterial Screening

The antibacterial potential of methanolic crude extract, purified fractions (A, B, C, and D), and oil fraction was assessed against *Shigella dysentery* (*S. dynasties*), *Escherichia coli* (*E. coli*), and *Salmonella Typhi* (*S. Typhi*) using the agar disc diffusion method. The strains were grown on nutrient broth, whereas the antibacterial spectrum of the extracts and fractions was assessed by the agar disc diffusion method, as mentioned before. A control (ampicillin) disc was also placed. The Clinical Laboratory Standards Institute guidelines (CLSI 2012) were followed [3] while determining the zone of inhibition (ZI) encountered by each extract and fractions against selected bacterial strains. The activity for each extract, purified fractions, and oil fraction was performed in triplicate and presented as the mean values.

2.9. Antioxidant Activities

2.9.1. DPPH Assay

A slightly modified DPPH assay as used before by Brand William was followed [21]. The absorbance of 3 mL from stock DPPH (20 mg in 100 mL of methanol) was adjusted to 0.75 at 517 nm. The DPPH stock solution was covered and kept in the dark overnight to generate free radicals. About 2 mL of each dilution (1000, 500, 250, 125, 62.5 μg/mL) prepared from methanolic extract stock (5 mg/5 mL of methanol) were mixed with 2 mL of pre-incubated DPPH stock solution and incubated for 15 min. Ascorbic acid was used as a standard. The absorbance of the reaction mixtures was recorded at 517 nm and the % inhibition was calculated as (Equation (2)):

$$\%\text{inhibition} = \frac{\text{A} - \text{B}}{\text{A}} \times 100 \tag{2}$$

where A = absorbance of pure DPPH in oxidized form, B = absorbance of the sample, which was measured after 15 min of reaction with DPPH.

2.9.2. ABTS Assay

A standard protocol for ABTS free radical scavenging potential of extracts was followed [22]. ABTS (7 mM) and $K_2S_2O_8$ (2.45 mM) were mixed (in methanol) and put in the dark for 24 h for free radicals' formation, which was then used as stock solution. The absorbance of 3 mL from the stock ABTS was adjusted to 0.75 at 745 nm, which was considered a control. About 300 μl of each of the serial dilutions of methanolic extract (1000, 500, 250, 125, 62.5 μg/mL) and 3 mL of stock ABTS were mixed and incubated for 15 min at 25 °C, and their absorbance was measured at 745 nm. Ascorbic acid was used as a control. The scavenging activity was calculated by Equation (2).

2.10. *In Vitro Antidiabetic Activities*

2.10.1. Inhibition of α-Amylase

The extracts were assessed as inhibitors of α-amylase following a standard protocol reported in the literature with some modifications [23]. In distilled water, an alpha-amylase stock solution (10 mg/100 mL) was prepared. About 10 μL of alpha-amylase stock solution was mixed with 30 μL of each sample dilution and 40 mL of starch solution, and these were kept at 37 °C for 30 min. After incubation, 20 μL of HCl (1M) was added to the reaction mixture, and its absorbance was measured at 580 nm. Acarbose was used as the reference standard. Alpha-amylase % inhibition was calculated as (Equation (3)):

$$\%\alpha - \text{amylase inhibition} = \frac{\text{control absorbance} - \text{sample absorbance}}{\text{control absorbance}} \times 100 \tag{3}$$

2.10.2. Inhibition of α-Glucosidase

The reported protocol was followed to assess the α-glucosidase inhibitory potential of the extracts [24]. To 100 μL α-glucosidase (0.5 units/mL), 50 μL of each sample dilutions and 600 μL of 0.1 M phosphate buffer (pH 6.9) were mixed. From the substrate (*p*-nitrophenyl-α-D-glucopyranoside) prepared as a 5 Mm solution in 0.1 M phosphate buffer, 100 μL of the substrate was added to the reaction mixture and kept for 15 min at 37 °C. The absorbance was recorded at 405 nm. The reaction mixture without α-glucosidase was labeled as blank, whereas the reaction mixture without the sample was taken as the control. The degree of the enzyme's activity inhibition was measured as:

$$\%\alpha - \text{glucosidase inhibition} = \frac{\text{control absorbance} - \text{sample absorbance}}{\text{control absorbance}} \times 100 \tag{4}$$

2.11. *Statistical Analysis*

All in vitro experiments were performed in three replicates. All results have been presented as mean ± SEM. The Student's *t*-test and one-way ANOVA followed by Dunnett's post hoc multiple comparison test was used to evaluate the significance of the data obtained. $p \leq 0.05$ was considered significant.

3. Results

3.1. Yield from Fractionation

Crude extract: 10 g, fraction A: 200 mg, B: 150 mg, C: 120 mg, and D: 100 mg were obtained. About 80 mg of purified hexane fraction was obtained in semi-solid form. In the subsequent studies, these components have been tested.

3.2. Qualitative Phytochemical Screening

Plants contain millions of compounds that are classified into roader groups of phytochemicals. Such constituents are determined as preliminary evaluations to decide the medicinal value of a plant. Table 1 represents the presence of different phytochemical groups showing that this plant is worthy of being investigated for its medicinal profile, being a rich source of phytochemicals.

Table 1. Phytochemical screening (qualitative) of *A. articulata* crude extract.

Phytochemical	Reagent	Observation	Result
Alkaloids	Dragendroff's	Reddish-orange precipitate	+
Tannins	Gelatin	Dirty (brownish) green precipitates	+
Flavonoids	Ferric chloride	The yellowish appearance that becomes clear after acid (HCL) addition	+
Triterpenoids	Liebermann Burchard	Brown ring	+
Glycosides	Keller Killiani	Reddish-brown layer	+

3.3. TPC and TFC in Resultant Extracts

A standard Gallic acid curve was constructed by preparing the dilutions 20, 40, 60, 80 and 100 mg/mL to estimate the TPC in different tested samples of *A. articulata* using a graphical regression method (Figure 2A). Comparatively higher TPC contents were estimated in almost all extracts as compared to TFC. Results show that the highest TPC (Figure 3) values were observed for crude extract and then oil and fraction A (72.1 ± 0.2, 69.0 ± 1.1 and 68.0 ± 0.4 mg GAE/g of dry sample, respectively).

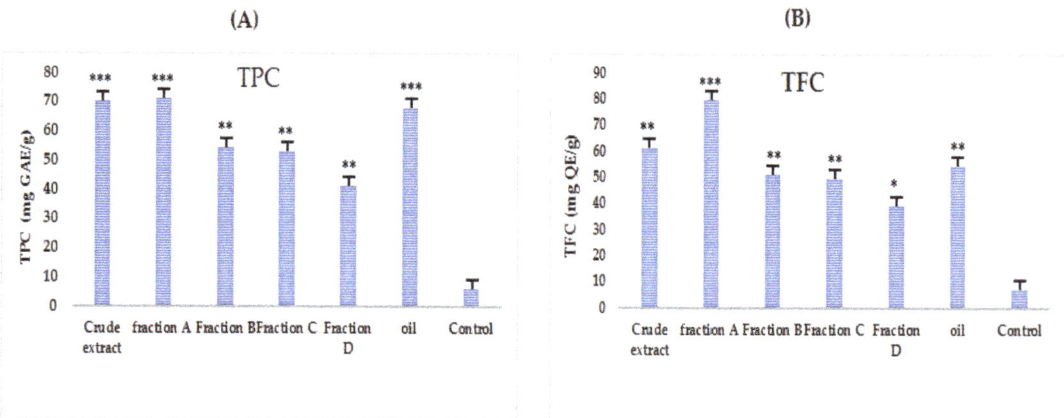

Figure 2. Total phenolic (**A**) and total flavonoids (**B**) contents in different tested samples (crude extract, purified fractions, and oil) of *A. articulata*. (**A**) TPC expressed as gallic acid equivalents (mg GAE)/g dry plant sample; (**B**) TFC expressed as quercetin equivalents (mg QE)/g dry plant sample. The data is represented as Mean ± SEM, n = 3. Values are significantly different as compared to positive control * $p < 0.05$, ** $p < 0.01$, *** $p < 0.001$.

To estimate the TFC in different tested samples of *A. articulata*, a regression curve of standard quercetin was constructed by preparing the dilutions 20, 40, 60, 80 and 100 mg/mL. The estimated contents are graphically presented in Figure 2B. Fraction A/crude extract followed by the oil fraction has shown the highest total flavonoid contents (62.0 ± 0.1, 62.3 ± 1.2, and 55.1 ± 0.3 mg QE/g of dry sample, respectively).

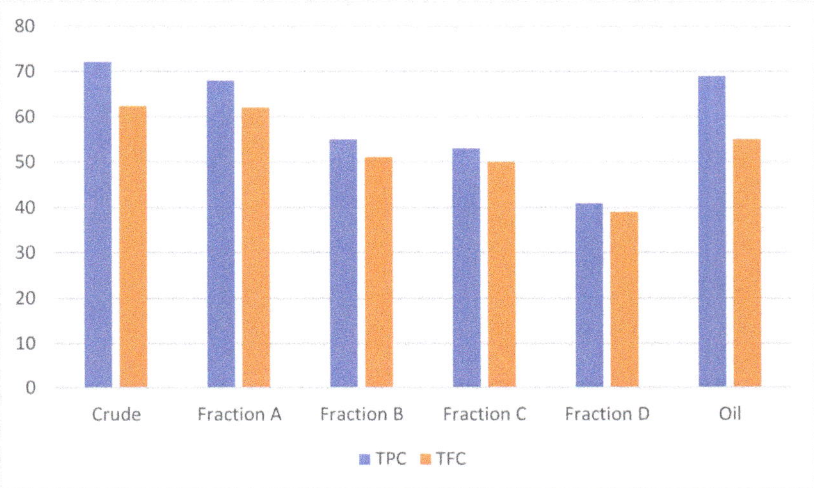

Figure 3. Total phenolic and total flavonoid content in different tested samples (crude extract, purified fractions, and oil) of *A. articulata*. TPC expressed as gallic acid equivalents (mg GAE)/g of dry plant sample, and TFC expressed as quercetin equivalents (mg QE)/g of dry plant sample. The data are represented as mean ± SEM, *n* = 3.

3.4. HPLC-UV Analysis

The HPLC chromatograms of the crude extract are presented in Figure 4A, purified fractions in Figure 4B–E, and the *n*-hexane fraction in Figure 4F. The compounds identified were: malic acid, gallic acid, chlorogenic acid, epigallocatechin gallate, Bis-HHDP-hex (pedunculagin), morin, 3-0-caffcoylquinic acid, Ellagic acid, kaempferol-3-(p-coumaroyl-diglucoside)-7-glucoside, catechin hydrate, rutin, syringic acid, quercetin-7-O-sophoroside, kaempferol-3-(caffeoyl-diglucoside)-7-rhamnosyl, mannose, pyrogallol, caffeic acid, mandelic acid, quercetin-3-(caffeoyldiglucoside)-7-glucoside, p-Coumaric acid, galactose, vitamin C, 3-0-caffeoylquinic acid, apigenin-7-O-rutinoside, quercetin 3,7-di-O-glucoside, rhamnose, 3,5-dicaffeoylquinic acid, mandelic acid, xylulose, quercetin 3,7-O-glucoside, glucose, quercitin-3-0-glysides, quercitin-3-O-rutinoside, quercitin-3-O-glycosides, quercetin, quercetin-3-(caffeoyldiglucoside)7-glucoside, quercitin-3-0-glycosides, in crude extract, various fraction (A, B, C, D, and *n*-hexane fraction. Each peak in the given chromatograms represents a phytoconstituent. For the identification of such constituent retention times of each component were compared with that of external standards. The quantification of each phenolic compound with their particular peak position and retention time (Rt) in chromatogram is presented in Tables 2–7.

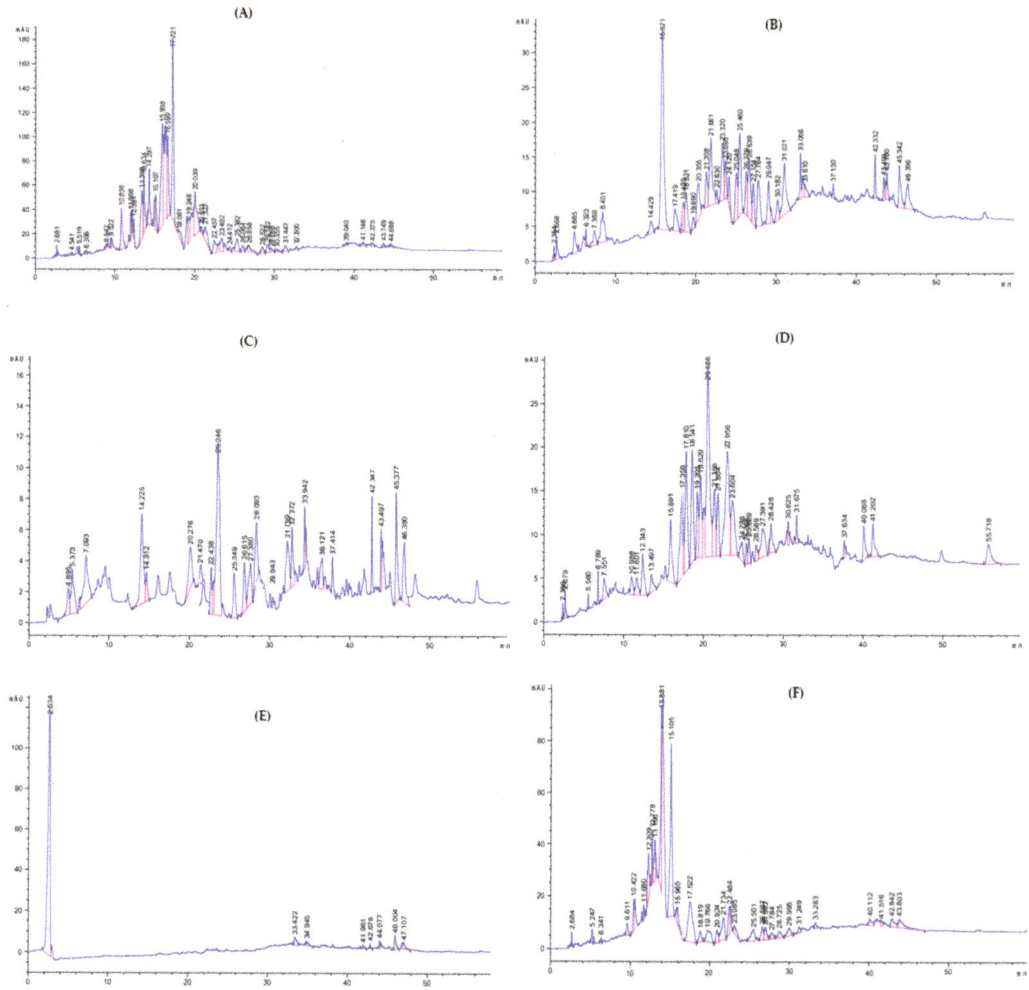

Figure 4. HPLC chromatograms of crude extract (**A**), various purified fractions (**B–E**), and *n*-hexane fraction (**F**) of *A. articulata*.

Table 2. Identified phytochemicals in crude extract of *A. articulata* through the HPLC-UV technique.

Retention Time (min)	Phytochemical Compounds	HPLC-UV λmax (nm)	Peak Area of Sample	Peak Area of Standard	Concentration (μg/mL)	Identification Reference
2	Malic acid	320	53.130	40.323	1.186	Ref. Stand
4	Gallic acid	320	41.239	195.40	0.189	Ref. Stand
6	Chlorogenic acid	320	32.966	12.929	2.295	Ref. Stand
8	Epigallocatechin gallate	320	44.782	7261.474	0.005	Ref. Stand
11	Bis-HHDP-hex(pedunculagin)	320	171.562	-	-	[25]
12	Morin	320	103.604	2.00	46.622	Ref. Stand

Table 2. *Cont.*

Retention Time (min)	Phytochemical Compounds	HPLC-UV λmax (nm)	Peak Area of Sample	Peak Area of Standard	Concentration (μg/mL)	Identification Reference
14	3-0-caffcoylquinic acid	320	514.593	-	-	[25]
16	Ellagic acid	320	912.321	319.242	2.572	Ref. Stand
18	Kaempferol-3-(p-coumaroyl-diglucoside)-7-glucoside	320	149.535	-	-	[25]
20	Catechin hydrate	320	810.747	78.00	9.355	Ref. Stand
22	Rutin	320	264.573	22.40	10.630	Ref. Stand
23	Syringic acid	320	254.546	-	-	[26]
24	Quercetin-7-O-sophoroside	320	50.8907	-	-	[26]
25	Kaempferol-3-(caffeoyl-diglucoside)-7-rhamnosyl	320	261.997	-	-	[26]
26	Mannose	320	85.1536	-	-	[25]
28	Pyrogallol	320	101.640	1.014	90.213	Ref. Stand
29	Caffeic Acid	320	129.708	-	-	[25]
30	Mandelic acid	320	45.405	72.00	0.567	Ref. Stand
31	Quercetin-3-(caffeoyldiglucoside)-7-glucoside	320	107.633	-	-	[27]
32	p-Coumaric acid	320	52.304	-	-	[27]
42	Galactose	320	57.017	-		[25]

Table 3. Identified phytochemical compounds in fraction A of *A. articulata* through the HPLC-UV technique.

Retention Time (min)	Phytochemical Compounds	HPLC-UV λmax (nm)	Peak Area of Sample	Peak Area of Standard	Concentration (μg/mL)	Identification Reference
4	Vitamin C	320	64.43	22.4	2.588	Ref. Stand
6	Chlorogenic acid	320	18.64	2.929	5.727	Ref. Stand
8	Epigallocatechin gallate	320	132.75	7261.474	0.0164	Ref. Stand
14	3-0-caffeoylquinic acid	320	45.345	-	-	[26]
18	Kaempferol-3-(p-coumaroyl-diglucoside)-7-glucoside	320	86.189	-	-	[26]
18	Apigenin-7-O-rutinoside	320	128.126	-	-	[26]
20	Catechin hydrate	320	134.971	78.00	1.557	Ref. Stand
22	Rutin	320	23.593	22.40	0.947	Ref. Stand
23	Syringic acid	320	176.298	-	-	[25]
23	Quercetin 3,7-di-O-glucoside	320	77.612	-	-	[25]
24	Quercetin-7-O-sophoroside	320	58.34	-	-	[25]

Table 3. *Cont.*

Retention Time (min)	Phytochemical Compounds	HPLC-UV λmax (nm)	Peak Area of Sample	Peak Area of Standard	Concentration (μg/mL)	Identification Reference
25	Kaempferol-3-(caffeoyl-diglucoside)-7-rhamnosyl	320	248.248	-	-	[26]
26	Rhamnose	320	103.713	-	-	[26]
27	3,5-dicaffeoylquinic acid	320	97.727	-	-	[26]
29	Caffeic acid	320	113.173	-	-	[27]
30	Mandellic acid	320	63.944	72.00	0.799	Ref. Stand
31	Quercetin-3-(caffeoyldiglucoside)-7-glucoside	320	223.444	-	-	[26]
42	Galactose	320	60.609	-	-	[27]
43	Xylulose	320	31.328	-	-	[27]

Table 4. Identified phytochemicals in purified fraction B of *A. articulata* through the HPLC-UV technique.

Retention Time (min)	Phytochemical Compounds	HPLC-UV λmax (nm)	Peak Area of Sample	Peak Area of Standard	Concentration (μg/mL)	Identification Reference
4	Gallic acid	320	36.4984	195.40	0.168	Ref. Stand
14	3-0-caffeoylquinic acid	320	160.818	-	-	[26]
20	Catechin hydrate	320	78.1380	78.00	0.902	Ref. Stand
22	Rutin	320	42.5292	22.40	1.709	Ref. Stand
23	Syringic acid	320	324.245	-	-	[26]
25	Kaempferol-3-(caffeoyl-diglucoside)-7-rhamnosyl	320	58.9445	-	-	[26]
27	3,5-dicaffeoylquinic acid	320	88.6597	-	-	[26]
28	Pyrogallol	320	85.8767	1.014	76.221	Ref. Stand
31	Quercetin-3-(caffeoyldiglucoside)-7-glucoside	320	88.6597	-	-	[26]
32	p-Coumaric acid	320	56.5793	-	-	[26]
43	Xylulose	320	36.3262	-	-	[27]

Table 5. Identified phytochemicals in fraction C of *A. articulata* through the HPLC-UV technique.

Retention Time (min)	Phytochemical Compounds	HPLC-UV λmax (nm)	Peak Area of Sample	Peak Area of Standard	Concentration (μg/mL)	Identification Reference
2	Malic acid	320	24.7858	40.323	0.554	Ref. Stand
11	Bis-HHDP-hex (pedunculagin	320	47.4527	-	-	[25]
12	Morin	320	100.496	2.00	45.223	Ref. Stand
18	Apigenin-7-O-rutinoside	320	246.3633	-	-	[26]
22	Rutin	320	424.706	22.40	17.064	Ref. Stand
23	Quercetin 3,7-O-glucoside	320	201.7868	-	-	[26]

Table 5. *Cont.*

Retention Time (min)	Phytochemical Compounds	HPLC-UV λmax (nm)	Peak Area of Sample	Peak Area of Standard	Concentration (μg/mL)	Identification Reference
25	Kaempferol-3-(caffeoyl-diglucoside)-7-rhamnosyl	320	30.1412	-	-	[26]
26	Rhamnose	320	32.8371	-	-	[27]
27	3,5-dicaffeoylquinic acid	320	120.040	-	-	[26]
28	Pyrogallol	320	66.255	1.014	58.806	Ref. Stand
30	Mandelic acid	320	29.289	72.00	0.366	Ref. Stand
31	Quercetin-3-(caffeoyldiglucoside)-7-glucoside	320	32.0566	-	-	[26]
37	Glucose	320	21.6425	-	-	[27]
41	Quercitin-3-0-glysides	320	78.9891	-	-	[26]

Table 6. Identified phytochemicals in purified fraction D of *A. articulata* through the HPLC-UV technique.

Retention Time (min)	Phytochemical Compounds	HPLC-UV λmax (nm)	Peak Area of Sample	Peak Area of Standard	Concentration (μg/mL)	Identification Reference
2	Malic acid	320	2514.20	40.323	56.116	Ref. Stand
34	Quercitin-3-O-rutinoside	320	41.0693	-	-	[26]
41	Quercitin-3-O-glycosides	320	30.249	-	-	[26]
42	Galactose	320	16.116	-	-	[27]

Table 7. Identified phytochemicals in *n*-hexane fraction of *A. articulata* through the HPLC-UV technique.

Retention Time (min)	Phytochemical Compounds	HPLC-UV λmax (nm)	Peak Area of Sample	Peak Area of Standard	Concentration (μg/mL)	Identification Reference
2	Malic acid	320	28.2673	40.323	0.631	Ref. Stand
6	Chlorogenic acid	320	14.9574	2.929	4.595	Ref. Stand
10	Quercetin	320	137.065	90.90	1.357	Ref. Stand
11	Bis-HHDP-hex (pedunculagin)	320	44.509	-	-	[26]
12	Morin	320	231.396	2.00	104.128	Ref. Stand
18	Apigenin-7-O-rutinoside	320	89.384	-	-	[26]
22	Rutin	320	159.626	22.40	6.414	Ref. Stand
23	Syringic acid	320	76.785	-	-	[26]
25	Kaempferol-3-(caffeoyl-diglucoside)-7-rhamnosyl	320	152.318	-	-	[26]
26	Rhamnose	320	100.004	-	-	[26]
31	Quercetin-3-(caffeoyldiglucoside) 7-glucoside	320	51.863	-	-	[26]
41	Quercitin-3-0-glycosides	320	31.616	-	-	[26]
42	Galactose	320	100.903	-	-	[27]

3.5. GC-MS Characterization of the Different Fractions

3.5.1. Purified Fraction A

The GC-MS chromatogram of fraction A is indicated in Figure 5A. Figure S1 represents the structural formulas of five phytochemical compounds identified from the given chromatogram, whereas Figure S2 represents their mass fragmentation pattern and Table S1 represents different parameters of the major phytochemical compounds identified.

3.5.2. Purified Fraction B

The GC-MS chromatogram of fraction B is indicated in Figure 5B. The technique confirmed the presence of 12 phytochemical compounds in fraction B, and their other parameters are presented in Table S2. Figures S3 and S4 represent the structural formulas of 12 phytochemical compounds and the pattern of their mass fragmentation, respectively.

3.5.3. Purified Fraction C

Figure 5C shows the GC-MS chromatogram of fraction C. Table S3 represents 13 phytochemical compounds identified along with some basic parameters related to the analysis performed. Figure S5 indicates the structural formulas of 13 phytochemical compounds, and Figure S6 indicates their mass fragmentation pattern.

3.5.4. Purified Fraction D

The GC-MS chromatogram of fraction D is presented in Figure 5D, where the presence of 13 phytochemical compounds was confirmed as presented in Table S4 along with analysis-related parameters. Figure S7 represents the structural formulas of phytochemical compounds, whereas Figure S8 represents the pattern of their mass fragmentation.

3.5.5. Oil Fraction

Figure 5E represents the GC-MS chromatogram of the purified oil fraction. Figure 5F shows the structural formulas of major phytochemical compounds (a and b), and Table S5 shows their different parameters.

3.6. Antibacterial Activity

The results of the antibacterial potential of the samples have been tabulated in Table S6 and graphically represented in Figure 6. The results depicted that all the samples except for B and C showed activity against the tested bacterial strains. The broad-spectrum antibiotic ampicillin was used as a positive control. The *n*-hexane (oil) fraction showed the highest ZI against all tested strains: *S. dysentery*, *E. coli*, and *S. Typhi* as 20, 24, and 16 mm respectively. An appreciable degree of antibacterial bacterial potential suggests the plant's possible usage as a source for isolating antibacterial compounds.

3.7. Antioxidant Activity of Crude and Purified Fractions of *A. articulata*

3.7.1. DPPH Assay

Almost all extracts inhibited DPPH free radicals; however, among them, *n*-hexane fraction, crude extract, and fraction B showed significant free radical inhibition with IC_{50} values of 45, 90, and 62 μg/mL, respectively, as presented in Table S7 and Figure 7A.

3.7.2. ABTS Assay

The ABTS scavenging potential of extracts is presented in Table S7 and Figure 7B. The results depict that fraction A and *n*-hexane extract possesses significant free radical inhibition with the lowest IC_{50} values of 75 and 71 μg/mL, respectively, as compared to the ascorbic acid used as the standard, which showed an IC_{50} value of 32 μg/mL.

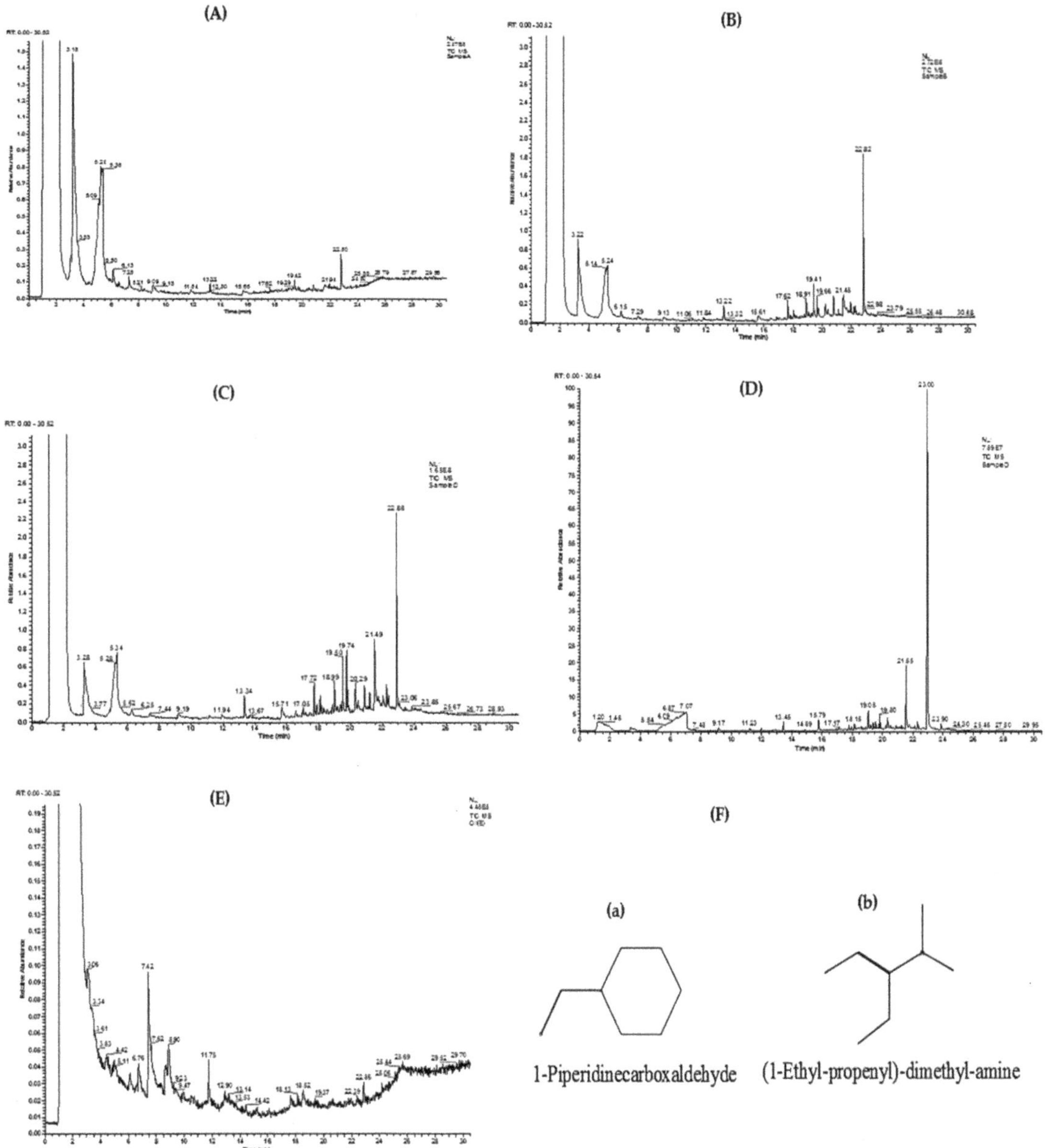

Figure 5. GC-MS chromatogram of *A. articulata* purified fractions (**A**–**D**), oil fraction (**E**), and major phytochemical compounds (**a**,**b**) identified in the oil fraction (**F**).

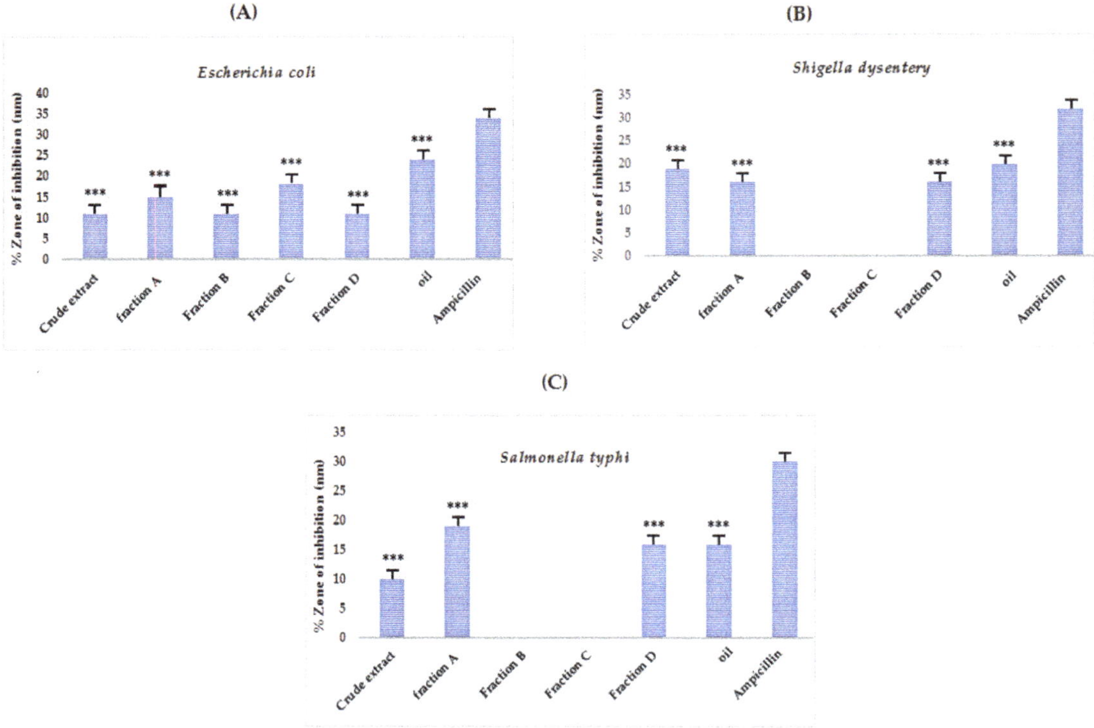

Figure 6. Antibacterial potential of crude extract, purified fractions (A, B, C, and D), and oil fraction of *A. articulata* against (**A**) *E. coli*, (**B**) *Shigella dysentery* and (**C**) *S. typhi*. The data are represented as mean ± SEM, $n = 3$. Values are significantly different as compared to positive control *** $p < 0.001$.

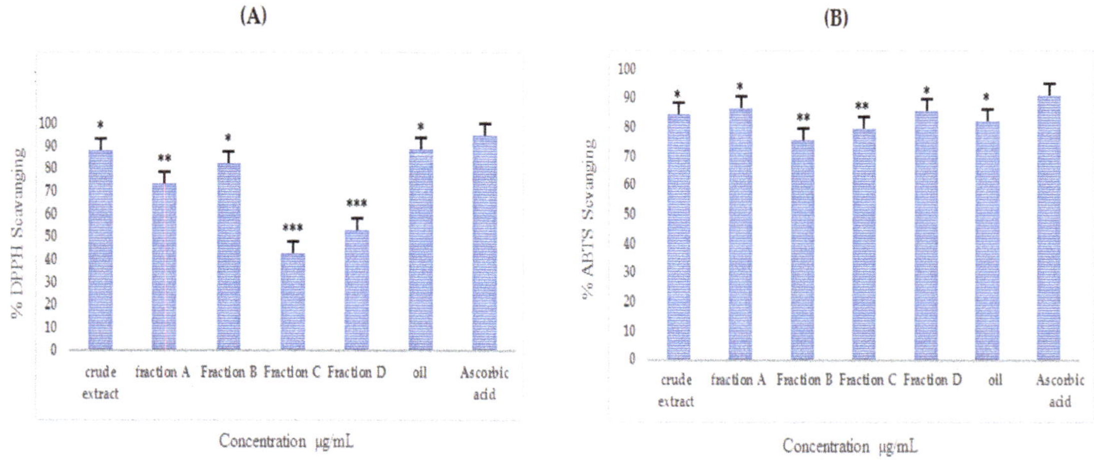

Figure 7. Antioxidant activity ((**A**) = DPPH; (**B**) = ABTS) of crude extract, purified fractions (A, B, C, and D), and oil fraction of *A. articulata*. The data plotted is mean ± SEM, $n = 3$. Values are significantly different as compared to positive control * $p < 0.05$, ** $p < 0.01$, *** $p < 0.001$.

3.8. In Vitro Antidiabetic Activities of Purified Fraction and Extract

3.8.1. In Vitro α-Glucosidase Inhibition

Figure 8A and Table S8 represent the %α-glucosidase inhibition of crude extract, oil, and the purified fractions. The crude extract showed the highest inhibition of the enzyme with an IC_{50} of 32 μg/mL followed by fraction B, which produced an IC_{50} of 60 μg/mL.

3.8.2. In Vitro α Amylase Inhibition

As shown in Figure 8B and Table S8, the %α-amylase inhibition is appreciable, and the highest inhibition was caused by crude extract with an IC_{50} of 34 μg/mL followed by fraction B, which produced an IC_{50} of 58 μg/mL.

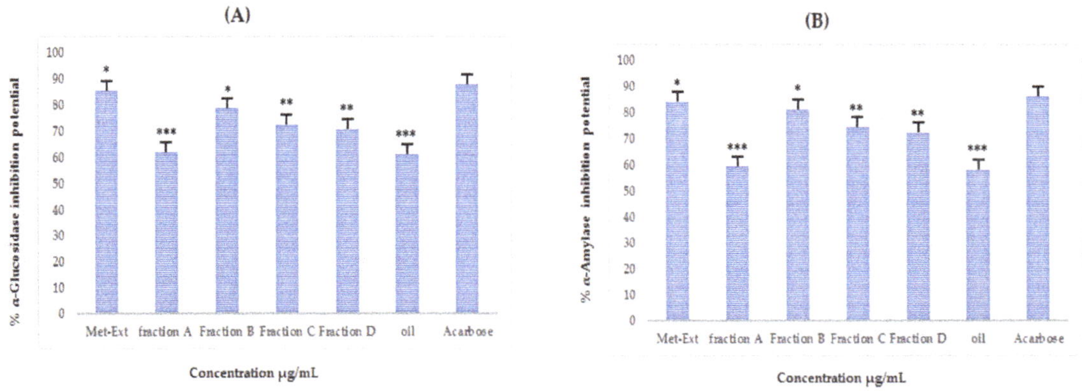

Figure 8. (**A**) %α-glucosidase inhibition potential and (**B**) %α-amylase inhibition potential of crude extract, purified fractions (A, B, C, and D), and oil fraction of *A. articulata* at different concentrations. The data are represented as mean ± SEM, $n = 3$. Values are significantly different as compared to positive control * $p < 0.05$, ** $p < 0.01$, *** $p < 0.001$.

4. Discussion

Presently, insulin therapies are the treatment of choice to control hyperglycemia in diabetes mellitus. Other strategies are the inhibition of alpha-amylase and glucosidase through different inhibitors, as both enzymes are responsible for releasing glucose from starch taken in food [28,29]. In this context, an attempt has been made in this study to identify the possible antidiabetic phytochemical that could inhibit the activity of carbohydrate digesting enzymes (α-amylase and α-glucosidase). The study revealed that the plant could be a potential candidate for isolating antidiabetic compounds.

With the increasing reports about the side effects of synthetic drugs, researchers have focused on plants to isolate effective therapeutic precursors with low or no side effects. Drug resistance is the other overwhelming problem in the modern era, and the search for new antibiotics of plant origin is in progress. The plant crude extract and purified fractions showed appreciable antibacterial activity, which is evident from the zones of inhibitions against selected bacterial strains, as shown in Figure 5.

Oxidative stress caused by free radicals that are constantly produced during normal metabolic processes is a serious health threat. Although these are constantly deactivated by the human defense system, in the modern era, humans have started relying on processed foods, which have given rise to the overproduction of free radicals. Research shows that plants could neutralize the free radicals because of their constituent phenolics [30], as benzene rings in such compounds can stabilize the singlet electron of the free radicals. Collectively, such phytoconstituents are named antioxidant compounds, which play an important role in human health by combating reactive oxygen species and, in turn, is the main contributor to a number of human diseases, including insulin resistance, cardiovascular diseases, atherosclerosis, and coronary heart disease. Butylated hydroxytoluene and butylated hydroxy anisole are strong synthetic antioxidant agents, but they are carcinogenic and toxic to humans. Therefore, plant-based phenolic compounds can be used as antioxidants to scavenge or stabilize free radicals involved in oxidative stress generated in human bodies as a result of oxidation of certain substances. It is found that the use of synthetic antioxidants is injurious to human health, and individuals taking them are at risk of cancer and other liver disorders. The antioxidants in plants have become a hotspot for researchers in recent decades due to the mentioned fact of low or no side effects. Studies have indicated that the use of natural antioxidants can reduce oxidative stress and reduce the risk of major human diseases, including oxidative stress [3,6]. The *n*-hexane fraction, crude extract, and fraction B were more potent against DPPH radicals, whereas against ABTS, the *n*-hexane fraction and fraction A were more potent, indicating that these extracts contained certain phytoconstituents capable of scavenging free radicals, which could thus be further investigated for the isolation of responsible compounds. The DPPH radicals in acholic medium undergo a reduction in the presence of hydrogen donating antioxidants. Phytochemicals such as flavonoids and phenolics are good antioxidants and play a vital role in scavenging the free radicals due to the presence of benzene rings in their structures [6–13]. The observed antioxidant potential can be correlated with the estimated TFC and TPC values, as these are the responsible scavengers in the extracts. The total polyphenol and flavonoid content in the fractions increased in the following order: crude extract, fraction A, and oil fraction. The crude extract has the highest TPC and TFC, i.e., (TPC = 72.1 mg GAE/g and TFC = 62 mg QE/g) followed by purified fraction A, which has the highest TPC and TFC, i.e., 68 GAE/g and 62 mg QE/g, inferring the plant is a good source of flavonoids and phenolics. As mentioned before, due to the presence of benzene rings in the structure, flavonoids and phenolics have been found to be excellent scavengers of the free radicals, which is why the tested radicals, ABTS and DPPH, were potently scavenged by the extracts, i.e., the total phenolic and flavonoid contents in the extracts and purified fractions were positively proportional to the antioxidant activities. The current results were in line with the previously reported studies [6,14]. The study of Kim et al. [31] showed plants that contained high TFC and TPC, and by virtue of these components, they exhibit various biological potentials. Their conclusion was based on findings of extracts from 40 plant species in Korea. As mentioned, phenolic and flavonoid compounds are strong antioxidants that can deactivate free radicals by offering their hydrogen atoms and electrons [32], which is the reason that plants with high TFC and TPC inhibit DPPH and ABTS radicals more potently in laboratory-scale experiments. The positive correlation between the total phenolic content and flavonoid content in the plant extracts and the antioxidant activities have been observed by other researchers as well [32]. The plants in the form of extracts could, therefore, offer strong activity against a wide range of oxidants and thus would have great medicinal applications. It can be seen from Table S7 that the crude extract and fractions exhibited significant activities against the DPPH and ABTS tested radicals, which needs to be further investigated. Furthermore, for the crude extract, the preliminary phytochemical tests (Table 1) were positive, indicating the presence of broad phytochemical groups and, consequently, the wider range of their therapeutic action.

The HPLC analysis of crude and purified fractions of *A. articulata* showed the presence of several possible compounds that might be responsible for antioxidant and antidiabetic activities. The antidiabetic properties of *A. articulata* crude extracts and fractions (Figure 7 and Table S8) were determined based on the inhibitory effect against two carbohydrate hydrolyzing enzymes, namely α-amylase and α-glucosidase. As mentioned before, starch is converted into disaccharides and oligosaccharides by pancreatic α-amylase, while disaccharides are broken down into glucose by intestinal α-glucosidase [3,6] and, thus, if inhibited, will lessen the glucose burden in diabetic patients as their inhibition could retard the breakdown of starch in the gastrointestinal tract and, therefore, would ameliorate hyperglycemia in human. The detected compounds are known to be antioxidant and antidiabetic agents, as indicated in the previously reported studies [3,6,9,14]. The current results of the screening are in close accordance with the already reported study of Nazir et al., where they confirmed the presence of quercetin, morin, and rutin in the methanolic extract of *Silybum marianum* (L.) seeds [33] and in the methanolic extracts of the fruit of *Elaeagnus umbellata* Thunb. [6]. The results of this study are in agreement with the findings of other studies where strong antioxidant activities were observed along with strong α-glucosidase and α-amylase inhibitions [3,6,9].

The medicinal plant has become a vital source of antioxidants in the last few decades. Literature surveys have shown that the ingestion of natural antioxidants can reduce oxidative stress-related diseases. Various studies have shown that the presence of malic acid, gallic acid, quercetin, morin, ellagic acid, rutin, chlorogenic acid, and epigallocatechin gallate can be liable for the antioxidant capacity observed [14,34,35]. It is evident from the literature that gallic acid, chlorogenic acid, epigallocatechin gallate, and morin have strong antioxidants and antidiabetic potentials [36,37].

The GC-MS analysis of the purified fraction also confirmed the presence of certain valuable phytochemical compounds: Acetdimethylamide, *N*-Nitrosomorpholine, 1,2-Benzenedicarboxylic acid, Mono(2-ethylhexyl) phthalate, Bis(2-ethylhexyl) phthalate, N-Acetyl-l-methioninamide, 2-Propanamine, Phenol, 2,4-di-tert-butyl, Benzene, (1-dodecyltridecyl)-, Benzene, (1-hexyltetradecyl)-, Benzene, (1-hexylheptyl)-, Isopropyl Palmitate, 10-Octadecenoic acid, methyl ester, 1-Docosene, Methyl ricinoleate, Oleic acid, tetradecyl ester, Diisooctyl phthalate, Asparagine, entacosane, 13-phenyl Eicosane, 7-phenyl, Dodecane, 6-phenyl, Palmitic acid, methyl ester, tert-Hexadecanethiol, Decyl oleate, octadecyl ester, Elaidic acid, isopropyl ester, Phenethyl alcohol, á-methyl, Benzyl-3-hydroxypyrrolidine, Diethyl Phthalate, 2,6-Dimethyl-pyridine-3,5-dicarboxylic acid, dihydrazide, Methoxycarbonyl-2-methoxyphenyl isothiocyanate, Phosphoric acid, dibutyl 3-trifluoromethyl-3-pentyl ester, 4-Acetylaminophthalic acid, dimethyl ester, Benzene-1,3-dicarboxylic acid, 5-acetylamino-, (2-Phenyl-1,3-dioxolan-4-yl) methyl (9E)-9-octadecenoate, 1-Heneicosyl formate, 18,19-Secoyohimban-19-oic acid, Cleavamine, 18á-carboxy-3,4à-dihydro-, 1-Piperidinecarboxaldehyde, and (1-Ethyl-propenyl)-dimethyl-amine, which could possibly have their share in the observed biological potentials. The findings of the present study could be correlated with the reported studies [19,38]. From the rich phytochemical composition of the selected plant, we hypothesized that the different levels of antidiabetic activity of the extract and different fractions of *A. articulata* are due to the varying levels of various phytochemicals in each extract/fraction. The purified fraction A followed by crude extract *A. articulata* exhibited higher levels of TPC and TFC, together with antioxidant and antidiabetic activity as compared with the other extracts/fractions. This indicates that phenolic compounds, including flavonoids, are key active compounds found in these extracts, and the plant could thus be a good candidate for further studies to isolate inhibitors of the tested radicals and enzymes.

5. Conclusions

From the current study results, it can be concluded that *A. articulata* in extract form should be considered an effective source to relieve oxidative stress and health complications associated with diabetes. This plant also has the potential to be used as an antimicrobial agent as it effectively inhibited the growth of selected bacterial strains. The α-amylase and α-glucosidase enzymes were inhibited by extracts to an appreciable extent suggesting that this plant could be used as a potential candidate to isolate effective antidiabetic drugs. The observed biological activities were at the expense of its rich phytochemical composition, confirmed through HPLC-UV and GC-MS techniques in this study. Further studies are needed to isolate pure compounds responsible for the observed biological potentials.

Supplementary Materials: The following supporting information can be downloaded at: https://www.mdpi.com/article/10.3390/molecules27113526/s1, Figure S1. Major phytochemical compounds identified in fraction A of *A. articulata*. Figure S2. Individual mass fragmentation patterns of each compound: (A) Acetdimethylamide (B) N-Nitrosomorpholine (C) 1,2-Benzenedicarboxylic acid (D) Mono(2-ethylhexyl) phthalate (E) Bis(2-ethylhexyl) phthalate. Figure S3. Major phytochemical compounds identified in fraction B of *A. articulata*. Figure S4. Individual mass fragmentation patterns of each compound: (A) N-Acetyl-l-methioninamide (B) 2-Propanamine (C) Phenol, 2,4-di-tert-butyl (D) Benzene, (1-dodecyltridecyl) (E) Benzene, (1-hexyltetradecyl) (F) Benzene, (1-hexylheptyl) (G) Isopropyl Palmitate (H) 10-Octadecenoic acid, methyl ester (I) 1-Docosene (J) Methyl ricinoleate (K) Oleic acid, tetradecyl ester (L) Diisooctyl phthalate. Figure S5. Major phytochemical compounds identified in fraction C of *A. articulata*. Figure S6. Individual mass fragmentation patterns of each compound: (A) Asparagine (B) entacosane, 13-phenyl (C) Eicosane, 7-phenyl (D) Dodecane, 6-phenyl (E) Palmitic acid, methyl ester (F) Isopropyl Palmitate (G) tert-Hexadecanethiol (H) Methyl ricinoleate (I) Decyl oleate (J) Oleic acid, octadecyl ester (K) Elaidic acid, isopropyl ester (L) Diisooctyl phthalate (M) Mono(2-ethylhexyl) phthalate. Figure S7. Major phytochemical compounds identified in fraction D of *A. articulata*. Figure S8. Individual mass fragmentation patterns of each compound: (A) Phenethyl alcohol, á-methyl (B) Benzyl-3-hydroxypyrrolidine (C) Diethyl Phthalate (D) 2,6-Dimethyl-pyridine-3,5-dicarboxylic acid, dihydrazide (E) Methoxycarbonyl-2-methoxyphenyl isothiocyanate (F) Phosphoric acid, dibutyl 3-trifluoromethyl-3-pentyl ester (G) Oleic Acid (H) 4-Acetylaminophthalic acid, dimethyl ester (I) Benzene-1,3-dicarboxylic acid (J) (2-Phenyl-1,3-dioxolan-4-yl) methyl (9E)-9-octadecenoate (K) 1-Heneicosyl formate (L) 18,19-Secoyohimban-19-oic acid (M) Cleavamine, 18á-carboxy-3,4à-dihydro-, methyl ester. Table S1. Major phytochemical compounds identified in fraction A of *A. articulata* and their various parameters. Table S2. Major phytochemical compounds and their parameters identified in fraction B of *A. articulata*. Table S3. List of major components and their parameters identified in fraction C of *A. articulata*. Table S4. Major phytochemical compounds and their parameters identified in fraction D of *A. articulata*. Table S5. Major phytochemical compounds and their parameters identified in the oil fraction of *A. articulata*. Table S6. Antibacterial potential of crude extract, oil, and purified fraction of *A. articulata*. Table S7. Antioxidant potential (DPPH and ABTS) of *A. articulata* crude extract, hexane, and subsequent fractions. Table S8. α-glucosidase and α-amylase inhibition of *A. articulata* crude extract and subsequent purified fractions at various concentrations.

Author Contributions: F.A.A.-J. and M.Z.: conceptualization; F.A.A.-J. and M.J.: methodology and formal analysis; N.N. and S.N.: writing—original draft; M.Z.: writing—review and editing; F.A.K. and M.T.: visualization; M.Z.: supervision and project administration, and validation; S.N.: formal analysis and data curation; A.N. and A.S. helped in the GC-MS analysis and biological activities. All authors have read and agreed to the published version of the manuscript.

Funding: This research received no external funding.

Institutional Review Board Statement: Not applicable.

Informed Consent Statement: Not applicable.

Data Availability Statement: The data presented in this manuscript belong to the research work supervized under Muhammad Zahoor and have not been deposited in any repository yet. However, the data are available to the researchers upon request.

Acknowledgments: The authors are thankful to Aljouf and Malakand Universities for providing research facilities.

Conflicts of Interest: The authors declare that they have no conflict of interest.

Abbreviations

WHO	World Health Organization
UV	Ultraviolet
HPLC	High performance liquid chromatography
GC-MS	Gas chromatography-mass spectroscopy
DPPH	2,2-Diphenyle-1 picrylhydrazyl
ABTS	2,2′-Azino-Bis-3 Ethylbenzothiazoline-6-Sulfonic Acid
TLC	Thin layer chromatography
TPC	Total phenolic content
TFC	Total Flavonoid Content
GAE	Gallic acid equivalent

QE	Quercetin equivalent
FID	Flame ionization detector
NIST	National Institute of Standards and Technology

References

1. Patel, S.B.; Ghane, S.G. Phyto-constituents profiling of Luffa echinata and in vitro assessment of antioxidant, antidiabetic, anticancer and anti-acetylcholine esterase activities. *Saudi J. Biol. Sci.* **2021**, *28*, 3835–3846. [CrossRef] [PubMed]
2. Beshah, F.; Hunde, Y.; Getachew, M.; Bachheti, R.K.; Husen, A.; Bachheti, A. Ethnopharmacological, phytochemistry and other potential applications of Dodonaea genus: A comprehensive review. *Curr. Res. Biotechnol.* **2020**, *2*, 103–119. [CrossRef]
3. Nazir, N.; Zahoor, M.; Nisar, M.; Khan, I.; Ullah, R.; Alotaibi, A. Antioxidants Isolated from Elaeagnus umbellata (Thunb.) Protect against Bacterial Infections and Diabetes in Streptozotocin-Induced Diabetic Rat Model. *Molecules* **2021**, *26*, 4464. [CrossRef] [PubMed]
4. Nazir, N.; Zahoor, M.; Nisar, M.; Karim, N.; Latif, A.; Ahmad, S.; Uddin, Z. Evaluation of neuroprotective and anti-amnesic effects of Elaeagnus umbellata Thunb. On scopolamine-induced memory impairment in mice. *BMC Complement. Med. Ther.* **2020**, *20*, 143. [CrossRef]
5. Juan, C.A.; Pérez de la Lastra, J.M.; Plou, F.J.; Pérez-Lebeña, E. The Chemistry of Reactive Oxygen Species (ROS) Revisited: Outlining Their Role in Biological Macromolecules (DNA, Lipids and Proteins) and Induced Pathologies. *Int. J. Mol. Sci.* **2021**, *22*, 4642. [CrossRef]
6. Nazir, N.; Zahoor, M.; Nisar, M.; Khan, I.; Karim, N.; Abdel-Halim, H.; Ali, A. Phytochemical analysis and antidiabetic potential of Elaeagnus umbellata (Thunb.) in streptozotocin-induced diabetic rats: Pharmacological and computational approach. *BMC Complement. Altern. Med.* **2018**, *18*, 332. [CrossRef]
7. Li, Z.; Cheng, Y.; Wang, D.; Chen, H.; Chen, H.; Ming, W.K.; Wang, Z. Incidence rate of type 2 diabetes mellitus after gestational diabetes mellitus: A systematic review and meta-analysis of 170,139 women. *J. Diabetes Res.* **2020**, *2020*, 3076463. [CrossRef]
8. Kambouche, N.; Merah, B.; Derdour, A.; Bellahouel, S.; Bouayed, J.; Dicko, A.; Younos, C.; Soulimani, R. Hypoglycemic and anti-hyperglycemic effects of Anabasis articulata (Forssk) Moq (Chenopodiaceae), an Algerian medicinal plant. *Afr. J. Biotechnol.* **2009**, *8*, 5589–5594.
9. Nazir, N.; Zahoor, M.; Ullah, R.; Ezzeldin, E.; Mostafa, G.A. Curative effect of catechin isolated from Elaeagnus umbellata Thunb. berries for diabetes and related complications in streptozotocin-induced diabetic rats model. *Molecules* **2021**, *26*, 137. [CrossRef]
10. Zafar, R.; Ullah, H.; Zahoor, M.; Sadiq, A. Isolation of bioactive compounds from Bergenia ciliata (haw.) Sternb rhizome and their antioxidant and anticholinesterase activities. *BMC Complement. Altern. Med.* **2019**, *19*, 296. [CrossRef]
11. Bari, W.U.; Zahoor, M.; Zeb, A.; Sahibzada, M.U.K.; Ullah, R.; Shahat, A.A.; Mahmood, H.M.; Khan, I. Isolation, pharmacological evaluation and molecular docking studies of bioactive compounds from Grewia optiva. *Drug Des. Dev.* **2019**, *13*, 3029.
12. Nazir, N.; Zahoor, M.; Nisar, M. A review on traditional uses and pharmacological importance of genus Elaeagnus species. *Bot. Rev.* **2020**, *86*, 247–280. [CrossRef]
13. Nazir, N.; Nisar, M.; Ahmad, S.; Wadood, S.F.; Jan, T.; Zahoor, M.; Ahmad, M.; Ullah, A. Characterization of phenolic compounds in two novel lines of Pisum sativum L. along with their in vitro antioxidant potential. *Environ. Sci. Pollut. Res.* **2020**, *27*, 7639–7646. [CrossRef]
14. Nazir, N.; Khalil, A.A.K.; Nisar, M.; Zahoor, M.; Ahmad, S. HPLC-UV characterisation, anticholinesterase, and free radical-scavenging activities of Rosa moschata Herrm. leaves and fruits methanolic extracts. *Braz. J. Bot.* **2020**, *43*, 523–530. [CrossRef]
15. Barros, L.; Oliveira, S.; Carvalho, A.M.; Ferreira, I.C. In vitro antioxidant properties and characterisation in nutrients and phytochemicals of six medicinal plants from the Portuguese folk medicine. *Ind. Crops Prod.* **2010**, *32*, 572–579. [CrossRef]
16. Khan, I.; Abbas, T.; Anjum, K.; Abbas, S.Q.; Shagufta, B.I.; Shah, S.A.A.; Akhter, N.; ul Hassan, S.S. Antimicrobial potential of aqueous extract of Camellia sinensis against representative microbes. *Pak. J. Pharm. Sci.* **2019**, *32*, 631–636.
17. Xiao, Y.; Zhu, S.; Wu, G.; ul Hassan, S.S.; Xie, Y.; Ishaq, M.; Sun, Y.; Yan, S.K.; Qian, X.P.; Jin, H.Z. Chemical Constituents of Vernonia parishii. *Chem. Nat. Compd.* **2020**, *56*, 134–136. [CrossRef]
18. Zeb, A. A reversed phase HPLC-DAD method for the determination of phenolic compounds in plant leaves. *Anal. Methods* **2015**, *7*, 7753–7757. [CrossRef]
19. Nazir, N.; Zahoor, M.; Uddin, F.; Nisar, M. Chemical composition, in vitro antioxidant, anticholinesterase, and antidiabetic potential of essential oil of Elaeagnus umbellata Thunb. *BMC Complement. Med. Ther.* **2021**, *21*, 73. [CrossRef]
20. Xie, Y.G.; Zhao, X.C.; ul Hassan, S.S.; Zhen, X.Y.; Muhammad, I.; Yan, S.K.; Yuan, X.; Li, H.L.; Jin, H.Z. One new sesquiterpene and one new iridoid derivative from Valeriana amurensis. *Phytochem. Lett.* **2019**, *32*, 6–9. [CrossRef]
21. Shams ul Hassan, S.; Ishaq, M.; Zhang, W.; Jin, H.-Z. An overview of the mechanisms of marine fungi-derived antiinflammatory and anti-tumor agents and their novel role in drug targeting. *Curr. Pharm. Des.* **2021**, *27*, 2605–2614. [CrossRef] [PubMed]
22. Shams, S.; Zhang, W.; Jin, H.; Basha, S.H.; Priya, S.V.S.S. In-silico anti-inflammatory potential of guaiane dimers from Xylopia vielana targeting COX-2. *J. Biomol. Struct. Dyn.* **2020**, *40*, 484–498. [CrossRef]
23. Miller, G.L. Use of dinitrosalicylic acid reagent for determination of reducing sugar. *Anal. Chem.* **1959**, *31*, 426–428. [CrossRef]
24. Ranilla, L.G.; Kwon, Y.I.; Apostolidis, E.; Shetty, K. Phenolic compounds, antioxidant activity and in vitro inhibitory potential against key enzymes relevant for hyperglycemia and hypertension of commonly used medicinal plants, herbs and spices in Latin America. *Bioresour. Technol.* **2010**, *101*, 4676–4689. [CrossRef]

25. Suryanti, V.; Marliyana, S.D.; Putri, H.E. Effect of germination on antioxidant activity, total phenolics, β-carotene, ascorbic acid and α-tocopherol contents of lead tree sprouts (*Leucaena leucocephala* (lmk.) de Wit). *Int. Food Res. J.* **2016**, *23*, 167–172.
26. Hamid, A.A.; Aiyelaagbe, O.O.; Usman, L.A.; Ameen, O.M.; Lawal, A. Antioxidants: Its medicinal and pharmacological applications. *Afr. J. Pure Appl. Chem.* **2010**, *4*, 142–151.
27. Ngueyem, T.A.; Brusotti, G.; Caccialanza, G.; Finzi, P.V. The genus Bridelia: A phytochemical and ethnopharmacological review. *J. Ethnopharmacol.* **2009**, *124*, 339–349. [CrossRef] [PubMed]
28. Buttermore, E.; Campanella, V.; Priefer, R. The increasing trend of Type 2 diabetes in youth: An overview. *Diabetes Metab. Syndr. Clin. Res. Rev.* **2021**, *15*, 102253. [CrossRef] [PubMed]
29. Tiji, S.; Bouhrim, M.; Addi, M.; Drouet, S.; Lorenzo, J.M.; Hano, C.; Bnouham, M.; Mimouni, M. Linking the phytochemicals and the α-glucosidase and α-amylase enzyme inhibitory effects of Nigella sativa seed extracts. *Foods* **2021**, *10*, 1818. [CrossRef] [PubMed]
30. Gani, M.A.; Shama, M. Phenolic Compounds. In *Bioactive Compounds-Biosynthesis, Characterisation and Applications*; Intechopen: London, UK, 2021.
31. Kim, E.J.; Choi, J.Y.; Yu, M.R.; Kim, M.Y.; Lee, S.H.; Lee, B.H. Total polyphenols, total flavonoid contents, and antioxidant activity of Korean natural and medicinal plants. *Korean J. Food Sci. Technol.* **2012**, *44*, 337–342. [CrossRef]
32. Aryal, S.; Baniya, M.K.; Danekhu, K.; Kunwar, P.; Gurung, R.; Koirala, N. Total phenolic content, flavonoid content and antioxidant potential of wild vegetables fromWestern Nepal. *Plants* **2019**, *8*, 96. [CrossRef]
33. Nazir, N.; Karim, N.; Abdel-Halim, H.; Khan, I.; Wadood, S.F.; Nisar, M. Phytochemical analysis, molecular docking and antiamnesic effects of methanolic extract of *Silybum marianum* (L.) Gaertn seeds in scopolamine induced memory impairment in mice. *J. Ethnopharmacol.* **2018**, *210*, 198–208. [CrossRef]
34. Ihsan, M.; Nisar, M.; Nazir, N.; Zahoor, M.; Khalil, A.A.K.; Ghafoor, A.; Khan, A.; Mothana, R.A.; Ullah, R.; Ahmad, N. Genetic diversity in nutritional composition of oat (Avena sativa L.) germplasm reported from Pakistan. *Saudi J. Biol. Sci.* **2022**, *29*, 1487–1500. [CrossRef]
35. Khan, J.; Majid, A.; Nazir, N.; Nisar, M.; Khan Khalil, A.A.; Zahoor, M.; Ihsan, M.; Ullah, R.; Bari, A.; Shah, A.B. HPLC Characterization of Phytochemicals and Antioxidant Potential of Alnus nitida (Spach) Endl. *Horticulturae* **2021**, *7*, 232. [CrossRef]
36. Qin, L.; Zang, M.; Xu, Y.; Zhao, R.; Wang, Y.; Mi, Y.; Mei, Y. Chlorogenic Acid Alleviates Hyperglycemia-Induced Cardiac Fibrosis through Activation of the NO/cGMP/PKG Pathway in Cardiac Fibroblasts. *Mol. Nutr. Food Res.* **2021**, *65*, 2000810. [CrossRef]
37. Liu, B.; Kang, Z.; Yan, W. Synthesis, Stability, and Antidiabetic Activity Evaluation of (−)-Epigallocatechin Gallate (EGCG) Palmitate Derived from Natural Tea Polyphenols. *Molecules* **2021**, *26*, 393. [CrossRef]
38. Nwosu, L.C.; Edo, G.I.; Onyibe, P.N.; Ozgor, E. The Phytochemical, Proximate, Pharmacological, FTIR, GC-MS Analysis of Cyperus Esculentus (Tiger Nut): A Fully Validated Approach in Health, Food and Nutrition. *Food Nutr.* **2022**, *46*, 101551. [CrossRef]

molecules

MDPI

Article

α-Glucosidase, α-Amylase and Antioxidant Evaluations of Isolated Bioactives from Wild Strawberry

Mohammed A. Huneif [1], Seham M. Alqahtani [1], Alqahtani Abdulwahab [1], Sultan A. Almedhesh [1], Mater H. Mahnashi [2,*], Muhammad Riaz [3], Najm Ur-Rahman [3], Muhammad Saeed Jan [4], Farhat Ullah [5], Muhammad Aasim [6] and Abdul Sadiq [5,*]

[1] Pediatric Department, Medical College, Najran University, Najran 55461, Saudi Arabia; huneif@hotmail.com (M.A.H.); drseham2015@gmail.com (S.M.A.); aaalsharih@nu.edu.sa (A.A.); almedhesh31@hotmail.com (S.A.A.)
[2] Department of Pharmaceutical Chemistry, College of Pharmacy, Najran University, Najran 55461, Saudi Arabia
[3] Department of Pharmacy, Shaheed Benazir Bhutto University, Sheringal 18050, Pakistan; pharmariaz@gmail.com (M.R.); najm@sbbu.edu.pk (N.U.-R.)
[4] Department of Pharmacy, University of Swabi, Swabi 94640, Pakistan; saeedjanpharmacist@gmail.com
[5] Department of Pharmacy, Faculty of Biological Sciences, University of Malakand, Chakdara 18000, Pakistan; farhataziz80@hotmail.com
[6] Department of Biotechnology, Faculty of Biological Sciences, University of Malakand, Chakdara 18000, Pakistan; draasim@uom.edu.pk
* Correspondence: mhmahneshi@nu.edu.sa (M.H.M.); sadiquom@yahoo.com (A.S.); Tel.: +96-65-0873-4539 (M.H.M.); +92-03-01-229-7102 (A.S.)

Citation: Huneif, M.A.; Alqahtani, S.M.; Abdulwahab, A.; Almedhesh, S.A.; Mahnashi, M.H.; Riaz, M.; Ur-Rahman, N.; Jan, M.S.; Ullah, F.; Aasim, M.; et al. α-Glucosidase, α-Amylase and Antioxidant Evaluations of Isolated Bioactives from Wild Strawberry. *Molecules* **2022**, 27, 3444. https://doi.org/10.3390/molecules27113444

Academic Editors: Syed Shams ul Hassan, Mohamed M. Abdel-Daim, Tapan Behl and Simona Bungau

Received: 14 April 2022
Accepted: 24 May 2022
Published: 26 May 2022

Publisher's Note: MDPI stays neutral with regard to jurisdictional claims in published maps and institutional affiliations.

Abstract: Diabetes mellitus is a metabolic disorder and is a global challenge to the current medicinal chemists and pharmacologists. This research has been designed to isolate and evaluate antidiabetic bioactives from *Fragaria indica*. The crude extracts, semi-purified and pure bioactives have been used in all in vitro assays. The in vitro α-glucosidase, α-amylase and DPPH free radical activities have been performed on all plant samples. The initial activities showed that ethyl acetate (**Fi.EtAc**) was the potent fraction in all the assays. This fraction was initially semi-purified to obtain **Fi.EtAc 1–3**. Among the semi-purified fractions, **Fi.EtAc 2** was dominant, exhibiting potent IC_{50} values in all the in vitro assays. Based on the potency and availability of materials, **Fi.EtAc 2** was subjected to further purification to obtain compounds **1** (2,4-dichloro-6-hydroxy-3,5-dimethoxytoluene) and **2** (2-methyl-6-(4-methylphenyl)-2-hepten-4-one). The two isolated compounds were characterized by mass and NMR analyses. The compounds **1** and **2** showed excellent inhibitions against α-glucosidase (21.45 for **1** and 15.03 for **2** μg/mL), α-amylase (17.65 and 16.56 μg/mL) and DPPH free radicals (7.62 and 14.30 μg/mL). Our study provides baseline research for the antidiabetic bioactives exploration from *Fragaria indica*. The bioactive compounds can be evaluated in animals-based antidiabetic activity in future.

Keywords: *Fragaria indica*; α-glucosidase; α-amylase; antioxidant; bioactives; antidiabetic

1. Introduction

The use of plants or their derivatives to treat various ailments is a practice as old as human civilization [1,2]. These plants not only produce primary metabolites for their own existence but also release secondary metabolites to interact with other organisms. The secondary metabolites are usually investigated for various bioactives [3,4]. Bioactives are chemicals produced by a living being to treat various diseases or alter biological functions [5]. Over a span of decades, there have been attempts to explore the use of plants to treat various disorders, including diabetes [6,7].

Diabetes is a metabolic disorder characterized by an elevated level of blood glucose with common symptoms of polyphagia, polydipsia and polyurea [8]. There are several biochemical pathways which can be targeted for the management of diabetes [6]. The more trivial and vital enzymatic targets of diabetes are α-glucosidase, α-amylase, protein tyrosine phosphatase. The free radicals play a vital role in implications of several disorders [9–11]. In DM, the free radicals within the body increase due to auto-oxidation of glucose, which further complicates the situation. The free radicals damage the β cells, which are majorly responsible for the synthesis of insulin, and thus diminish the insulin synthesis [12]. The human body has the capability to combat the free radicals due to an auto-immune response [13]. However, at certain stages, when the free radicals are at uncontrolled level, the auto-immune system fails to control it [14]. At this stage of high free radicals within the body, the antioxidant treatment is necessary [15]. Reactive oxygen species and pro-inflammatory cytokines and chemokines cause the activation of JNK and NF-κB pathways that promote the development of diabetes [16]. There are 537 million adults suffering from diabetes and the number is expected to reach 783 million by 2045. The tremendous rise in diabetics confirms that it is a global challenge for policymakers and researchers to take necessary steps to overcome this challenge.

As the previous molecules are losing effectiveness, the result is certain effects, such as weight gain, heart problems, infections and weakened kidney. Despite significant additions to the list antidiabetics, researchers have to do more in search of safe, effective and efficient drug. Natural bioactives might be an effective therapeutic intervention, as studies have shown that phytochemicals are efficient agents in controlling diabetes via different mechanisms [17,18].

Fragaria indica Andrews, commonly known as wild strawberry, belongs to the family Rosaceae. It has scientific synonym *Duchesnea indica* (Andrews) and *Potentilla indica* (Andrews). It is naturalized in Africa, Europe and North America and distributed in Asian countries such as Pakistan, Kashmir, Afghanistan, Bhutan, China, India, Indonesia, Japan, Korea and Nepal [19]. The plant is known locally as *the Zmaki toot, Khunmurch,* and *Blmngye.* It is used traditionally for sore throat, tuberculosis, [20,21] nutrient, anthelmintic, and diabetic patients [22]. The plant has been explored for antioxidant, anti-inflammatory [23], and immunomodulatory potential [24,25]. Until now, other species of Fragaria such as *Fragaria* × *ananassa* and *Fragaria nilgerrensis* have been evaluated for their antidiabetic potentials [26,27]. However, to the best of our literature survey, we noticed that the antidiabetic potential of *Fragaria indica* is still unexplored. Keeping in view the global burden of diabetes and weaknesses of the available synthetic antidiabetic molecules and unexplored nature of *F. indica,* we plan the current work to investigate the antidiabetic and antioxidant potentials of the *F. indica* crude extract, semi-purified extracts and pure bioactives.

2. Results

2.1. Isolation of Bioactive Compounds

In our designed study, we initially screened the crude extract and different solvent fractions of *F. indica* for in vitro α-glucosidase, α-amylase and DPPH activities. Based on the potency of ethyl acetate fraction (**Fi.EtAc**), we subjected the ethyl acetate fraction to normal gravity column chromatography with slow elution of solvent system. The solvent system was started with 100% *n*-hexane and was gradually increased in polarity by adding small percentages of ethyl acetate. Initially, we obtained three semi-purified phytocomponents **Fi.EtAc 1**, **Fi.EtAc 2** and **Fi.EtAc 3**. The semi-purified phytochemicals were further subjected to in vitro α-glucosidase, α-amylase and DPPH activities. The semi-purified components were isolated based on TLC analysis. Among the three semi-purified ethyl acetate fractions, **Fi.EtAc 2** was the potent one based on its IC_{50} values obtained in the in vitro assays. The **Fi.EtAc 2** was further purified to obtain two bioactives **1** and **2** (as shown in Figure 1).

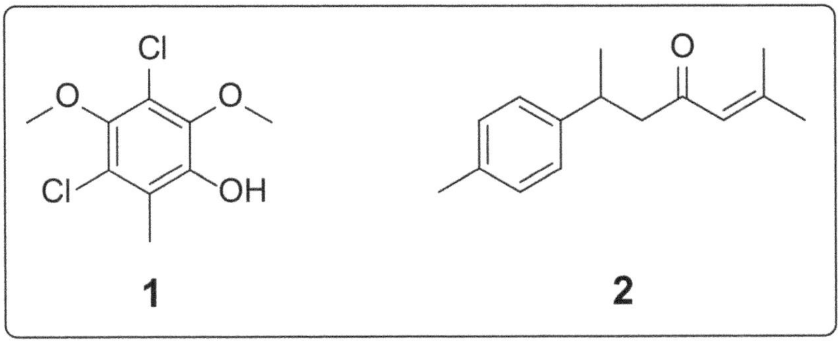

Figure 1. Isolated bioactive compounds from *Fragaria indica*.

2.2. Structure Confirmations of the Isolated Bioactives

The compound **1** is chemically 2,4-dichloro-6-hydroxy-3,5-dimethoxytoluene. The molecular weight of compound **1** was confirmed as 236 $[M-1]^+$, and its fragmentation pattern was 236 (68%, molecular ion peak), 221 (100%, the base peak), 206 (2%), 193 (25%), 178 (38%), 171 (5%), 158 (15%), 143 (14%), 127 (5%), 103 (11%), 87 (19%) and 67 (18%). The ^{1}H-NMR spectrum of compound **1** showed three distinct singlets, each of 3H. The toluoyl methyl group (Ar-CH_3) appeared at 2.314 chemical shift. The two methoxy groups (CH_3O-Ar-OCH_3) appeared as two singlets at 3.870 and 3.877 chemical shift. The phenolic hydrogen (HO-Ar) appeared as singlet of one proton at 6.041. The ^{13}C-NMR of compound **1** showed signals at chemical shift values of 15.38, 55.03, 112.18, 115.70, 119.75, 133.11, 137.39 and 139.59 ppm.

Chemically, compound **2** is 2-methyl-6-(4-methylphenyl)-2-hepten-4-one. The molecular weight of compound **2** was confirmed as 216, and its fragmentation pattern was 216 (28%, molecular ion peak), 201 (10%), 132 (16%), 119 (46%), 83 (100%, base peak) and 55 (25%). The ^{1}H-NMR spectrum also showed all the proton patterns of the compound **2**. The four aromatic protons appeared in two distinct doublets (each having two protons integration) at 7.12 and 7.07 chemical shift values. The observed coupling constant values (J) were 8.36 and 8.05 Hz, respectively for the two doublets. The hydrogen atom on the alkene unit of the compound **2** appeared at 6.151 as a sharp singlet. One of the methylene protons of compound **2** (-CH_2-) gave a multiplet between chemical shift of 3.463 and 3.373. The second methylene proton (-CH_2-) merged with methyne proton (-CH-) in the multiplet range of 3.054 to 2.963. The aromatic methyl group (Ar-CH_3) appeared as a singlet of three protons at 2.238. The two methyl groups attached with the alkene unit gave two distinct singlets at 2.082 and 1.994, with integration values of 3H for each one. The methyl group attached at α-position to the aromatic ring gave a doublet at 1.28 with a J value of 7.07 Hz. The ^{13}C-NMR of compound **2** showed signals at chemical shift values of 18.03, 19.45, 21.78, 26.21, 37.67, 45.83, 126.92, 130.60, 132.20, 137.16, 140.14, 156.05 and 201.24 ppm. The mass and NMR spectra of the isolated bioactive compounds are provided in the Supplementary Materials.

2.3. In Vitro Activities on Crude Extracts of Fragaria indica

2.3.1. α-Glucosidase and α-Amylase

The α-glucosidase and α-amylase activities of the crude extracts and subfractions of *Fragaria indica* are summarized in Table 1. The tested concentrations of the plant's samples and standard drug used were 1000, 500, 250, 125 and 62.5 μg/mL. At maximum concentrations, the crude extract (**Fi.Cr**), *n*-hexane (**Fi.Hex**), ethyl acetate (**Fi.EtAc**), chloroform (**Fi.Chf**) and aqueous (**Fi.Aq**) fractions gave percent inhibitions of 64.44, 61.50, 71.69, 61.23 and 57.85%. The standard drug acarbose at the same concentration (1000 μg/mL) gave a percent inhibition of 96.00%. Overall in our fractions, **Fi.EtAc** was the potent one in inhibition of α-glucosidase enzyme with the observation of an IC_{50} value of 117.54 μg/mL.

Table 1. α-Glucosidase and α-amylase results of *Fragaria indica* crude extracts.

Sample	Concentration (μg/mL)	α-Glucosidase		α-Amylase	
		Percent Inhibition	IC_{50} (μg/mL)	Percent Inhibition	IC_{50} (μg/mL)
Fi.Cr	1000	64.44 ± 0.09 ***	232.10	66.90 ± 0.72 ***	218.19
	500	56.87 ± 0.39 ***		57.12 ± 0.89 ***	
	250	51.83 ± 1.07 ***		52.64 ± 1.38 ***	
	125	44.23 ± 0.44 ***		45.40 ± 0.93 ***	
	62.5	36.29 ± 0.43 ***		35.22 ± 0.94 ***	
Fi.Hex	1000	61.50 ± 2.26 ***	340.56	65.50 ± 2.26 ***	249.85
	500	54.01 ± 0.42 ***		56.01 ± 0.42 ***	
	250	45.07 ± 0.62 ***		49.07 ± 0.62 ***	
	125	39.70 ± 0.35 ***		43.70 ± 0.35 ***	
	62.5	34.73 ± 0.66 ***		35.73 ± 0.66 ***	
Fi.EtAc	1000	71.69 ± 0.77 ***	117.54	73.60 ± 1.63 ***	96.82
	500	63.67 ± 0.61 ***		66.82 ± 0.85 ***	
	250	56.44 ± 0.51 ***		61.25 ± 1.40 ***	
	125	51.76 ± 0.58 ***		53.10 ± 0.60 ***	
	62.5	43.54 ± 0.50 ***		44.61 ± 0.43 ***	
Fi.Chf	1000	61.23 ± 0.22 ***	296.86	63.48 ± 0.25 ***	259.11
	500	55.45 ± 0.90 ***		56.47 ± 0.04 ***	
	250	46.90 ± 0.60 ***		48.47 ± 0.44 ***	
	125	41.00 ± 0.30 ***		42.44 ± 0.09 ***	
	62.5	37.90 ± 0.45 ***		37.43 ± 1.39 ***	
Fi.Aq	1000	57.85 ± 0.56 ***	429.39	57.79 ± 0.62 ***	398.46
	500	51.64 ± 0.75 ***		52.45 ± 0.49 ***	
	250	44.58 ± 0.77 ***		46.75 ± 0.58 ***	
	125	38.75 ± 0.63 ***		38.73 ± 0.64 ***	
	62.5	32.58 ± 0.70 ***		33.47 ± 0.56 ***	
Acarbose	1000	96.00 ± 0.30	17.28	82.43 ± 0.52	12.84
	500	90.61 ± 0.43		74.03 ± 0.64	
	250	84.03 ± 0.86		71.56 ± 0.49	
	125	77.58 ± 0.77		67.05 ± 0.49	
	62.5	71.48 ± 0.74		63.26 ± 0.93	

All values are taken as mean ± SEM (n = 3). Two-way ANOVA followed by Bonferoni test were followed. Values significantly different in comparison to standard drug, i.e., *** = $p < 0.001$.

Likewise, the α-amylase inhibitory potential was also in a similar pattern to that of α-glucosidase, as shown in Table 1. The observed IC_{50} values for **Fi.Cr**, **Fi.Hex**, **Fi.EtAc**, **Fi.Chf** and **Fi.Aq** fractions were 218.19, 249.85, 96.82, 259.11 and 398.46 μg/mL, respectively in comparison to the standard acarbose (12.84 μg/mL). In α-glucosidase and α-amylase activities, we observed that ethyl acetate fraction of *Fragaria indica* (**Fi.EtAc**) was the potent inhibitor fraction.

2.3.2. DPPH Results

The antioxidant potentials of crude extracts and different solvent fractions of *Fragaria inidica* were assessed by DPPH assay and the results are shown in Table 2. Based on Table 2, we can clearly observe that overall, the plant contains potential antioxidant properties. At 1000 μg/mL, **Fi.Cr**, **Fi.Hex**, **Fi.EtAc**, **Fi.Chf** and **Fi.Aq** fractions gave 65.66, 63.44, 78.81, 67.85 and 60.54% inhibitions. Ascorbic acid gave 91.90% inhibition at the highest tested concentration under the same set of experiments. In the DPPH assay, the **Fi.EtAc** was the fraction with potent IC_{50} value of 59.55 μg/mL in comparison to the standard drug IC_{50} value, which was noted as 4.98 μg/mL. The other fractions showed IC_{50} values of 200.89 (**Fi.Cr**), 236.91 (**Fi.Hex**), 142.39 (**Fi.Chf**) and 349.35 μg/mL (**Fi.Aq**).

Table 2. DPPH free radicals scavenging results of crude extracts of *Fragaria indica*.

Sample	Concentration (μg/mL)	Percent Inhibition	IC_{50} (μg/mL)
Fi.Cr	1000	65.66 ± 0.78 ***	200.89
	500	60.62 ± 0.74 ***	
	250	52.62 ± 0.74 ***	
	125	44.86 ± 0.60 ***	
	62.5	37.48 ± 0.64 ***	
Fi.Hex	1000	63.44 ± 0.09 ***	236.91
	500	57.87 ± 0.39 ***	
	250	51.83 ± 1.07 ***	
	125	43.23 ± 0.44 ***	
	62.5	36.29 ± 0.43 ***	
Fi.EtAc	1000	76.81 ± 0.60 ***	59.55
	500	70.74 ± 0.61 ***	
	250	64.68 ± 0.60 ***	
	125	58.63 ± 0.76 ***	
	62.5	49.79 ± 0.63 ***	
Fi.Chf	1000	67.85 ± 0.56 ***	142.39
	500	62.64 ± 0.75 ***	
	250	55.58 ± 0.77 ***	
	125	47.75 ± 0.63 ***	
	62.5	42.58 ± 0.70 ***	
Fi.Aq	1000	60.54 ± 0.48 ***	349.35
	500	52.30 ± 0.66 ***	
	250	45.58 ± 0.59 ***	
	125	41.52 ± 0.62 ***	
	62.5	35.45 ± 0.57 ***	
Ascorbic acid	1000	91.90 ± 0.96	4.98
	500	87.08 ± 0.47	
	250	82.40 ± 0.20	
	125	77.61 ± 0.43	
	62.5	75.45 ± 0.90	

All values are taken as mean ± SEM (n = 3). Two-way ANOVA followed by Bonferoni test were followed. Values significantly different in comparison to standard drug, i.e., *** = $p < 0.001$.

2.4. *In Vitro Activities on Semi-Purified Ethyl Acetate Fractions of* Fragaria indica

2.4.1. In Vitro α-Glucosidase and Amylase Results

The α-glucosidase and α-inhibitory potentials of semi-purified ethyl acetate fractions (**Fi.EtAc 1, Fi.EtAc 2** and **Fi.EtAc 3**) are shown in Table 3. As obvious from Table 3 results, the activities profile enhanced with the semi-purification. Specifically, the semi-purified ethyl acetate fraction **Fi.EtAc 2** exhibited IC_{50} values in parallel comparison to the standard drug. The fraction **Fi.EtAc 2** exhibited IC_{50} values of 14.81 and 14.54 μg/mL against α-glucosidase and α-amylase enzymes, respectively. The standard drug gave IC_{50} values of 17.28 and 12.84 μg/mL against α-glucosidase and α-amylase enzymes, respectively. Similarly, the potencies of other semi-purified fractions were also excellent. The fraction **Fi.EtAc 1** gave IC_{50} values of 63.85 and 38.60 μg/mL against α-glucosidase and α-amylase enzymes, respectively. Similarly, the semi-purified fraction **Fi.EtAc 3** gave IC_{50} values of 20.56 μg/mL (α-glucosidase) and 19.62 μg/mL (α-amylase).

Table 3. α-Glucosidase and α-amylase results of semi-purified ethyl acetate fraction of *Fragaria indica*.

Sample	Concentration (μg/mL)	α-Glucosidase		α-Amylase	
		Percent Inhibition	IC_{50} (μg/mL)	Percent Inhibition	IC_{50} (μg/mL)
Fi.EtAc 1	1000	81.81 ± 0.60 ***	63.85	82.45 ± 0.55 ns	38.60
	500	76.74 ± 0.61 ***		76.53 ± 0.41 ns	
	250	67.68 ± 0.60 ***		71.42 ± 0.46 ns	
	125	61.63 ± 0.76 ***		65.68 ± 0.64 *	
	62.5	47.79 ± 0.63 ***		53.63 ± 0.64 ***	
Fi.EtAc 2	1000	87.63 ± 0.64 ***	14.81	89.37 ± 0.54 ns	14.54
	500	82.45 ± 0.55 ***		84.44 ± 0.50 ns	
	250	76.53 ± 0.41 ***		77.51 ± 0.72 ns	
	125	71.42 ± 0.46 ***		72.28 ± 0.61 ns	
	62.5	65.68 ± 0.64 ***		67.46 ± 0.62 ns	
Fi.EtAc 3	1000	83.08 ± 1.04 ***	20.56	85.43 ± 1.26 ns	19.62
	500	76.45 ± 0.90 ***		78.83 ± 0.66 ns	
	250	70.58 ± 0.63 ***		72.93 ± 0.90 ns	
	125	65.40 ± 0.20 ***		67.26 ± 0.77 ns	
	62.5	61.80 ± 0.90 ***		63.10 ± 0.95 ns	
Acarbose	1000	96.00 ± 0.30	17.28	82.43 ± 0.52	12.84
	500	90.61 ± 0.43		74.03 ± 0.64	
	250	84.03 ± 0.86		71.56 ± 0.49	
	125	77.58 ± 0.77		67.05 ± 0.49	
	62.5	71.48 ± 0.74		63.26 ± 0.93	

All values are taken as mean ± SEM ($n = 3$). Two-way ANOVA followed by Bonferoni test were followed. Values significantly different in comparison to standard drug, i.e., * = $p < 0.05$, *** = $p < 0.001$ and ns: not significant.

2.4.2. DPPH Results

The three semi-purified ethyl acetate fractions of *Fragaria indica* were also found to possess strong antioxidant properties, as can be seen in Table 4. With semi-purification, the fractions showed excellent antioxidant activity profiles. The observed IC_{50} values for the three fractions **Fi.EtAc 1**, **Fi.EtAc 2** and **Fi.EtAc 3** were 14.95, 20.59 and 26.25 μg/mL, respectively in DPPH free radicals scavenging activity. Based on the relative potencies and amount of each semi-purified fraction, we selected **Fi.EtAc 2** for further purification to obtain the bioactive compounds.

Table 4. DPPH results of semi-purified ethyl acetate fractions of *Fragaria indica*.

Sample	Concentration (μg/mL)	Percent Inhibition	IC_{50} (μg/mL)
Fi.EtAc 1	1000	95.00 ± 0.32 ns	14.95
	500	90.63 ± 0.45 ns	
	250	84.05 ± 0.88 ns	
	125	78.56 ± 0.79 ns	
	62.5	71.46 ± 0.76 **	
Fi.EtAc 2	1000	85.72 ± 0.79 ***	20.59
	500	81.68 ± 0.63 ***	
	250	76.46 ± 0.53 ***	
	125	69.78 ± 0.60 ***	
	62.5	61.56 ± 0.52 ***	
Fi.EtAc 3	1000	84.83 ± 0.62 ***	26.25
	500	80.76 ± 0.63 ***	
	250	75.70 ± 0.62 ***	
	125	66.65 ± 0.78 ***	
	62.5	59.81 ± 0.65 ***	

Table 4. *Cont.*

Sample	Concentration (μg/mL)	Percent Inhibition	IC_{50} (μg/mL)
Ascorbic acid	1000	91.90 ± 0.96	4.98
	500	87.08 ± 0.47	
	250	82.40 ± 0.20	
	125	77.61 ± 0.43	
	62.5	75.45 ± 0.90	

All values are taken as mean ± SEM (n = 3). Two-way ANOVA followed by Bonferoni test were followed. Values significantly different in comparison to standard drug, i.e., ** = $p < 0.01$, *** = $p < 0.001$ and ns: not significant.

2.5. In Vitro Activities on Isolated Compounds of *Fragaria indica*

The two isolated compounds were characterized and were then subjected to in vitro antidiabetic activities and results are summarized in Table 5. The compound **1** (2,4-dichloro-6-hydroxy-3,5-dimethoxytoluene) and **2** (2-methyl-6-(4-methylphenyl)-2-hepten-4-one) showed excellent activities profiles exhibiting IC_{50} values of 21.45 and 17.65 (α-glucosidase) and 15.03 and 16.56 μg/mL (α-amylase), respectively. Similarly, of the two compounds, specifically, the phenolic type of derivative (**1**) was the more potent antioxidant. The observed IC_{50} values for compounds **1** and **2** were 7.62 and 14.30 μg/mL, respectively against the DPPH free radicals. The standard drug IC_{50} under the same set of condition was 4.98 μg/mL.

Table 5. α-Glucosidase, α-amylase and DPPH results of isolated compounds from *Fragaria indica*.

Sample	α-Glucosidase IC_{50} (μg/mL)	α-Amylase IC_{50} (μg/mL)	DPPH IC_{50} (μg/mL)
Compound **1**	21.45	17.65	7.62
Compound **2**	15.03	16.56	14.30
Acarbose	17.28	12.84	-
Ascorbic acid	-	-	4.98

3. Discussion

Researchers, specifically the medicinal chemists and pharmacologists are constantly searching for new drug molecules [28–30]. The organic chemists design and explore new methods for the synthesis of valuable molecules which can be important drug or natural products [31,32]. On the other hand, the medicinal chemists are in constant evaluation of medicinal compounds for various vital targets [33–35]. The pharmacological researchers use various in vitro and in vivo methods to confirm the possible use of a compound for a specific disease [36]. One of the major tools is the computational approach, which allows researchers to obtain the binding energies for a specific biological target [37]. The medicinal values of compounds can be explored from natural and synthetic origins [38,39]. Among the major health issues, diabetes is a global challenge to the researchers. Various natural and synthetic origins have been employed for the management of diabetes [40]. However, due to the emerging increasing number of diabetes patients, more natural sources can be explored to treat diabetes. Poly- and oligosaccharides are converted into monosaccharides by the action of intestinal α-glucosidase and α-amylase, and the activity significantly contributes to postprandial hyperglycemia. Synthetic agents such as acarbose can inhibit these enzymes, but the side effects, such as diarrhea, flatulence, and stomach pain, make them unsuitable [41]. The present study was designed to develop antidiabetics from natural products based on this strategy with lesser side effects [42].

The ethyl acetate fraction of *F. indica* showed maximum inhibition of α-glucosidase, α-amylase enzyme and DPPH free radical among the tested fractions and crude extract. The inhibitory activity was enhanced with semi-purification of the ethyl acetate fraction. The isolated compounds showed equivalent potent activities to the standard used.

Compound **1** is chemically 2,4-dichloro-6-hydroxy-3,5-dimethoxytoluene, while compound **2** is 2-methyl-6-(4-methylphenyl)-2-hepten-4-one. Compound **1** seems to be the derivative of 3,5-dimethoxytoluene, an important aromatic compound of roses with sedative action and used in aromatherapy as a relaxing fragrance [43]. Compound **2**, commonly named ar-turmerone, was isolated from *Curcuma longa* with proven antivenom and anti-inflammatory potentials [44,45]. It is a major component of turmeric oil with anti-dermatophytic activity [46] and a neuroprotective effect [47]. However, the first time it was reported was in *F. indica*.

The wild-type strawberry, i.e., *F. indica* herb was investigated for the first time for in vitro antidiabetic and antioxidant activities. However, other species, especially fruits, are extensively investigated for such activities. The enzyme α-amylase and α-glucosidase activity were inhibited by ellagic acid or derivatives in aqueous fruit extract of *Fragaria* × *ananassa* [26]. Similarly, another specie *F. vesca* leaf extract was reported to inhibit α-glucosidase and α-amylase enzyme activity [41]. Flavonoids in *n*-butanol extract of *F. nilgerrensis* Schlecht produced an antidiabetic effect [27].

The antioxidant properties of strawberries have been well documented, both in vitro and in vivo [48–50]. A study on aqueous extract of *Fragaria vesca* L. collected in Bulgaria showed strong ABTS inhibitory activity due to more polyphenols than those with less ABTS inhibitory potentials [51].

Strawberries and strawberry-based phytochemicals are potential dietary sources for managing hyperglycemia [52]. The leaves of the *Fragaria* × *ananassa* Duch. cultivars have also been reported for antihyperglycemic and antioxidant potentials [53].

The beneficial effects of the different strawberry extracts on blood glucose levels are popularizing its role as functional food [54,55]. The role was further supported by a clinical trial where a diet supplemented with strawberry fruit extract lowered glycohemoglobin (HbA1c) levels after six weeks of treatment [56].

Our experiments revealed that polyphenolic compounds of ethyl acetate fractions, e.g., compounds **1** and **2**, are potential major components possessing anti-diabetic and antioxidant potentials; however, further analytical and structure-activity studies are required to identify and synthesize more active components in the field.

4. Materials and Methods

4.1. Collection of Medicinal Plant and Extraction

The plant was collected from Laram Qilla Talash (Latitude 45.990530, Longitude 12.673570), District Dir Lower of Khyber Pakhtunkhwa (KPK) Pakistan in August 2020 and was identified as *Fragaria indica* Andrews by Nasrullah, Plant Taxonomist at the department of Botany University of Malakand, Chakdara, Pakistan. A sample specimen was deposited at the herbarium of the same department, voucher no. H.UOM.BG.551. The extraction was carried out as per our previous mentioned procedure [57]. The crude extract was subjected to different solvents fractions based on the polarity basis as per our previously reported procedure [58].

4.2. Phytochemistry and Bioactives Isolation

In our initial in vitro studies, we observed that among all the fractions, ethyl acetate fraction (**Fi.EtAc**) was the most prominent. Based on this, the ethyl acetate fraction was subjected to gravity column. The column elution was started with 100% of pure *n*-hexane as non-polar component. The polarity of solvent was gradually increased with the addition of ethyl acetate as polar modifier solvent. After the column chromatography, we were able to collect three different ethyl acetate fractions (**Fi.EtAc 1–3**) as visualized on pre-coated TLC plate under UV lamp. The three semi-purified ethyl acetate fractions were also subjected to all in vitro assays. Among the semi-purified ethyl acetate fractions, **Fi.EtAc 2** was observed with potent IC_{50} values in our in vitro experiments. Based on the potency, **Fi.EtAc 2** was further subjected to purification to obtain the bioactive compounds. At the end of purification process, we were able to obtain two bioactive compounds (**1** and **2**). The isolated compounds were characterized by their masses and also their structures were confirmed by proton NMR [6].

4.3. In Vitro α-Glucosidase Inhibition

α-Glucosidase inhibitory assay on *F. indica* extracts, solvent fractions and isolated compounds were performed according to the reported procedure [59] using acarbose as standard drug. The α-glucosidase solution (120 µL) of 0.50 U/mL in 0.10 M phosphate buffer (pH 6.90) was prepared and *p*-nitrophenyl-α-D-glucopyranoside (5 mM) was added as substrate solution. Different concentrations ranging from 62.50 µg/mL to 1000 µg/mL of the crude extracts, semi-purified fractions and isolated compounds of *F. indica* were prepared accordingly. The plant's samples (extracts, semi-purified and purified compounds) were mixed with the solutions of enzyme as per the experiment and were incubated at 37 °C.

After initial incubation for 15 min, 20 µL *p*-nitrophenyl-α-D-glucopyranoside solution was mixed and was again incubated in the same way. Finally, 80 µL solution of sod. carbonate (0.20 M) was added to finish up the reaction. The solution containing all the substrates except α-glucosidase served as a blank. The absorbances of experiments were measured at 405 nm. The experiments were repeated 3 times and the calculations were made as per the standard protocols.

4.4. In Vitro α-Amylase Inhibition

The α-amylase inhibitory activity was carried out according to the established protocols for *F. indica* extracts and isolated compounds [60] The amylase solution was prepared in phosphate buffer. The solutions of the *F. indica* and isolated compounds (62.50 to 1000 µg/mL) were added to the amylase solution and allowed to react. After reaction completion, starch solution was added and incubated as per the standard method. Afterwards, a solution of dinitro-salicylic acid was added to both test and control groups. The final solutions were heated for 3 min in boiling water and the absorbances were measured at 656 nm. The experiments were performed in triplicates and percent inhibitions were calculated.

4.5. DPPH Free Radicals Scavenging Assay

The antioxidant activity was carried out for *F. indica* extracts, semi-purified ethyl acetate fractions and isolated compounds using DPPH free radicals. DPPH solution was prepared (24 mg/100 mL of methanol). Stock solutions (1 mg/mL) of *F. indica* extracts and isolated compounds were prepared in methanol and then diluted to 1000, 500, 250, 125, 62.5 µg/mL. Sample and DPPH solutions were mixed in a ratio of 1:1 and were incubated at 23 °C for 30 min. Finally, absorbance was measured at 517 nm using UV spectrophotometer (Thermo electron corporation, Waltham, MA, USA). Ascorbic acid was used as positive control. Percent radical scavenging activity was measured using the reported equations [61].

4.6. Estimation of IC_{50} Values

The half inhibitory concentration (IC_{50}) was calculated using Microsoft Excel program as per our previously reported method [62].

4.7. Statistical Data Analysis

The results in assays were expressed as mean ± S.E.M. GraphPad prism (GraphPad Software, San Diego, CA, USA) was used for one-way ANOVA followed by Dunnett's multiple comparison test with positive control and test groups. The p values less than 0.05 were considered as statistically significant. The level of significance was expressed as $* = p < 0.05$, $** = p < 0.01$ and $*** = p < 0.001$.

5. Conclusions

It can be concluded from our current research results that *Fragaria indica* is a rich source of bioactives. Herein, we have explored the *Fragaria indica* for its antidiabetic potential. We initially confirmed from crude extracts' in vitro assays that ethyl acetate fraction of the wild strawberry was potent among all. Then, we semi-purified ethyl acetate fractions **Fi.EtAc**

1 to **3**. The **Fi.EtAc 2** was the fraction with dominant activities. We were able to isolate two compounds, i.e., **1** (2,4-dichloro-6-hydroxy-3,5-dimethoxytoluene) and **2** (2-methyl-6-(4-methylphenyl)-2-hepten-4-one). The two compounds showed overwhelming activity profile against α-glucosidase, α-amylase and DPPH free radicals. The two compounds can be further evaluated for antidiabetic targets using in vivo models to obtain more fruitful results.

Supplementary Materials: The following are available online at https://www.mdpi.com/article/10.3390/molecules27113444/s1. The NMR and MS spectra are provided in the supporting information. Figure S1: GC-MS chromatogram of Compound **1**; Figure S2: ^{1}H-NMR spectrum of Compound **1**; Figure S3: ^{13}C-NMR spectrum of Compound **1**; Figure S4: GC-MS chromatogram of Compound **2**; Figure S5: ^{1}H-NMR spectrum of Compound **2**; Figure S6: ^{13}C-NMR spectrum of Compound **2**.

Author Contributions: M.R., N.U.-R. and M.S.J. helped in plant collection, extraction and fractionation. M.A.H., S.M.A., A.A., S.A.A. and M.H.M. helped in phytochemistry part. F.U., M.A. and M.S.J. helped in the assays. A.S. supervised the whole project and refined the manuscript for publication. All authors have read and agreed to the published version of the manuscript.

Funding: Authors would like to acknowledge the support of the Deputy for Research and Innovation-Ministry of Education, Kingdom of Saudi Arabia for this research through a grant (NU/IFC/ENT/01/010) under the institutional Funding Committee at Najran University, Kingdom of Saudi Arabia.

Institutional Review Board Statement: Not applicable.

Informed Consent Statement: Not applicable.

Data Availability Statement: The whole data is available within the manuscript and the Supplementary Materials.

Acknowledgments: We would like to acknowledge Nasrullah Khan, Department of Botany, University of Malakand, Chakdara, Pakistan for plant identification.

Conflicts of Interest: The authors declare no conflict of interest.

Sample Availability: Samples of the plant are available from the authors.

References

1. Süntar, I. Importance of ethnopharmacological studies in drug discovery: Role of medicinal plants. *Phytochem. Rev.* **2019**, *19*, 1199–1209. [CrossRef]
2. Akram, M.; Tahir, I.M.; Shah, S.M.A.; Mahmood, Z.; Altaf, A.; Ahmad, K.; Munir, N.; Daniyal, M.; Nasir, S.; Mehboob, H. Antiviral potential of medicinal plants against HIV, HSV, influenza, hepatitis, and coxsackievirus: A systematic review. *Phytother. Res.* **2018**, *32*, 811–822. [CrossRef] [PubMed]
3. Verpoorte, R. Exploration of nature's chemodiversity: The role of secondary metabolites as leads in drug development. *Drug Discov. Today* **1998**, *3*, 232–238. [CrossRef]
4. Zafar, R.; Ullah, H.; Zahoor, M.; Sadiq, A. Isolation of bioactive compounds from *Bergenia ciliata* (haw.) Sternb rhizome and their antioxidant and anticholinesterase activities. *BMC Complement. Altern. Med.* **2019**, *19*, 296. [CrossRef] [PubMed]
5. Ksean, M. Natural products research. *Nat. Prod. Chem. Res.* **2012**, *1*, e101. [CrossRef]
6. Mahnashi, M.H.; Alqahtani, Y.S.; Alqarni, A.O.; Alyami, B.A.; Jan, M.S.; Ayaz, M.; Ullah, F.; Rashid, U.; Sadiq, A. Crude extract and isolated bioactive compounds from *Notholirion thomsonianum* (Royale) Stapf as multitargets antidiabetic agents: In-vitro and molecular docking approaches. *BMC Complement. Med. Ther.* **2021**, *21*, 270. [CrossRef]
7. Sadiq, A.; Rashid, U.; Ahmad, S.; Zahoor, M.; Alajmi, M.F.; Ullah, R.; Noman, O.M.; Ullah, F.; Ayaz, M.; Khan, I.; et al. Treating Hyperglycemia From *Eryngium caeruleum* M. Bieb: In-vitro α-Glucosidase, Antioxidant, in-vivo Antidiabetic and Molecular Docking-Based Approaches. *Front. Chem.* **2020**, *8*, 558641. [CrossRef]
8. Care, D. Economic Costs of Diabetes in the US in 2017. *Diabetes Care* **2018**, *41*, 917. [CrossRef]
9. Shah, S.M.M.; Sadiq, A.; Ullah, F.; Shah, S.M.H. Antioxidant, total phenolic contents and antinociceptive potential of *Teucrium stocksianum* methanolic extract in different animal models. *BMC Complement. Altern. Med.* **2014**, *14*, 181. [CrossRef]
10. Sadiq, A.; Mahmood, F.; Ullah, F.; Ayaz, M.; Ahmad, S.; Haq, F.U.; Khan, G.; Jan, M.S. Synthesis, anticholinesterase and antioxidant potentials of ketoesters derivatives of succinimides: A possible role in the management of Alzheimer's. *Chem. Central J.* **2015**, *9*, 31. [CrossRef]
11. Zafar, R.; Zubair, M.; Ali, S.; Shahid, K.; Waseem, W.; Naureen, H.; Haider, A.; Jan, M.S.; Ullah, F.; Sirajuddin, M.; et al. Zinc metal carboxylates as potential anti-Alzheimer's candidate: In vitro anticholinesterase, antioxidant and molecular docking studies. *J. Biomol. Struct. Dyn.* **2021**, *39*, 1044–1054. [CrossRef] [PubMed]

12. Hussain, F.; Khan, Z.; Jan, M.S.; Ahmad, S.; Ahmad, A.; Rashid, U.; Ullah, F.; Ayaz, M.; Sadiq, A. Synthesis, in-vitro α-glucosidase inhibition, antioxidant, in-vivo antidiabetic and molecular docking studies of pyrrolidine-2,5-dione and thiazolidine-2,4-dione derivatives. *Bioorg. Chem.* **2019**, *91*, 103128. [CrossRef] [PubMed]
13. Puertollano, M.A.; Puertollano, E.; De Cienfuegos, G.A.; De Pablo, M.A. Dietary Antioxidants: Immunity and Host Defense. *Curr. Top. Med. Chem.* **2011**, *11*, 1752–1766. [CrossRef]
14. Zeb, A.; Sadiq, A.; Ullah, F.; Ahmad, S.; Ayaz, M. Investigations of anticholinestrase and antioxidant potentials of methanolic extract, subsequent fractions, crude saponins and flavonoids isolated from *Isodon rugosus*. *Biol. Res.* **2014**, *47*, 76. [CrossRef]
15. Bibi, A.; Shah, T.; Sadiq, A.; Khalid, N.; Ullah, F.; Iqbal, A. L-Isoleucine-catalyzed Michael Synthesis of *N*-Alkylsuccinimide Derivatives and Their Antioxidant Activity Assessment. *Russ. J. Org. Chem.* **2019**, *55*, 1749–1754. [CrossRef]
16. Shoelson, S.E.; Lee, J.; Goldfine, A.B. Inflammation and insulin resistance. *J. Clin. Investig.* **2006**, *116*, 1793–1801. [CrossRef] [PubMed]
17. Dembinska-Kiec, A.; Mykkänen, O.; Kiec-Wilk, B.; Mykkänen, H. Antioxidant phytochemicals against type 2 diabetes. *Br. J. Nutr.* **2008**, *99*, ES109–ES117. [CrossRef]
18. Leiherer, A.; Mündlein, A.; Drexel, H. Phytochemicals and their impact on adipose tissue inflammation and diabetes. *Vasc. Pharmacol.* **2013**, *58*, 3–20. [CrossRef]
19. Qureshi, H.; Arshad, M.; Bibi, Y. Invasive flora of Pakistan: A critical analysis. *Int. J. Biosci.* **2014**, *4*, 407–424.
20. Kayani, S.; Ahmad, M.; Zafar, M.; Sultana, S.; Khan, M.P.Z.; Ashraf, M.A.; Hussain, J.; Yaseen, G. Ethnobotanical uses of medicinal plants for respiratory disorders among the inhabitants of Gallies—Abbottabad, Northern Pakistan. *J. Ethnopharmacol.* **2014**, *156*, 47–60. [CrossRef]
21. Hamayun, M.; Khan, M.A.; Begum, S. Marketing of medicinal plants of Utror-Gabral valleys, Swat, Pakistan. *Ethnobot Leafl.* **2003**, *2005*, 44.
22. Saqib, A.A.; Gul, S. Traditional knowledge of medicinal herbs among indigenous communities in Maidan Valley, Lower Dir, Pakistan. *Bull. Environ. Pharmacol. Life Sci.* **2018**, *7*, 1–23.
23. Sereno-Villaseñor, L.; Hernández-García, A.; Torres-Martínez, R.; Meléndez-Herrera, E.; Manzo-Avalos, S.; Martínez-Flores, H.E.; Saavedra-Molina, A.; Salgado-Garciglia, R. Antioxidant and Anti-inflammatory Effects of the Methanolic Extract of *Potentilla indica* Fruits. *FASEB J.* **2020**, *34*, 1. [CrossRef]
24. Ang, H.Y.; Subramani, T.; Yeap, S.K.; Omar, A.R.; Ho, W.Y.; Abdullah, M.P.; Alitheen, N.B. Immunomodulatory effects of *Potentilla indica* and *Dendrophthoe pentandra* on mice splenocytes and thymocytes. *Exp. Ther. Med.* **2014**, *7*, 1733–1737. [CrossRef] [PubMed]
25. Long, M.; Yu, X.; Li, B.; Xiong, Y.; Xiang, B.; He, Q. Characterisation of antioxidant, anti-inflammatory, and immunomodulatory activities of polysaccharides derived from *Duchesnea indica* (Andrews) Focke. *Int. Food Res. J.* **2020**, *27*, 76–384.
26. Pinto, M.D.S.; de Carvalho, J.E.; Lajolo, F.M.; Genovese, M.I.; Shetty, K. Evaluation of Antiproliferative, Anti-Type 2 Diabetes, and Antihypertension Potentials of Ellagitannins from Strawberries (*Fragaria* × *ananassa* Duch.) Using In Vitro Models. *J. Med. Food* **2010**, *13*, 1027–1035. [CrossRef]
27. Gao, L.; Wang, X.; Lin, Z.; Song, N.; Liu, X.; Chi, X.; Shao, T. Antidiabetic and Neuroprotective Effect of the N-Butanol Extract of *Fragaria nilgerrensis* Schlecht. in STZ-Induced Diabetic Mice. *Evid.-Based Complement. Altern. Med.* **2018**, *2018*, 6938370. [CrossRef]
28. Nugent, T.C.; Bibi, A.; Sadiq, A.; Shoaib, M.; Umar, M.N.; Tehrani, F.N. Chiral picolylamines for Michael and aldol reactions: Probing substrate boundaries. *Org. Biomol. Chem.* **2012**, *10*, 9287–9294. [CrossRef]
29. Nugent, T.C.; Negru, D.E.; El-Shazly, M.; Hu, D.; Sadiq, A.; Bibi, A.; Umar, M.N. Sequential Reductive Amination-Hydrogenolysis: A One-Pot Synthesis of Challenging Chiral Primary Amines. *Adv. Synth. Catal.* **2011**, *353*, 2085–2092. [CrossRef]
30. Sadiq, A.; Mahnashi, M.H.; Alyami, B.A.; Alqahtani, Y.S.; Alqarni, A.O.; Rashid, U. Tailoring the substitution pattern of pyrrolidine-2,5-dione for discovery of new structural template for dual COX/LOX inhibition. *Bioorg. Chem.* **2021**, *112*, 104969. [CrossRef]
31. Nugent, T.C.; Sadiq, A.; Bibi, A.; Heine, T.; Zeonjuk, L.L.; Vankova, N.; Bassil, B.S. Noncovalent Bifunctional Organocatalysts: Powerful Tools for Contiguous Quaternary-Tertiary Stereogenic Carbon Formation, Scope, and Origin of Enantioselectivity. *Chem. Eur. J.* **2012**, *18*, 4088–4098. [CrossRef]
32. Jabeen, M.; Choudhry, M.I.; Miana, G.A.; Rahman, K.M.; Rashid, U.; Khan, H.-U.; Arshia; Sadiq, A. Synthesis, pharmacological evaluation and docking studies of progesterone and testosterone derivatives as anticancer agents. *Steroids* **2018**, *136*, 22–31. [CrossRef] [PubMed]
33. Jan, M.S.; Ahmad, S.; Hussain, F.; Ahmad, A.; Mahmood, F.; Rashid, U.; Abid, O.-U.; Ullah, F.; Ayaz, M.; Sadiq, A. Design, synthesis, in-vitro, in-vivo and in-silico studies of pyrrolidine-2,5-dione derivatives as multitarget anti-inflammatory agents. *Eur. J. Med. Chem.* **2019**, *186*, 111863. [CrossRef] [PubMed]
34. Ahmad, S.; Iftikhar, F.; Ullah, F.; Sadiq, A.; Rashid, U. Rational design and synthesis of dihydropyrimidine based dual binding site acetylcholinesterase inhibitors. *Bioorg. Chem.* **2016**, *69*, 91–101. [CrossRef] [PubMed]
35. Ahmad, G.; Rasool, N.; Rizwan, K.; Imran, I.; Zahoor, A.F.; Zubair, M.; Sadiq, A.; Rashid, U. Synthesis, in-vitro cholinesterase inhibition, in-vivo anticonvulsant activity and in-silico exploration of *N*-(4-methylpyridin-2-yl)thiophene-2-carboxamide analogs. *Bioorg. Chem.* **2019**, *92*, 103216. [CrossRef]
36. Sadiq, A.; Zeb, A.; Ullah, F.; Ahmad, S.; Ayaz, M.; Rashid, U.; Muhammad, N. Chemical Characterization, Analgesic, Antioxidant, and Anticholinesterase Potentials of Essential Oils from *Isodon rugosus* Wall. ex. Benth. *Front. Pharmacol.* **2018**, *9*, 623. [CrossRef]

37. Munir, A.; Khushal, A.; Saeed, K.; Sadiq, A.; Ullah, R.; Ali, G.; Ashraf, Z.; Mughal, E.U.; Jan, M.S.; Rashid, U.; et al. Synthesis, in-vitro, in-vivo anti-inflammatory activities and molecular docking studies of acyl and salicylic acid hydrazide derivatives. *Bioorg. Chem.* **2020**, *104*, 104168. [CrossRef]
38. Farooq, U.; Naz, S.; Shams, A.; Raza, Y.; Ahmed, A.; Rashid, U.; Sadiq, A. Isolation of dihydrobenzofuran derivatives from ethnomedicinal species *Polygonum barbatum* as anticancer compounds. *Biol. Res.* **2019**, *52*, 1. [CrossRef]
39. Sultana, N.; Sarfraz, M.; Tanoli, S.T.; Akram, M.S.; Sadiq, A.; Rashid, U.; Tariq, M.I. Synthesis, crystal structure determination, biological screening and docking studies of N^1-substituted derivatives of 2,3-dihydroquinazolin-4(1*H*)-one as inhibitors of cholinesterases. *Bioorg. Chem.* **2017**, *72*, 256–267. [CrossRef]
40. American Diabetes Association. Introduction: Standards of Medical Care in Diabetes—2022. *Diabetes Care* **2022**, *45*, S1–S2. [CrossRef]
41. Takács, I.; Szekeres, A.; Takács, Á.; Rakk, D.; Mézes, M.; Polyák, Á.; Lakatos, L.; Gyémánt, G.; Csupor, D.; Kovács, K.J.; et al. Wild Strawberry, Blackberry, and Blueberry Leaf Extracts Alleviate Starch-Induced Hyperglycemia in Prediabetic and Diabetic Mice. *Planta Med.* **2020**, *86*, 790–799. [CrossRef] [PubMed]
42. Kumar, S.; Narwal, S.; Kumar, V.; Prakash, O. α-glucosidase inhibitors from plants: A natural approach to treat diabetes. *Pharmacogn. Rev.* **2011**, *5*, 19–29. [CrossRef]
43. Scalliet, G.; Journot, N.; Jullien, F.; Baudino, S.; Magnard, J.-L.; Channelière, S.; Vergne, P.; Dumas, C.; Bendahmane, M.; Cock, J.; et al. Biosynthesis of the major scent components 3,5-dimethoxytoluene and 1,3,5-trimethoxybenzene by novel rose *O*-methyltransferases. *FEBS Lett.* **2002**, *523*, 113–118. [CrossRef]
44. Ferreira, L.A.; Henriques, O.B.; Andreoni, A.A.; Vital, G.R.; Campos, M.M.; Habermehl, G.G.; de Moraes, V.L. Antivenom and biological effects of ar-turmerone isolated from *Curcuma longa* (Zingiberaceae). *Toxicon* **1992**, *30*, 1211–1218. [CrossRef]
45. Golding, B.T.; Pombo, E.; Samuel, C.J. Turmerones: Isolation from turmeric and their structure determination. *J. Chem. Soc. Chem. Commun.* **1982**, 363–364. [CrossRef]
46. Jankasem, M.; Wuthi-Udomlert, M.; Gritsanapan, W. Antidermatophytic Properties of *Ar*-Turmerone, Turmeric Oil, and *Curcuma longa* Preparations. *ISRN Dermatol.* **2013**, *2013*, 250597. [CrossRef]
47. Saga, Y.; Hatakenaka, Y.; Matsumoto, M.; Yoshioka, Y.; Matsumura, S.; Zaima, N.; Konishi, Y. Neuroprotective effects of aromatic turmerone on activity deprivation-induced apoptosis in cerebellar granule neurons. *NeuroReport* **2020**, *31*, 1302–1307. [CrossRef]
48. Wang, S.Y.; Lin, H.-S. Antioxidant Activity in Fruits and Leaves of Blackberry, Raspberry, and Strawberry Varies with Cultivar and Developmental Stage. *J. Agric. Food Chem.* **2000**, *48*, 140–146. [CrossRef]
49. Scalzo, J.; Mezzetti, B.; Battino, M. Total antioxidant capacity evaluation: Critical steps for assaying berry antioxidant features. *BioFactors* **2005**, *23*, 221–227. [CrossRef]
50. Tulipani, S.; Romandini, S.; Busco, F.; Bompadre, S.; Mezzetti, B.; Battino, M. Ascorbate, not urate, modulates the plasma antioxidant capacity after strawberry intake. *Food Chem.* **2009**, *117*, 181–188. [CrossRef]
51. Kiselova, Y.; Ivanova, D.; Chervenkov, T.; Gerova, D.; Galunska, B.; Yankova, T. Correlation between the in vitro antioxidant activity and polyphenol content of aqueous extracts from bulgarian herbs. *Phytother. Res.* **2006**, *20*, 961–965. [CrossRef] [PubMed]
52. Cheplick, S.; Kwon, Y.-I.; Bhowmik, P.; Shetty, K. Phenolic-linked variation in strawberry cultivars for potential dietary management of hyperglycemia and related complications of hypertension. *Bioresour. Technol.* **2010**, *101*, 404–413. [CrossRef] [PubMed]
53. El-Hawary, S.S.; Mohammed, R.; El-Din, M.E.; Hassan, H.M.; Ali, Z.Y.; Rateb, M.E.; El Naggar, E.M.B.; Othman, E.M.; Abdelmohsen, U.R. Comparative phytochemical analysis of five Egyptian strawberry cultivars (*Fragaria* × *ananassa* Duch.) and antidiabetic potential of Festival and Red Merlin cultivars. *RSC Adv.* **2021**, *11*, 16755–16767. [CrossRef] [PubMed]
54. Mandave, P.; Khadke, S.; Karandikar, M.; Pandit, V.; Ranjekar, P.; Kuvalekar, A.; Mantri, N. Antidiabetic, Lipid Normalizing, and Nephroprotective Actions of the Strawberry: A Potent Supplementary Fruit. *Int. J. Mol. Sci.* **2017**, *18*, 124. [CrossRef]
55. Ibrahim, D.S.; El-Maksoud, M.A.E.A. Effect of strawberry (*Fragaria* × *ananassa*) leaf extract on diabetic nephropathy in rats. *Int. J. Exp. Pathol.* **2015**, *96*, 87–93. [CrossRef]
56. Moazen, S.; Amani, R.; Rad, A.H.; Shahbazian, H.; Ahmadi, K.; Jalali, M.-T. Effects of Freeze-Dried Strawberry Supplementation on Metabolic Biomarkers of Atherosclerosis in Subjects with Type 2 Diabetes: A Randomized Double-Blind Controlled Trial. *Ann. Nutr. Metab.* **2013**, *63*, 256–264. [CrossRef]
57. Shah, S.M.M.; Ahmad, Z.; Yaseen, M.; Shah, R.; Khan, S.; Khan, B. Phytochemicals, in vitro antioxidant, total phenolic contents and phytotoxic activity of *Cornus macrophylla* Wall bark collected from the North-West of Pakistan. *Pak. J. Pharm. Sci.* **2015**, *28*, 23–28.
58. Shah, S.M.M.; Ullah, F.; Shah, S.M.H.; Zahoor, M.; Sadiq, A. Analysis of chemical constituents and antinociceptive potential of essential oil of *Teucrium Stocksianum* bioss collected from the North West of Pakistan. *BMC Complement. Altern. Med.* **2012**, *12*, 244. [CrossRef]
59. Aslam, H.; Khan, A.-U.; Naureen, H.; Ali, F.; Ullah, F.; Sadiq, A. Potential application of *Conyza canadensis* (L) Cronquist in the management of diabetes: In vitro and in vivo evaluation. *Trop. J. Pharm. Res.* **2018**, *17*, 1287. [CrossRef]
60. Sadiq, A.; Mahnashi, M.H.; Rashid, U.; Jan, M.S.; Alshahrani, M.A.; Huneif, M.A. 3-(((1*S*,3*S*)-3-((*R*)-Hydroxy(4-(trifluoromethyl)phenyl)methyl)-4-oxocyclohexyl)methyl)pentane-2,4-dione: Design and Synthesis of New Stereopure Multi-Target Antidiabetic Agent. *Molecules* **2022**, *27*, 3265. [CrossRef]

61. Huneif, M.A.; Alshehri, D.B.; Alshaibari, K.S.; Dammaj, M.Z.; Mahnashi, M.H.; Majid, S.U.; Javed, M.A.; Ahmad, S.; Rashid, U.; Sadiq, A. Design, synthesis and bioevaluation of new vanillin hybrid as multitarget inhibitor of α-glucosidase, α-amylase, PTP-1B and DPP4 for the treatment of type-II diabetes. *Biomed. Pharmacother.* **2022**, *150*, 113038. [CrossRef]
62. Mahnashi, M.H.; Alyami, B.A.; Alqahtani, Y.S.; Jan, M.S.; Rashid, U.; Sadiq, A.; Alqarni, A.O. Phytochemical profiling of bioactive compounds, anti-inflammatory and analgesic potentials of *Habenaria digitata* Lindl.: Molecular docking based synergistic effect of the identified compounds. *J. Ethnopharmacol.* **2021**, *273*, 113976. [CrossRef] [PubMed]

molecules

MDPI

Article

Isolation of Echimidine and Its C-7 Isomers from *Echium plantagineum* L. and Their Hepatotoxic Effect on Rat Hepatocytes

Michał Gleńsk [1,*], Marta K. Dudek [2], Peter Kinkade [3], Evelyn C. S. Santos [4], Vitold B. Glinski [3], Daneel Ferreira [5], Ewa Seweryn [6], Sławomir Kaźmierski [2], Joao B. Calixto [4] and Jan A. Glinski [3,*]

1 Department of Pharmacognosy and Herbal Drugs, Wroclaw Medical University, Borowska 211A, 50556 Wrocław, Poland
2 Centre of Molecular and Macromolecular Studies, Polish Academy of Sciences, Sienkiewicza 112, 90363 Łódź, Poland; mdudek@cbmm.lodz.pl (M.K.D.); kaslawek@cbmm.lodz.pl (S.K.)
3 Planta Analytica LLC, 461 Danbury Rd., New Milford, CT 06776, USA; kinkadeanalytical@gmail.com (P.K.); v.glinski@plantaanalytica.com (V.B.G.)
4 Centro de Inovação e Ensaios Pré-Clínicos (CIEnP), Av. Luiz Boiteux Piazza, 1302-Cachoeira do Bom Jesus, Florianópolis 88056000, SC, Brazil; evelyn.santos@cienp.org.br (E.C.S.S.); joao.calixto@cienp.org.br (J.B.C.)
5 Department of BioMolecular Sciences, Division of Pharmacognosy, Research Institute of Pharmaceutical Sciences, School of Pharmacy, University of Mississippi, Oxford, MI 38677, USA; dferreir@olemiss.edu
6 Department of Chemistry and Immunochemistry, Wroclaw Medical University, M. Skłodowskiej-Curie 48/50, 50369 Wrocław, Poland; ewa.seweryn@umw.edu.pl
* Correspondence: michal.glensk@umw.edu.pl (M.G.); jan@plantaanalytica.com (J.A.G.)

Citation: Gleńsk, M.; Dudek, M.K.; Kinkade, P.; Santos, E.C.S.; Glinski, V.B.; Ferreira, D.; Seweryn, E.; Kaźmierski, S.; Calixto, J.B.; Glinski, J.A. Isolation of Echimidine and Its C-7 Isomers from *Echium plantagineum* L. and Their Hepatotoxic Effect on Rat Hepatocytes. *Molecules* **2022**, *27*, 2869. https://doi.org/10.3390/molecules27092869

Academic Editors: Syed Shams ul Hassan, Mohamed M. Abdel-Daim, Tapan Behl and Simona Bungau

Received: 25 March 2022
Accepted: 28 April 2022
Published: 30 April 2022

Abstract: Echimidine is the main pyrrolizidine alkaloid of *Echium plantagineum* L., a plant domesticated in many countries. Because of echimidine's toxicity, this alkaloid has become a target of the European Food Safety Authority regulations, especially in regard to honey contamination. In this study, we determined by NMR spectroscopy that the main HPLC peak purified from zinc reduced plant extract with an MS $[M + H]^+$ signal at m/z 398 corresponding to echimidine (**1**), and in fact also represents an isomeric echihumiline (**2**). A third isomer present in the smallest amount and barely resolved by HPLC from co-eluting (**1**) and (**2**) was identified as hydroxymyoscorpine (**3**). Before the zinc reduction, alkaloids (**1**) and (**2**) were present mostly (90%) in the form of an *N*-oxide, which formed a single peak in HPLC. This is the first report of finding echihumiline and hydroxymyoscorpine in *E. plantagineum*. Retroanalysis of our samples of *E. plantagineum* collected in New Zealand, Argentina and the USA confirmed similar co-occurrence of the three isomeric alkaloids. In rat hepatocyte primary culture cells, the alkaloids at 3 to 300 μg/mL caused concentration-dependent inhibition of hepatocyte viability with mean IC_{50} values ranging from 9.26 to 14.14 μg/mL. Our discovery revealed that under standard HPLC acidic conditions, echimidine co-elutes with its isomers, echihumiline and to a lesser degree with hydroxymyoscorpine, obscuring real alkaloidal composition, which may have implications for human toxicity.

Keywords: echimidine; echimidine isomers; hepatotoxicity; *Echium plantagineum* L.; rat hepatocytes primary culture

1. Introduction

Pyrrolizidine alkaloids (PAs) are toxic compounds widespread throughout the plant kingdom, occurring in about 3% of flowering plants [1]. Over 350 PAs have been identified so far [2]. The plants containing PAs belong mainly to the Asteraceae (Senecioneae and Eupatorieae tribes), Boraginaceae (all genera), and Fabaceae (genus Crotalaria) families [3–6]. PAs occur as free-base/tertiary forms or their *N*-oxides. Both forms are hepatotoxic and genotoxic [7,8]. Often, the *N*-oxides are found in higher concentrations than the corresponding free bases (tertiary PAs) [9,10].

The well-documented liver toxicity of 1,2-unsaturated PAs results in hepatic veno-occlusive disease (HVOD). Although chronic exposure to PAs may not produce readily recognizable symptoms, their hepatotoxicity can lead to irreversible liver damage and death [5,6].

Because PAs may enter into the food chain by consumption of teas and herbal products, eggs, milk, cereals and honey, awareness was recently raised by the European Food Safety Authority (EFSA) [5,6,11–13]. Although legal limits for PAs in food have not been yet established in the European Union, the German Federal Institute for Risk Assessment (BfR) recommended a limit for intake of no higher than 0.007 μg of 1,2-unsaturated PAs per day per kg body weight [14].

Certain PAs-containing plants such as *Echium plantagineum* and *E. vulgare* are important foraging sources for honeybees. The Food Standards Australia New Zealand (FSANZ) recognizes that honey produced from *E. plantagineum* will contain PAs and suggests blending it with other kinds of honey to reduce PAs concentration or that consumers should not exclusively eat this honey [10,13].

The main pyrrolizidine alkaloid of *E. plantagineum* L., known as Paterson's curse or Salvation Jane in Australia, is echimidine [15–17]. This plant is native to the Mediterranean basin in Spain and Portugal and is naturalized across much of southern Australia. It was originally introduced as a potherb in the mid-1800s and later spread as an unintended contaminant of pasture seed and hay. *E. plantagineum* is estimated to have invaded over 30 million ha of grazing land in Australia [18,19]. In addition to the fact that *E. plantagineum* contains pyrrolizidine alkaloids, its traditional uses as diaphoretic diuretic, cough and wound healing agent is mentioned. Moreover, echium oil also has many potential uses in cosmetic and personal care product industries [20].

The presence of PAs in honey is well documented [10,13], but at what concentration level could they be considered dangerous is still being debated. Currently, there are several research programs underway, initiated by countries that either export or import the honey, aimed at the determination of toxic limits for pyrrolizidine alkaloids. For most of the studies, echimidine has been chosen as one of the representative alkaloids.

In this paper, we describe the isolation and determination of three isomeric pyrrolizidine alkaloids from *E. plantagineum*, including echimidine, echihumiline and hydroxymyoscorpine, the latter two being isolated from this source for the first time. The latter compounds have been previously isolated from other plants [4,17]. Considering that these alkaloids may contribute unequally to the hepatotoxicity effects, the "echimidine" peak was further resolved into true echimidine (**1**) and the mixture of echihumiline (**2**) and hydroxymyoscorpine (**3**) and their cytotoxicity were assessed in rat primary cell hepatocytes. Additionally, the hepatotoxic activity of echiumine (**4**), the alkaloid isolated from the same plant material as the three aforementioned isomeric alkaloids, was also evaluated.

2. Results and Discussion

2.1. Chromatography Investigation

Echimidine is currently used as a standard for the determination of toxic limits for pyrrolizidine alkaloids in many products; therefore, its purity is of great importance. In our study, we noticed that the methanol extract contained nearly 90% of echimidine, as the *N*-oxide represented by a sharp RP-HPLC peak. The *N*-oxides were reduced with zinc dust in the presence of sulfuric acid. After the reduction, the alkaloidal fraction was collected by extraction with ethyl acetate. The alkaloid mixture was subjected to a CPC purification step that produced a fraction composed of echimidine plus its isomers. An RP HPLC analysis of this fraction under acidic conditions (0.05% TFA) consistently produced a sharp peak plus a small 1–2% area partially resolved peak eluting ahead of the main peak showing an MS ion at m/z 398 $[M + H]^+$ corresponding to echimidine (Figure 1). Considering the HPLC profile and keeping the MS result in mind, the sample may be regarded as an echimidine of good purity.

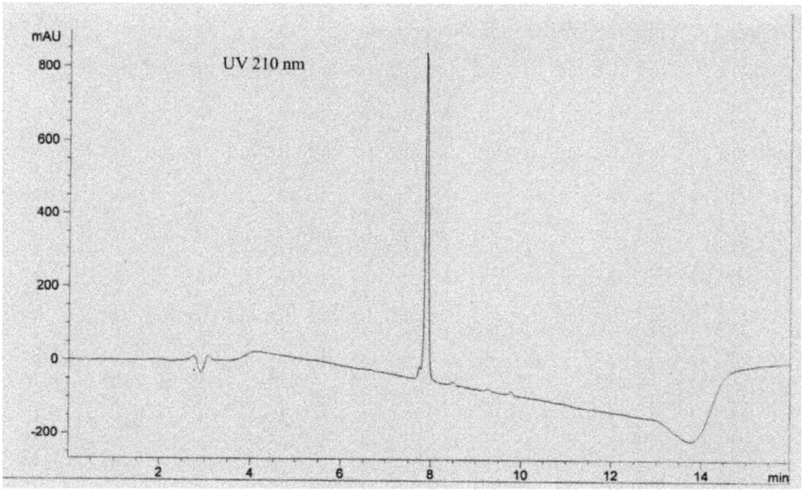

Figure 1. An RP HPLC analysis of the "echimidine" peak from *E. plantagineum* under acidic conditions (0.05% TFA). This peak in fact consists of Echimidine, Echihumiline and Hydroxymyoscorpine.

However, further NMR and HPLC data analysis indicated that this single peak comprises a mixture of alkaloids. Using the "core-shell" RP HPLC column (Kinetex EVO C18, Phenomenex) in a buffer system at pH 6.8 (or higher), the peak attributed to "echimidine" was resolved into two well-separated peaks (Figure 2). Peak A consists of echihumiline and hydroxymyoscorpine, whereas peak B represents echimidine.

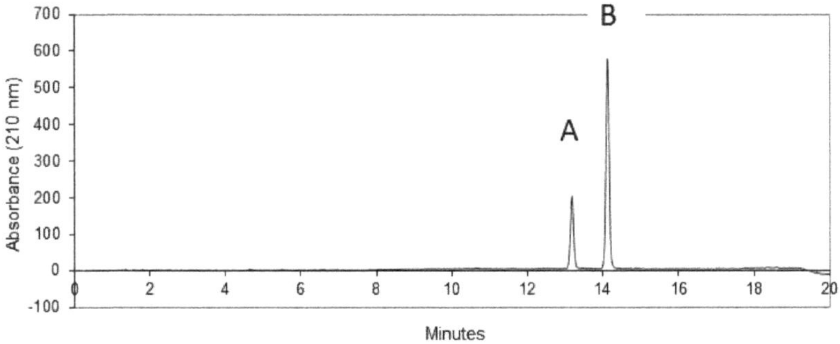

Figure 2. An RP HPLC analysis of the "echimidine" peak from *E. plantagineum* under buffer conditions (pH 6.8). Peak A—mixture of echihumiline and hydroxymyoscorpine. Peak B—echimidine.

Additionally, with the aim of centrifugal partition chromatography (CPC) in a biphasic system, consisting of chloroform as the mobile phase and citrate buffer at pH 5.6 as the stationary phase, we were able to simultaneously collect a fraction that contained mainly echiumine (**4**) (Figure 3), an alkaloid closely related to echimidine.

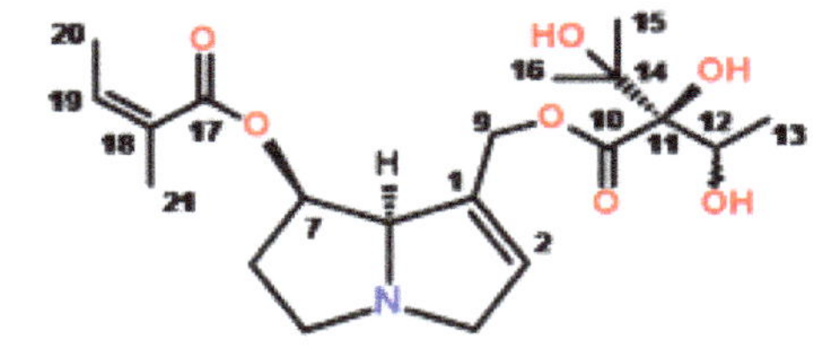

Echimidine (**1**)

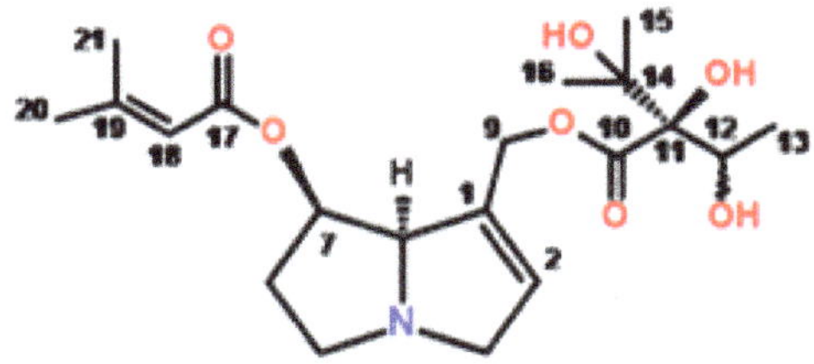

Echihumiline (**2**)

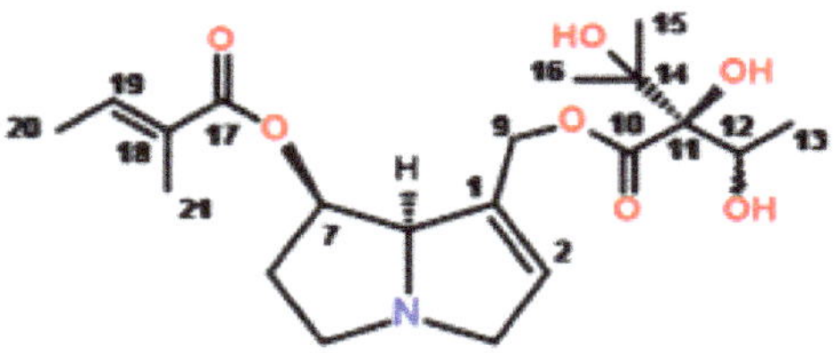

Hydroxymyoscorpine (**3**)

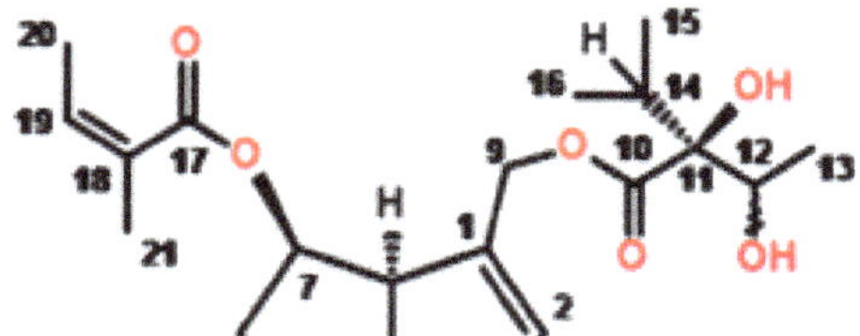

Echiumine (**4**)

Figure 3. Structures of the investigated alkaloids (Stereochemistry at position C-12 not established).

2.2. NMR Analysis

The NMR data analysis proved that peak **B** is indeed echimidine, while peak **A** comprised of two largely unresolved alkaloids, echihumiline (major) and hydroxymyoscorpine (minor). Examination of several older (highly enriched in echimidine) comparable fractions derived from different *E. planatgineum* plant collections revealed that the ratio of echihumiline to echimidine varied from 0.13 to 0.43. The ratio of hydroxymyoscropine to echimidine varied from non-detectable to 0.02. Nevertheless, the earlier eluting peak **A** in Figure 2, which in fact is a 3:1 mixture consisting of two alkaloids by NMR data analysis, could not be fully resolved through pH-controlled systems or the use of various stationary phases (C_8, C_{12}, C_{18}).

A detailed analysis of the ^{1}H NMR spectrum of the constituent comprising peak **A** clearly indicates that two compounds, in the ratio of 3:1, are present. This was evaluated on the basis of integral values of the well-separated ^{1}H resonances, which are directly proportional to the number of protons giving rise to a given signal. The spectroscopic pattern of both compounds is characteristic of pyrrolizidine alkaloids, with many signals resonating at the same frequencies. However, broad singlets at δ_H 4.48 and 4.44, assignable to H-8 of the major and minor components (Table 1), respectively, are well separated. There is also a difference in the ^{1}H and ^{13}C chemical shifts of the C-18-C-21 moieties. In the minor component, the ^{1}H signals at δ_H 6.75, 1.75 and 1.76, together with their adjacent carbons, are characteristic of a tigloyl moiety and assignable to C-19, C-20 and C-21 sites (Table 1), respectively, implying that this component is hydroxymyoscorpine (Figure 3) [21]. In the ^{1}H NMR spectrum of the major component, the olefinic proton (H-18) signal is shielded to δ_H 5.57, in contrast to H-19 of echimidine and hydroxymyoscorpine, in which these signals resonate at 6.75 and 6.09, respectively. In addition, both methyl groups are deshielded to δ_H 1.88 and 2.12, in comparison to 1.75 and 1.76 ppm seen for hydroxymyoscorpine (see Table 1). These chemical shifts are characteristic of a senecioic acid ester moiety, which permits the identification of the major component represented by peak **A** as echihumiline (Figure 3) [22].

Table 1. NMR data for echihumiline, hydroxymyoscorpine and echimidine ($CDCl_3$, 500 MHz, T = 295 K).

Position	Peak (A)				Peak (B)	
	Major Component (1.00) Echihumiline		**Minor Component (0.33) Hydroxymyoscorpine**		**Echimidine**	
	δ (^{13}C), Multiplicity	**δ (^{1}H) *J* in Hz**	**δ (^{13}C), Multiplicity**	**δ (^{1}H) *J* in Hz**	**δ (^{13}C), Multiplicity**	**δ (^{1}H) *J* in Hz**
1	132.9, C		132.8, C		132.9, C	
2	128.1, CH	5.84 bs	128.2, CH	5.86 bs	128.2, CH	5.85 bs
3	62.4, CH_2	4.00 bd (15.2) 3.42 m	62.7, CH_2	4.00 bd (15.2) 3.41 m	62.5, CH_2	4.01 bd (15.2) 3.40 m
5	53.7, CH_2	3.39 m 2.70 m	53.8, CH_2	3.39 m 2.70 m	53.8, CH_2	3.42 m 2.71 m
6	34.3, CH_2	2.11 m (2H)	34.3, CH_2	2.11 m (2H)	34.4, CH_2	2.13 m (2H)
7	73.6, CH	5.40 m	73.6, CH	5.40 m	73.6, CH	5.45 m
8	75.8, CH	4.48 bs	75.9, CH	4.44 bs	75.9, CH	4.50 bs
9	62.3, CH_2	4.92 d (13.1) 4.64 d (13.1)	62.5, CH_2	4.89 (overlapped) 4.63 d (13.2)	62.3, CH_2	4.90 d (13.2) 4.65 d (13.2)
10	174.2, C		174.2, C		174.2, C	
11	82.9, C		82.9, C		83.1, C	
12	69.7, CH	4.18 q (6.3)	69.7, CH	4.18 q (6.4)	69.7, CH	4.17 q (6.4)
13	18.5, CH_3	1.24 d (6.3)	18.5, CH_3	1.24 d (6.4)	18.5, CH_3	1.24 d (6.4)
14	72.7, C		72.7, C		73.7, C	
15	25.9, CH_3	1.22 s	25.9, CH_3	1.21 s	26.0, CH_3	1.21 s
16	24.8, CH_3	1.29 s	24.8, CH_3	1.28 s	24.8, CH_3	1.28 s
17	165.7, C		167.0, C		166.8, C	
18	115.5, CH	5.57 m	128.4, C		127.2, C	
19	158.4, C		138.0, CH	6.75 qd (6.2; 1.5)	139.6, CH	6.09 qq (7.3; 1.6)
20	27.5, CH_3	1.88 d (0.9)	14.5, CH_3	1.76 dd	15.8, CH_3	1.94 dd (7.3; 1.6)
21	20.3, CH_3	2.12 d (0.9)	11.9, CH_3	1.75 bs	20.5, CH_3	1.79 d (1.6)

The ^{1}H NMR spectrum of the second compound comprising peak **B** exhibits five methyl, four methylene, two olefinic and three methine proton signals. The chemical shifts at δ_H 6.09 and 5.85 (each 1H, corresponding to H-19 and H-2, see Table 1) are characteristic of olefinic moieties. One of these signals (δ_H 6.09) correlates in the HMBC spectrum with two methyl signals at δ_C 20.48 and 15.78 (C-21 and C-20, respectively), and one carboxyl signal at δ_C 166.75 (C-17). The protons adjacent to the two methyl groups resonate at δ_H 1.79 (H-21) and 1.94 (H-20), respectively, and in conjunction with a carboxylic carbon at δ_C 166.75 (C-17), an sp^2 quaternary carbon signal at δ_C 127.21 (C-18), and a methine carbon signal at δ_C 139.63 (C-19) are all characteristic of an angeloyl moiety [23]. The location of this moiety is indicated by an HMBC correlation of the carboxylic carbon with a signal at δ_H 5.45 (H-7, Table 1) belonging to the pyrrolizidine moiety, as indicated by additional HMBC and COSY correlations. A second side-chain moiety is substituted with a carbonyl group resonating at δ_C 174.24 and exhibiting long-range correlation to a proton signal at δ_H 4.17. This signal shows further correlations to a methyl group at δ_C 18.48 and two oxygenated tertiary carbons at δ_C 73.73 and 83.07. Both these carbons give HMBC cross-peaks to two methyl signals at δ_C 24.77 and 25.98. As a result, the compound representing peak B was unambiguously identified as echimidine (Figure 3). All ^{1}H and ^{13}C chemical shifts for echimidine and its isomers are collated in Table 1 and are in perfect agreement with published data [4,17,24,25].

The apparent similarity of the ^{1}H and ^{13}C NMR spectra of these alkaloids (see Figure 4) may be a reason why their existence has not been noticed before in what was regarded previously as a pure "echimidine" peak. The presence of a nitrogen atom, which is easily protonated at lower pH, and as a result may cause significant changes to proton chemical shifts, may further interfere with the proper identification, especially when the spectra are often registered in $CDCl_3$, which may have a different degree of acidity.

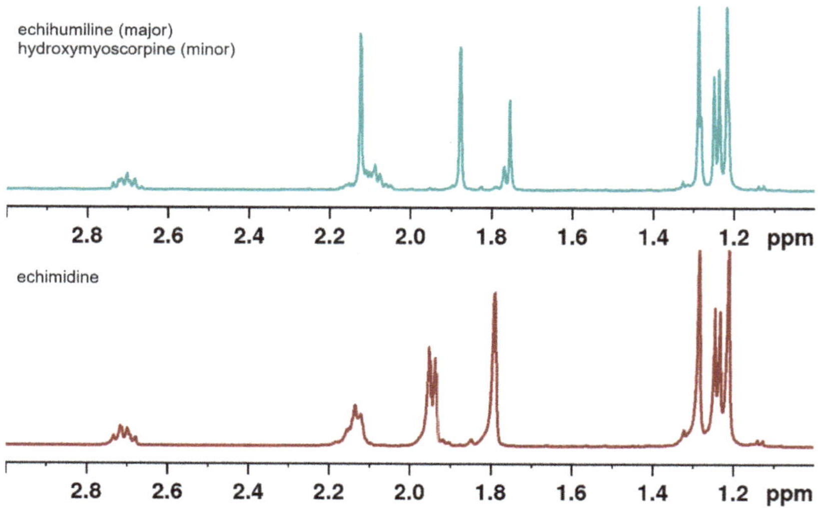

Figure 4. *Cont.*

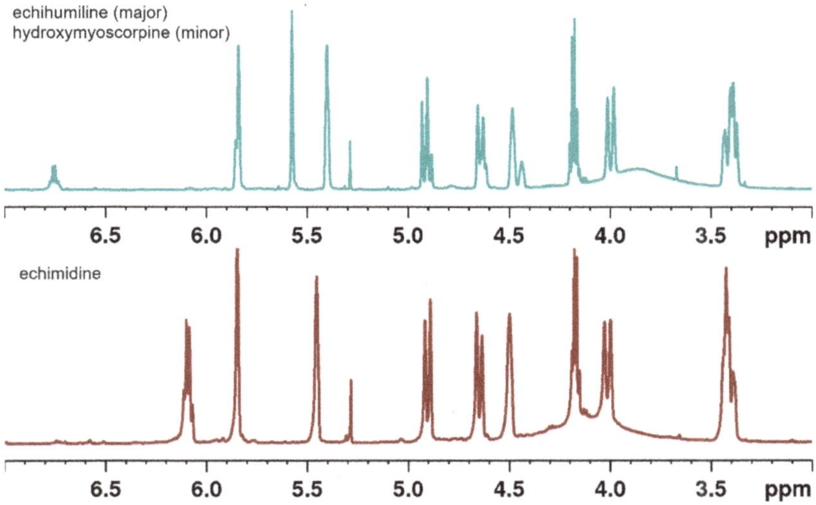

Figure 4. ^{1}H NMR spectra of echimidine and echihumiline-hydroxymyoscorpine mixture ($CDCl_3$, 500 MHz).

2.3. *Hepatotoxicity Assessment*

Figure 5A shows the percentage of viability of hepatocytes treated with different concentrations of echiumine, a fraction with an "echimidine" peak (being in fact a mixture of echimidine, echihumiline and hydroxymyoscorpine), echimidine and a mixture of echihumiline and hydroxymyoscorpine (3 to 300 μg/mL).

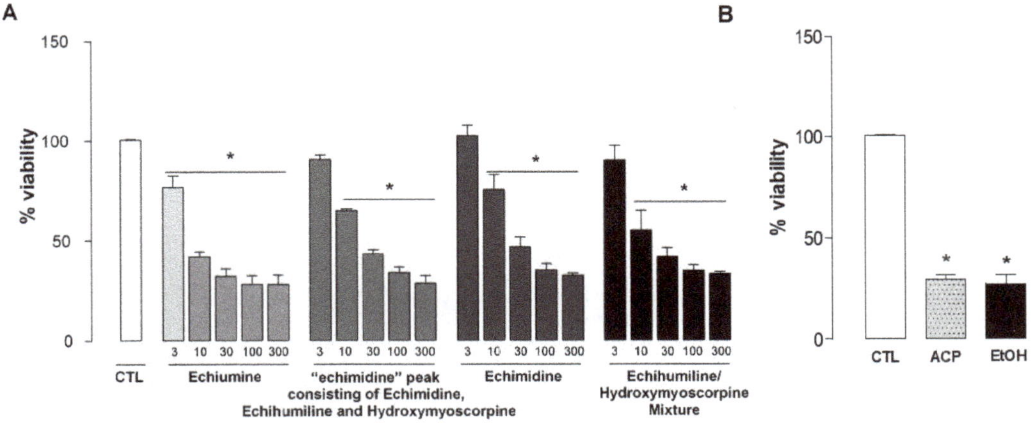

Figure 5. Effect of the compounds on cell viability in primary culture of rat hepatocytes. (**A**) Rat hepatocytes were treated with echiumine, a fraction with an "echimidine" peak, echimidine or echihumiline/hydroxymyoscorpine mixture (3–300 μg/mL) for 48 h. (**B**) Rat hepatocytes were treated with acetaminophen (ACP; 15.1 mg/mL) or ethanol (EtOH; 80 μL/mL) as a positive control. Cell viability was determined by MTT assay. The results are expressed in percentage of control (CTL) and each column presents the mean and the SEM of triplicate assays of two independent experiments. * $p < 0.05$ when compared with control (CTL) values (one-way ANOVA or *t*-test, when applicable).

All of them caused a concentration-dependent inhibition of rat hepatocyte primary cell culture viability. The calculated mean IC_{50} values for the four compounds ranged from 7.47 to 14.14 μg/mL and they do not differ significantly among them (Table 2). As a positive control (Figure 5B), we assessed the effect of two hepatotoxic agents, namely acetaminophen and ethanol at the concentrations of 15.1 and 80 μL/mL, respectively. Both agents inhibited hepatocyte viability in 70.68% and 72.51%, respectively, confirming that the bioassay worked properly.

Table 2. Mean IC_{50} and maximal inhibitions values on cell viability in primary culture hepatocytes for different compounds.

Compound	IC_{50}	95% CI	I_{max} (%)
"echimidine" peak consisitng of Echimidine, Echihumiline and Hydroxymyoscorpine	14.14 μg/mL	9.01 to 22.17	70.96
Echimidine	13.79 μg/mL	7.84 to 24.24	67.37
Echihumiline/Hydroxymyoscorpine Mixture	9.26 μg/mL	4.33 to 19.81	66.51
Echiumine	7.47 μg/mL	3.26 to 17.11	71.64
Acetaminophen	3.82 mg/mL	2.0 to 7.32	68.31

IC_{50} = concentration required to inhibit the cytotoxicity by 50%; CI = the 95% confidence interval, I_{max} = maximum inhibition of cell viability. Each group represents the triplicate of two independent animals.

Collectively, the results support the view that the alkaloids isolated from *Echium plantagineum* L. exhibit similar hepatotoxicity when assessed in rat hepatocyte primary culture cells. Importantly, the IC_{50} value of isolated echimidine (13.79 μg/mL) was similar to that of the fraction containing the three isomers (14.14 μg/mL). However, the highest toxicity was found for echiumine (7.47 μg/mL), which was isolated alongside with the three aforementioned alkaloids from the same plant sample. According to some authors, the pyrrolizidine alkaloids are converted in the liver into reactive metabolites. Therefore, chronic exposure to sub-lethal doses may cause cumulative damage or cancer. PAs intake can induce damage to liver cells, inducing hepato-sinusoidal obstruction syndrome or veno-occlusive disease. Testing of a range of PAs revealed that nuclear receptor PXR was exclusively activated by the open-chain diesters such as echimidine and lasiocarpine, suggesting that only open-chain diesters act as PXR agonists. This might imply that a PXR-dedicated mode of action may contribute to the hepatotoxicity of PAs that is dependent on PA structure [26].

3. Materials and Methods

3.1. General Experimental Procedures

Centrifugal Partition Chromatography was performed on the FCPC Kromaton A100 (Rousellet Robatel, Annonay, France) system. Analytical HPLC was carried out on an Agilent HP1100 model equipped with quaternary pump, DAD, and autosampler. Preparative HPLC was performed on a Gilson model 333/334 with H3M pump heads. A linear UV-VIS 200 detector with a preparative cell was used. The UV signal was monitored and integrated with a model N2000 Baseline Chromatography Data System from Baseline Chromatech Research Centre (www.qinhuan.com, accessed on 24 March 2022). Analytical HPLC analyses were performed on a Kinetex EVO C_{18}, 5 μm, 100 Å, 4.6 × 150 mm column, and preparative on a Kinetex EVO C_{18}, 5 μm, 100 Å, 21.2 × 150 mm column (Phenomenex, Torrance, CA, USA). The mobile phase buffer was the same as for the preparative HPLC (32 mM lithium phosphate, adjusted to pH 7.2 with phosphoric acid). The samples/compounds were eluted from the analytical column on a gradient from 5 to 16% acetonitrile in 15 min, followed by a column wash with 29% acetonitrile. The flow rate was 1.0 mL/min. The UV absorbance signal on the diode array detector was monitored at 210 nm.

3.2. Plant Material

The plant material was collected in New Zealand, and the voucher specimens (CHR 637254, CHR 637255) were deposited at Allan Herbarium (CHR), Landcare Research,

PO Box 69040, Lincoln 7640, NZ. The American Echium was propagated from the seeds sold by Outsidepride company, 915 N. Main St., Independence, OR 97351, in Bethel, Connecticut. Argentinian *E. plantagineum* was collected in B6600 Mercedes, Provincia de Buenos Aires, Argentina.

3.3. Extraction and Isolation

Collected in New Zealand, aerial parts of *E. plantagineum* were dried and then extracted with MeOH. After the removal of MeOH, the resulting oleoresin was defatted by partitioning between *n*-hexane-MeOH-water (10:9:1). The lower phase was concentrated to remove the organic solvents and diluted with water. The *N*-oxides of the Pas were reduced to free alkaloids by treating the aqueous phase with 50% H_2SO_4 till the pH adjusted to 2, and then an excess of zinc dust was added during stirring. In about 6 h the pH became nearly neutral, and another portion of sulfuric acid and zinc dust was added. After 20 h of stirring, the reaction mixture was made alkaline with Na_2CO_3, and the alkaloids were extracted several times with ethyl acetate containing 5% of isopropanol. These extracts were combined and quickly evaporated to dryness to avoid hydrolysis of ethyl acetate by the alkaloids. Next, the obtained solid was again partitioned between water acidified with sulfuric acid and ethyl acetate. The alkaloids migrated into the aqueous acidic phase, and the neutral compounds remained in the ethyl acetate phase. Finally, the aqueous layer was made alkaline with Na_2CO_3 and the alkaloids were re-extracted into ethyl acetate. Evaporation of the solvent produced a crude mixture of pyrrolizidine alkaloids. Centrifugal Partition Chromatography, using chloroform as the mobile phase and citrate buffer at pH 5.6 as the stationary phase, was used to produce the bands of echimidine fraction and echiumine. After this step, the echimidine fraction was purified by preparative HPLC. The mobile phase contained a buffer of 32 mM lithium phosphate, adjusted to pH 7.2 with phosphoric acid. The column was equilibrated using a 9:1 mixture of the buffered mobile phase and acetonitrile and run in a gradient from 10% to 20% MeCN over 14.4 min at 20 mL/min. For preparative runs, samples of 60 mg were injected in methanol/water/acetic acid (1:1:0.005) at a concentration of 100 mg/mL. A total of 1.3 g of CPC-fractionated echimidine fraction was loaded in 22 preparative HPLC runs. As a result, 630 mg of echimidine and 280 mg of echihumiline/hydroxymyoscorpine mixture were obtained.

3.4. NMR Spectroscopy

The 1D and 2D NMR spectra were recorded on a Bruker Avance III 500 spectrometer (Bruker BioSpin, Rheinstetten, Germany), operating at 500.13 MHz for ^1H and 125.76 MHz for ^{13}C. The spectrometer was equipped with a 5 mm BBI probe head with an actively shielded Z-gradient coil connected to a GAB/2 gradient unit capable of producing B_0 gradients with a maximum strength of 50 G/cm. During all measurements, the temperature was set at 295 K and was controlled and stabilized with a BCU 05 cooling unit controlled by a BVT3200 variable temperature unit. All spectra were recorded using 3 mm NMR tubes (Norell) and $CDCl_3$ (Armar Chemical, Döttingen, Switzerland) as a solvent. Chemical shifts were referenced to the residual solvent signals at 7.26 and 77.00 ppm for ^1H and ^{13}C, respectively. All spectra, except long-range ^1H-^{13}C correlation spectra, were acquired with the original Bruker pulse sequences. For observation of the long-range correlations, the Impact-HMBC pulse sequence was used. The spectra were acquired and processed using the TopSpin 3.1 program (Bruker BioSpin) running under Windows 7 (64 bit) OS on the HP Z700 workstation and used for operating and controlling the spectrometer.

3.5. Hepatotoxicity Assay

3.5.1. Animals

Male Sprague Dawley (250–350 g) rats were used throughout this study. Animals originated from Charles River Laboratories (Raleigh, NC, USA) were bred in the Centre for Innovation and Pre-Clinical Studies animal house facility (Florianópolis-S.C., Brazil).

All animals were Specific Pathogen Free (5 per cage) and were housed at 22 ± 1 °C in a light-controlled environment under a 12–12-h light-dark cycle (lights on at 07:00 AM). All experimental procedures were based on the "Principles of Laboratory Animal Care" from the NIH publication No. 85-23 [27], the international standards of animals recommended by Brazilian Law (# 11.794–10/08/2008) [28] and were approved by the Animal Ethics Committee of the CIEnP—Centre of Innovation and Pre-Clinical Studies (#243/00/CEUA/CIEnP).

3.5.2. Hepatocyte Isolation and Cultivation

Hepatocytes were isolated from male Sprague-Dawley rats by a modified two-step collagenase perfusion method [29]. The viability of hepatocytes was >80% (determined by trypan blue exclusion). Hepatocytes were cultured in a 48 well plate at a density of 2×10^5 cells/well in William's medium supplemented with 5% fetal bovine serum and 1% penicillin/streptomycin under standard conditions in a humidified atmosphere at 37 °C and 5% CO_2 for 4 h. The medium was removed and replaced by HepatoZyme. After 18 h of plating, cells were treated with the compounds echiumine, echimidine fraction, echimidine (isolated) or the echihumiline/hydroxymyoscorpine mixture at concentrations of 3, 10, 30, 100 and 300 μg/mL. Acetaminophen (15.1 mg/mL) and ethanol (63.2 μg/mL) were used as positive controls. The effects of the compounds were evaluated for 48 h.

3.5.3. MTT Assay

MTT assay viability was performed 48 h after the treatment of cells and was assessed by measuring the formation of formazan from the MTT spectrophotometric test, according to Mosmann (1983) [30]. At the end of the experiment, the hepatocytes were incubated with 0.5 mg/mL MTT for 4 h min at 37 °C. After the MTT was removed and the blue formazan was extracted from cells with DMSO (100 μL/well). The absorbance was read at 570 nm in a microplate reader (SpectraMax Plus).

3.5.4. Data Analysis

Graphic data were expressed as mean ± SEM. Statistical evaluation of the results was carried out using the appropriate one-way analysis of variance (ANOVA) followed by a post-hoc Dunnett's test, whereas control groups (ACP or EtOH) were analyzed by Student's *t*-test. p values lower than 0.05 (* $p < 0.05$) were considered statistically significant. Data were analyzed using GraphPad® Prism 5.0 software (San Diego, CA, USA).

4. Conclusions

For the first time, we have proven that the fraction with an "echimidine" peak in RP HPLC derived from *E. plantagineum* consists of three alkaloids: echimidine, echihumiline, and hydroxymyoscorpine. We believe that the two unreported alkaloids, echihumiline and hydroxymyoscorpine, have been overlooked in this species because MS does not differentiate isomers. Furthermore, as a result of significant structural similarity, most of the ^1H NMR signals are overlapping, and thus, signals of echihumiline and hydroxymyoscorpine, which are at lower concentrations, can be easily overlooked. We surmise that their presence varies and may be limited to some varieties, or they may occur at low concentrations compared to echimidine. Additionally, the present results support the view that the alkaloids isolated from *Echium plantagineum* L. exhibit similar hepatotoxicity when assessed in rat hepatocyte primary culture cells.

Author Contributions: Conceptualization, supervision and writing J.A.G. and J.B.C.; formal analysis writing and editing M.G. and M.K.D.; review and editing D.F. and E.S.; investigation and analysis., P.K., E.C.S.S., V.B.G. and S.K. All authors have read and agreed to the published version of the manuscript.

Funding: This research received no external funding.

Institutional Review Board Statement: The animal study protocol was approved by the Animal Ethics Committee of the CIEnP—Centre of Innovation and Pre-Clinical Studies (#243/00/CEUA/CIEnP).

Informed Consent Statement: Not applicable.

Data Availability Statement: Data sharing not applicable.

Acknowledgments: We are very grateful to John Proudfoot for reading the manuscript and the valuable suggestions. The publication (Article Processing Charge) is supported by the Wroclaw Medical University.

Conflicts of Interest: The authors declare no conflict of interest.

Sample Availability: Samples of the isolated alkaloids are available from the authors.

References

1. Smith, L.W.; Culvenor, C.C.J. Plant sources of hepatotoxic pyrrolizidine alkaloids. *J. Nat. Prod.* **1981**, *44*, 129–152. [CrossRef] [PubMed]
2. Stegelmeier, B.L.; Edgar, J.A.; Colegate, S.M.; Gardner, D.R.; Schoch, T.K.; Coulombe, R.A.; Molyneux, R.J. Pyrrolizidine alkaloid plants, metabolism and toxicity. *J. Nat. Toxins* **1999**, *8*, 95–116. [PubMed]
3. Coulombe, R.A. Pyrrolizidine alkaloids in foods. *Adv. Food Nutr. Res.* **2003**, *45*, 61–99. [PubMed]
4. El-Shazly, A.; Wink, M. Diversity of Pyrrolizidine Alkaloids in the Boraginaceae. Structures, Distribution, and Biological Properties, (review). *Diversity* **2014**, *6*, 188–282. [CrossRef]
5. Valese, A.C.; Molognoni, L.; de Sá Ploêncio, L.A.; de Lima, F.G.; Gonzaga, L.V.; Górniak, S.L.; Daguer, H.; Barreto, F.; Costa, A.C.O. A fast and simple LC-ESI-MS/MS method for detecting pyrrolizidine alkaloids in honey with full validation and measurement uncertainty. *Food Control* **2016**, *67*, 183–191. [CrossRef]
6. Martinello, M.; Borin, A.; Stella, R.; Bovo, D.; Biancotto, G.; Gallina, A.; Mutinelli, F. Development and validation of a QuEChERS method coupled to liquid chromatography and high-resolution mass spectrometry to determine pyrrolizidine and tropane alkaloids in honey. *Food Chem.* **2017**, *234*, 295–302. [CrossRef]
7. Fu, P.P.; Chou, M.W.; Xia, Q.; Yang, Y.C.; Yan, J.; Doerge, D.R.; Chan, P.C. Genotoxic pyrrolizidine alkaloids and pyrrolizidine alkaloid N-oxides-mechanisms leading to DNA adduct formation and tumorigenicity. *J. Environ. Sci. Health Part C* **2001**, *19*, 353–385. [CrossRef]
8. Chou, M.W.; Wang, Y.P.; Yan, J.; Yang, Y.C.; Berger, R.D.; Williams, L.D.; Doerge, D.R.; Fu, P.P. Riddelliine N-oxide is a phytochemical and mammalian metabolite with genotoxic activity that is comparable to the parent pyrrolizidine alkaloid riddelliine. *Toxicol. Lett.* **2003**, *145*, 239–247. [CrossRef]
9. Hartmann, T.; Witte, L. Chemistry, Biology and Chemoecology of the Pyrrolizidine Alkaloids. In *Alkaloids: Chemical and Biological Perspectives*; Pelletier, S.W., Ed.; Pergamon Press: Oxford, UK, 1995; Volume 9, pp. 155–233.
10. Lucchetti, M.A.; Glauser, G.; Kilchenmann, V.; Dübecke, A.; Beckh, G.; Praz, C.; Kast, C. Pyrrolizidine Alkaloids from *Echium vulgare* in Honey Originate Primarily from Floral Nectar. *J. Agric. Food Chem.* **2016**, *64*, 5267–5273. [CrossRef]
11. EFSA (European Food Safety Authority). Opinion of the scientific panel on contaminants in the food chain on a request from the European Commission related to pyrrolizidine alkaloids as undesirable substances in animal feed. *EFSA J.* **2007**, *447*, 1–51.
12. EFSA (European Food Safety Authority). Scientific opinion on pyrrolizidine alkaloids in food and feed. EFSA panel of contaminants in the food chain. *EFSA J.* **2011**, *9*, 2406.
13. Griffin, C.T.; Mitrovic, S.M.; Danaher, M.; Furey, A. Development of a fast isocratic LC-MS/MS method for the high-throughput analysis of pyrrolizidine alkaloids in Australian honey. *Food Addit. Contam. Part A* **2015**, *32*, 214–228. [CrossRef] [PubMed]
14. Bundesinstitut für Risikobewertung/The Federal Institute of Risk Assessment. Chemical Analysis and Toxicity of Pyrrolizidine Alkaloids and Assessment of the Health Risks Posed by Their Occurrence in Honey. BfR Opinion No. 038/2011. 2011. Available online: https://www.bfr.bund.de/cm/349/chemical-analysis-and-toxicity-of-pyrrolizidine-alkaloids-and-assessment-of-the-health-risks-posed-by-their-occurence-in-honey.pdf (accessed on 10 November 2021).
15. Culvenor, C.C.J.; Edgar, J.A.; Smith, L.W. Pyrrolizidine alkaloids in honey from *Echium plantagineum* L. *J. Agric. Food Chem.* **1981**, *29*, 958–960. [CrossRef] [PubMed]
16. Colegate, S.M.; Edgar, J.A.; Knill, A.M.; Lee, S.T. Solid-phase Extraction and HPLC-MS Profiling of Pyrrolizidine Alkaloids and their N-oxides: A Case Study of *Echium plantagineum*. *Phytochem. Anal.* **2005**, *16*, 108–119. [CrossRef]
17. Mehrabani, M.; Ghannadi, A.; Sajjadi, E.; Ghassemi, N.; Shams-Ardakani, M. Toxic pyrrolizidine alkaloids of *Echium amoenum* Fisch. & Mey. *DARU* **2006**, *14*, 122–127.
18. Piggin, C. The biology of Australian weeds: 8. *Echium plantagineum* L. *J. Aust. Inst. Agric. Sci.* **1982**, *48*, 3–16.
19. Weston, P.A.; Weston, L.A.; Hildebrand, S. Metabolic profiling in *Echium plantagineum*: Presence of bioactive pyrrolizidine alkaloids and napthoquinones from accessions across southeastern Australia. *Phytochem. Rev.* **2013**, *12*, 831–837. [CrossRef]
20. Wang, W.; Jin, J.; Xu, H.; Shi, Y.; Boersch, M.; Yin, Y. Comparative analysis of the main medicinal substances and applications of *Echium vulgare* L. and *Echium plantagineum* L.: A review. *J. Ethnopharmacol.* **2022**, *285*, 114894. [CrossRef]
21. Roeder, E.; Rengel, B. Pyrrolizidine alkaloids from *Lithospermum erythrorhizon*. *Phytochemistry* **1990**, *29*, 690–693. [CrossRef]
22. El-Shazly, A.; Sarg, T.; Ateya, A.; Abdel Aziz, E.; El-Dahmy, S.; Witte, L.; Wink, M. Pyrrolizidine and tetrahydroisoquinoline alkaloids from *Echium humile*. *Phytochemistry* **1996**, *42*, 225–230. [CrossRef]

23. El-Shazly, A.; Abdel-All, M.; Tei, A.; Wink, M. Pyrrolizidine alkaloids from *Echium rauwolfii* and *Echium horridum* (Boraginaceae). *Z. Naturforsch. C J. Biosci.* **1999**, *54*, 295–300. [CrossRef] [PubMed]
24. Roeder, E.; Liu, K.; Bourauel, T. Pyrrolizidine alkaloids from *Echium pininana. Phytochemistry* **1991**, *30*, 3107–3110. [CrossRef]
25. Carvalho, J.C.B.; Almeida, H.S.; Lobo, J.F.R.; Ferreira, J.L.P.; Oliveira, A.P.; Rocha, L. Pyrrolizidine alkaloids in two endemic capeverdian *Echium* species. *Biochem. Syst. Ecol.* **2013**, *50*, 1–6. [CrossRef]
26. Luckert, C.; Braeuning, A.; Lampen, A.; Hessel-Pras, S. PXR: Structure-specific activation by hepatotoxic pyrrolizidine alkaloids. *Chem. Biol. Interac.* **2018**, *288*, 38–48. [CrossRef]
27. National Research Council (US) Committee for the Update of the Guide for the Care and Use of Laboratory Animals. *Guide for the Care and Use of Laboratory Animals. The National Academies Collection: Reports Funded by National Institutes of Health*, 8th ed.; National Academies Press (US): Washington, DC, USA, 2011.
28. Brazil. Law n.° 11.794. 8 November 2008.
29. Seglen, P.O. Preparation of isolated rat liver cells. *Methods Cell Biol.* **1976**, *13*, 29–83.
30. Mosmann, T. Rapid colorimetric assay for cellular growth and survival: Application to proliferation and cytotoxicity assays. *J. Immunol. Methods* **1983**, *65*, 55–63. [CrossRef]

molecules

MDPI

Article

Effects of Shrimp Peptide Hydrolysate on Intestinal Microbiota Restoration and Immune Modulation in Cyclophosphamide-Treated Mice

Asif Iqbal Khan [1], Ata Ur Rehman [1], Nabeel Ahmed Farooqui [1], Nimra Zafar Siddiqui [1], Qamar Ayub [2], Muhammad Noman Ramzan [3], Liang Wang [4,*] and Yi Xin [1,*]

1 Department of Biotechnology, College of Basic Medical Science, Dalian Medical University, Dalian 116044, China; asif.iqbal@duhs.edu.pk (A.I.K.); ata_burraq@yahoo.com (A.U.R.); nabeel.farooqui99@yahoo.com (N.A.F.); nimra.siddiqui12@gmail.com (N.Z.S.)

2 College of Clinical Laboratory Sciences, Dalian Medical University, Dalian 116044, China; qamar_ayub@yahoo.com

3 Department of Biochemistry and Molecular Biology, College of Basic Medical Science, Dalian Medical University, Dalian 116044, China; mnomanr8894@gmail.com

4 Stem Cell Clinical Research Center, National Joint Engineering Laboratory, Regenerative Medicine Center, The First Affiliated Hospital of Dalian Medical University, No. 193, Lianhe Road, Shahekou District, Dalian 116011, China

* Correspondence: wangliang@dmu.edu.cn (L.W.); jimxin@hotmail.com (Y.X.); Tel.: +86-411-83635963-2169 (L.W.); +86-411-86110295 (Y.X.)

Citation: Khan, A.I.; Rehman, A.U.; Farooqui, N.A.; Siddiqui, N.Z.; Ayub, Q.; Ramzan, M.N.; Wang, L.; Xin, Y. Effects of Shrimp Peptide Hydrolysate on Intestinal Microbiota Restoration and Immune Modulation in Cyclophosphamide-Treated Mice. *Molecules* **2022**, *27*, 1720. https://doi.org/10.3390/molecules27051720

Academic Editors: Syed Shams ul Hassan, Mohamed M. Abdel-Daim, Tapan Behl, Simona Bungau and Masahide Hamaguchi

Received: 23 January 2022
Accepted: 3 March 2022
Published: 6 March 2022

Abstract: The gut microbiota is important in regulating host metabolism, maintaining physiology, and protecting immune homeostasis. Gut microbiota dysbiosis affects the development of the gut microenvironment, as well as the onset of various external systemic diseases and metabolic syndromes. Cyclophosphamide (CTX) is a commonly used chemotherapeutic drug that suppresses the host immune system, intestinal mucosa inflammation, and dysbiosis of the intestinal flora. Immunomodulators are necessary to enhance the immune system and prevent homeostasis disbalance and cytotoxicity caused by CTX. In this study, shrimp peptide hydrolysate (SPH) was evaluated for immunomodulation, intestinal integration, and microbiota in CTX-induced immunosuppressed mice. It was observed that SPH would significantly restore goblet cells and intestinal mucosa integrity, modulate the immune system, and increase relative expression of mRNA and tight-junction associated proteins (Occludin, Zo-1, Claudin-1, and Mucin-2). It also improved gut flora and restored the intestinal microbiota ecological balance by removing harmful microbes of various taxonomic groups. This would also increase the immune organs index, serum levels of cytokines (IFN-γ, IL1β, TNF-α, IL-6), and immunoglobin levels (IgA, IgM). The Firmicutes/Bacteroidetes proportion was decreased in CTX-induced mice. Finally, SPH would be recommended as a functional food source with a modulatory effect not only on intestinal microbiota, but also as a potential health-promoting immune function regulator.

Keywords: shrimp (*Penaeus Chinensis*); peptides; immunosuppression gut microbiota; cyclophosphamide; gut health

1. Introduction

The gut microbiome is a diversified community of trillions of bacteria (microorganisms) that live in the mammalian body and perform a variety of physiological functions in the host's gastrointestinal tract [1,2]. The host and the microbiota have common associations in a micro-ecosystem in the digestive tract. The microbiota is pivotal in regulating host metabolism, i.e., digesting the carbohydrate to supply the nutrients for maintenance of metabolic system, vitamin synthesis, maintain physiology, protect colonization of the pathogen by shielding the harmful microbe's invasion [3–5]. The intestinal barrier system consists of four layers that work together to keep the gastrointestinal tract in a state of homeostasis: immunological, mechanical, biological, and chemical barriers [6,7]. The mechanical barrier is composed of different types of cells: goblet cells, absorption cells, intestinal stem cells, endocrine cells, and Paneth cells, all of which collaborate to maintain the function and structure of the intestine [8]. Mucin-2 (Muc-2) is a component of chemical barriers, formed as an exudate of goblet cells and adheres to the surface of epithelial cells to prevent pathogens from attacking the mechanical barriers of the intestine [8,9]. In the intestine, the microbiota stimulates cellular proliferation, mucus production, villus thickening, vascularization, epithelial junction maintenance, and mucosal surface widening [10,11]. Gut microbiota dysbiosis plays a role in the development of the gut microenvironment, as well as the onset of various external systemic diseases and metabolic syndromes [5,12,13]. Nonetheless, commensal microbes play a vital role in gastrointestinal secretions, immunological response, and food metabolism [4,5,14].

In recent years, there has been a general understanding of the gut microbiota, with the rational use of probiotics as a functional food being well-established [2,15,16]. Pathogens are known to cause infections and are strongly associated with the development or enhancement of disease. While commensal gut microbiota modifies both genetic and epigenetic factors and modulates host health directly or indirectly [8,12], there is an established between immunity, the mucous membrane immune system, and the gut microbiome. Intestinal flora imbalance results in downregulation of immunological response, which in turn impacts immune system development, altering the immune response of distant mucous membrane tissue [13,14]. The nature of the intestinal flora is altered by an abnormal immune system, which in turn influences the immune response of the distant mucosa, leading to immune imbalance [14]. The host responds to the metabolites produced by microbes, altering the growth of immune cells and resulting in the formation of an immunological response at the limits of the body [13].

Since 1958, CTX has been one of the most used chemotherapeutic drugs to treat autoimmune disease and cancer, despite its dominating adverse effects of destroying and interfering with cells' DNA, folic acid, and thus causing immunosuppression [7,15–17]. A high dose of CTX may affect not only tumor cells, but also lead to downregulating the immune system while increasing intestinal mucosa inflammation, dysbiosis of the intestinal flora, and increased permeability of the intestine. All of them resulting in secondary infection by potential pathogens [7,18,19]. As a result, immunomodulators are important to stimulate the immune system to reduce the adverse effects of CTX and to prevent homeostasis disbalance and cytotoxicity [18,20].

A bioactive peptide is a group of polypeptides that produce in vitro enzymatic hydrolysis and fermentation of various marine or food resources employing appropriate proteolytic enzymes [17,21], structurally contain more than 2–20 residues of amino acid, due to hydrophobic characteristics and small size, bioactive compounds are easier to absorb [19]. Marine species contain physiologically bioactive components for active pharmaceutical ingredients and are an important compound for human nutrition. Marine peptides, particularly, have attracted a lot of interest because of their potential for improving health and lowering illness risk [22–24]. Peptides may play a significant role in the intestine by regulating digestive enzymes, stabilizing the intestinal epithelial tight junctions, and modulating dietary absorption [25,26]. Peptide hydrolysate from marine sources is safe, physiologically beneficial, improves immunity, and performs a variety of functions, i.e., anticancer, antibacterial, antiviral, antihypertensive, antithrombotic, hormonal, and cholesterol-lowering effects [20,22,23,27]. It has been reported that shark-derived hydrolysate upregulates the cytokines production and protects intestinal epithelial [24]. Oyster peptides stimulate cytokines, modify gut microbiota, and ameliorate intestinal damage [15].

However, the effect of shrimp peptide hydrolysate (SPH) on intestinal microbiota and mucosal integrity is less studied and their functional activity is not known. Therefore, the current study investigates the effect of SPH on immunomodulation, intestinal integration, intestinal microbiota in cyclophosphamide induces immunosuppressed mice.

2. Results

2.1. Protein Concentration of SPH

The concentration of SPH was determined by the Bradford method. The hydrolysate concentration increases hourly, and chymotrypsin (1%) concentration gives maximum yields of protein hydrolysate (Figure S1). The tris-tricine SDS gel electrophoresis analysis of SPH revealed that these proteins were of low molecular mass peptides, i.e., <10 kDa shown in Figure S2.

2.2. Molecular Mass Distribution of SPH

The MALDI-TOF-MS analysis revealed seven peaks in SPH (Figure 1). The analyzed molecular mass distribution was 0–8000 m/z. Five high-intensity peaks (3482.9, 2511.1, 1969.7, 1181.8, 223.5) were identified at <4000 m/z as major peptides.

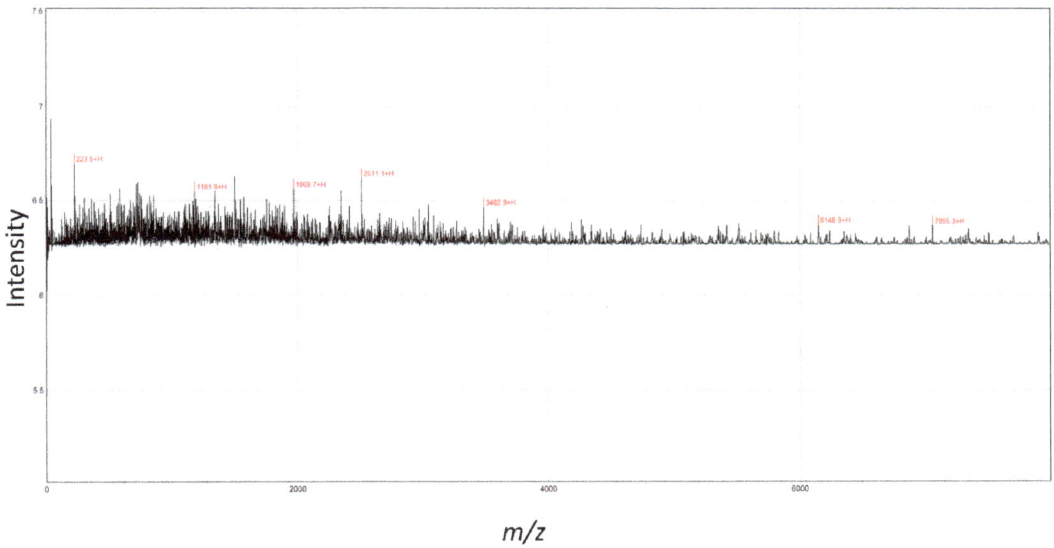

Figure 1. MALDI-TOF-MS analyses of peptides from shrimp protein hydrolysate. The mass spectrum was obtained from the peptide sample. Peaks with m/z values indicate the identified peaks.

2.3. SPH Ameliorative Effect on Immunosuppressed BALB/c Mice

CTX administration was significantly decreased food and water intake, bodyweight in the Model-CTX group, as shown in Figure 2a–c. In SPH treatment groups, mice had significantly improved body weight and food/water intake as compared with the model group. The spleen and thymus indices of the control group were significantly higher ($p < 0.05$) than the model group. Spleen and thymus indices of SPH treatment groups were significant compared with the model group ($p < 0.05$) while no significant differences were noted between the SPH treatment groups and the control groups (Figure 2d,e).

2.4. SPH Effect in Immunosuppressed Mice on Cytokine Levels

The serum concentration of cytokines, i.e., TNF-α, IFN-γ, IL-1β, IL-6, and immunoglobulins (IgA, IgM) in the model-CTX group were significantly lower ($p < 0.05$) than the control group. Serum cytokine (TNF-α, IFN-γ, IL-1β, IL-6) and immunoglobulins (IgA and IgM) concentration was significantly increased at post-treatment of SPH (Figure 3a–f). Colonic tissue mRNA expression of Mucin-2, Occludin, Zo-1, and claudin-1 was decreased in the Model-CTX group while SPH treatment groups showed a significant increase (Figure 4a–d).

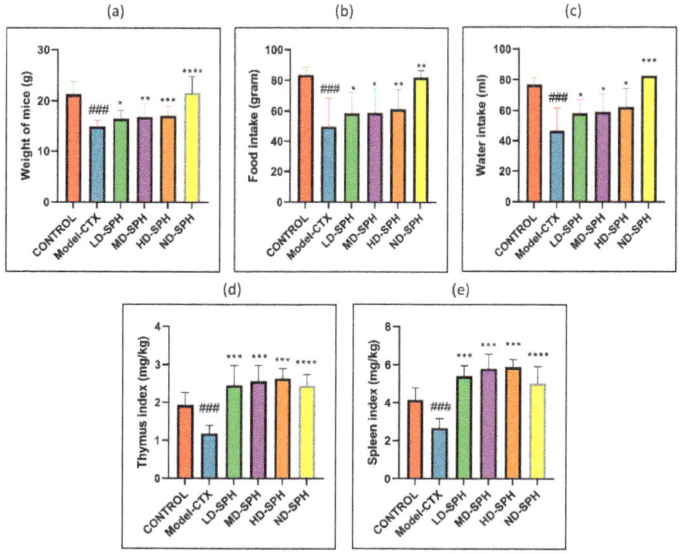

Figure 2. SPH effect on body weight, food and water intake, immune organ indices, and on proliferation response of splenocyte in CTX treated mice (6 groups, $n = 10$.) (**a**) Bodyweight. (**b**) Water intake. (**c**) Food Intake. (**d**) Spleen index. (**e**) Thymus index. ### $p < 0.001$ comparison to the Normal control group. * $p < 0.05$, ** $p < 0.01$, *** $p < 0.001$, and **** $p < 0.0001$, comparison to the Model-CTX group.

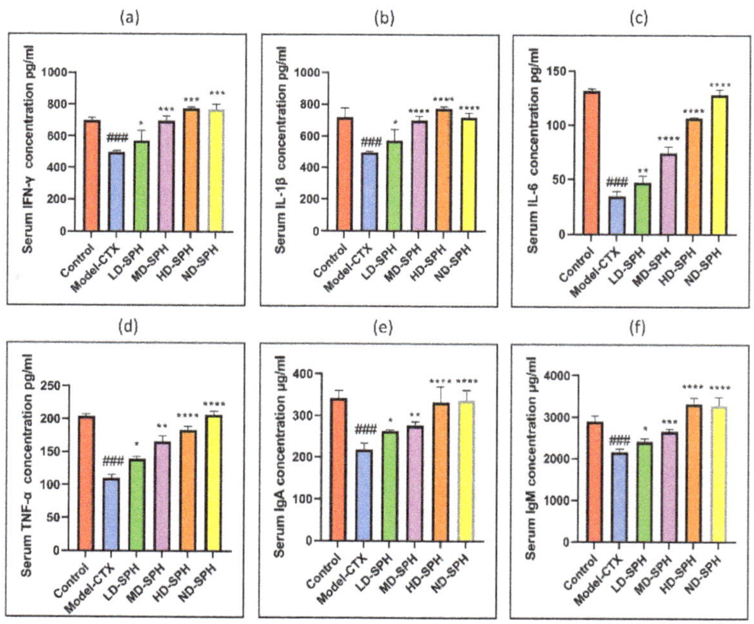

Figure 3. Effects of SPH on serum cytokines (**a**) IFN-γ, (**b**) IL-1β, (**c**) IL-6, (**d**) TNF-α, and immunoglobulin concentration (**e**) IgA, and (**f**) IgM. ### $p < 0.001$ comparison to the Normal control group. * $p < 0.05$, ** $p < 0.01$, *** $p < 0.001$, and **** $p < 0.0001$, comparison to the Model-CTX group.

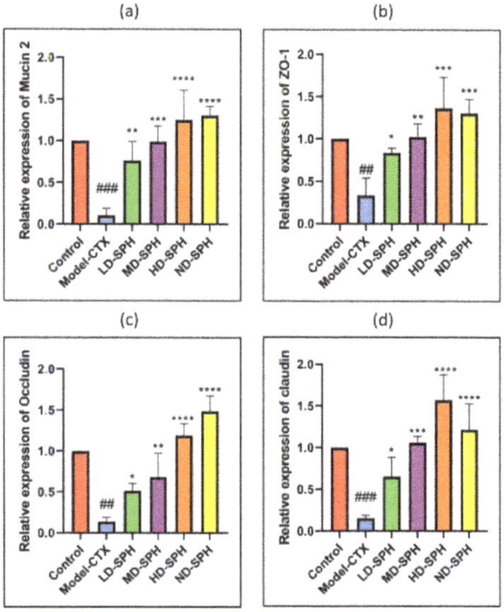

Figure 4. Relative expression of mRNA in colon tissues: (**a**) Mucin-2, (**b**) Zo-1, (**c**) Occludin, and (**d**) claudin-1, mRNA levels were standardized against Beta-actin expression, and as shown the mean ± SD fold increase compared with the control group. ### $p < 0.001$ comparison to the Normal control group. * $p < 0.05$, ** $p < 0.01$, *** $p < 0.001$, and **** $p < 0.0001$, comparison to the Model-CTX group.

2.5. Effects of SPH on Mouse Spleen, Thymus, and Colon Histomorphology

The histomorphological changes in the spleen and thymus as well as the effects of SPH are shown in Figure 5. The morphology of the spleen's white and red pulp was observed normal through H&E staining in the Normal control group. The lymphoid nodules were clear, with lymphocytes clustered together. The spleens in the Model CTX group had a disrupted structure, with an unclear margin between the white and red pulps. Furthermore, there is an ambiguous border between the medulla and the thymic cortex. The thymocyte estimation was decreased in the Model CTX group. While the SPH treatment groups (LD.SPH, MD.SPH, HD.SPH, and ND.SPH) have had similar results as a control group, with tightly organized thymocytes comprising a clear nucleus and fewer intracellular spaces in a dose-dependent manner. Histomorphological findings showed that SPH reserved the induced CTX impairment in the thymus and spleen in a dose-dependent manner.

The histomorphology of the colon shows that permeability increased after the administration of CTX, villi breakage, elongation of shallow crypts, loss in goblet cells as shown in Figure 6. The goblet cells proportion was evaluated in the colon after cyclophosphamide treatment (Figure 7a,b). The Model-CTX had a significantly lower relative number of goblet cells when compared with the control group. Periodic-acid Schiff (PAS) was utilized to stain neutral glycoprotein or mucus (purple-stained) containing goblet cells in the colon tissues. PAS and Alcian blue periodic-acid Schiff (AB-PAS) staining revealed that the glycoprotein content was significantly decreased in the Model-CTX group compared with the control group. Moreover, SPH treatment increased and recovered the enormous amount of goblet cells and mucin expression in LD.SPH, MD.SPH, HD.SPH, and ND.SPH in a dose-dependent manner. The expression of tight junction proteins in the colonic epithelium is affected by morphological changes. To further understand the impact of CTX on a mucosal barrier, Mucin2, Occludin, Claudin, and ZO-1, three of the most well-studied tight junction proteins were analyzed. Treatment with CTX reduced the expression of Mucin2, Occludin, Claudin, and ZO-1, while SPH treatment improved the interstitial barrier and increased the expression of tight junction protein as depicted in Figures 8 and 9.

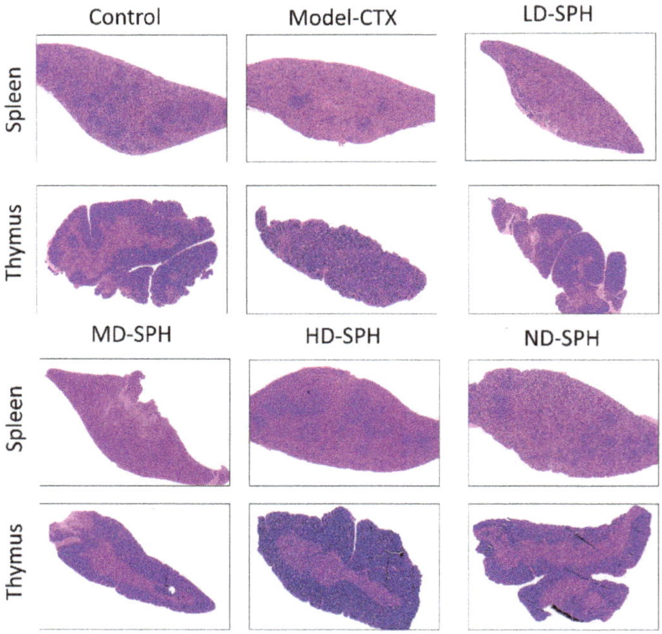

Figure 5. SPH impacts on spleen and thymus tissues histology. Magnification 10×, Scale bar: 100 μm.

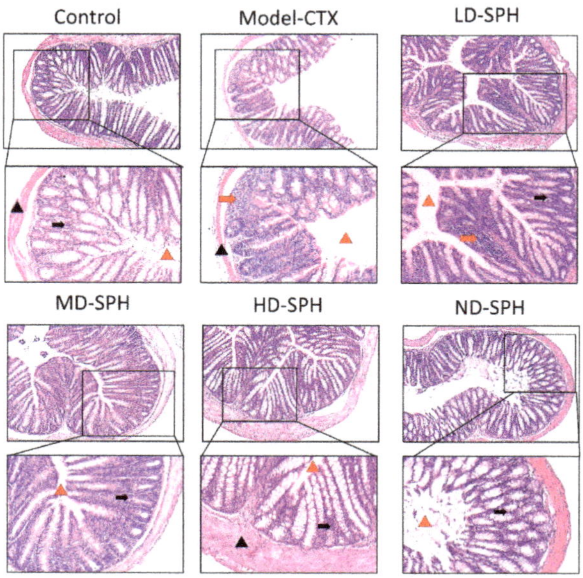

Figure 6. Effect of SPH on colon histologic modifications in mice with dysregulation caused by CTX. Colon sections stained with hematoxylin and eosin (H&E). Inflammatory cells (represented by red arrows), the mucosal space (represented by a red arrowhead), epithelial cells (represented by black arrows), and the epithelial surface (represented by black arrowhead). Magnification 20×, scale bar: 100 μm.

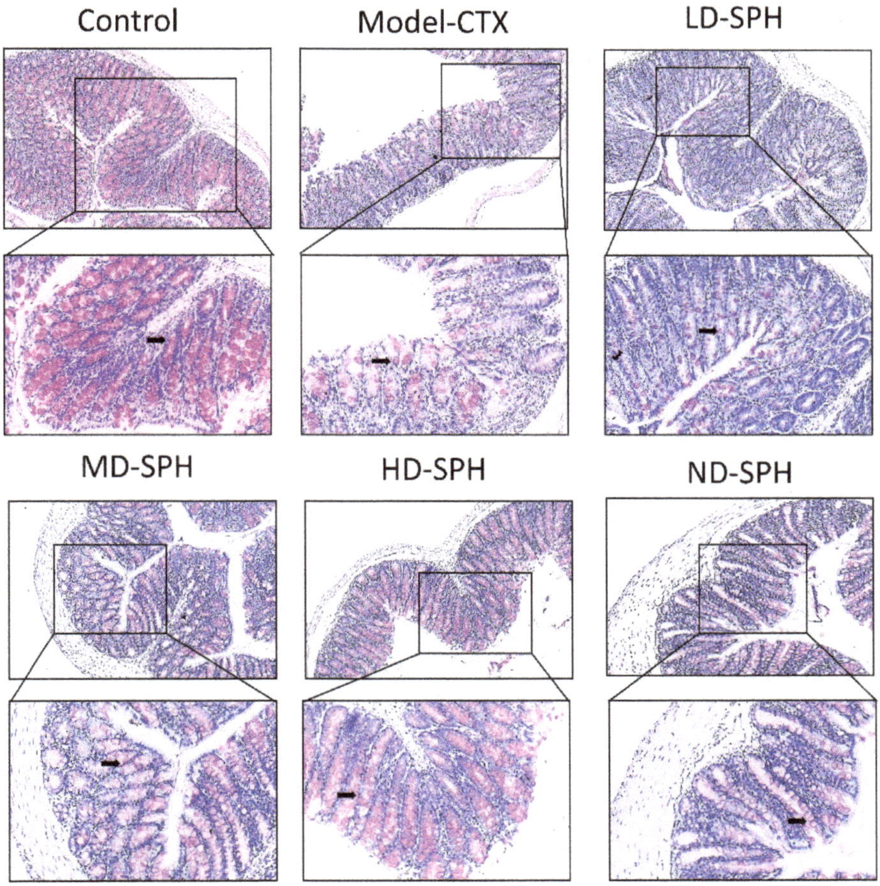

(a)

Figure 7. *Cont.*

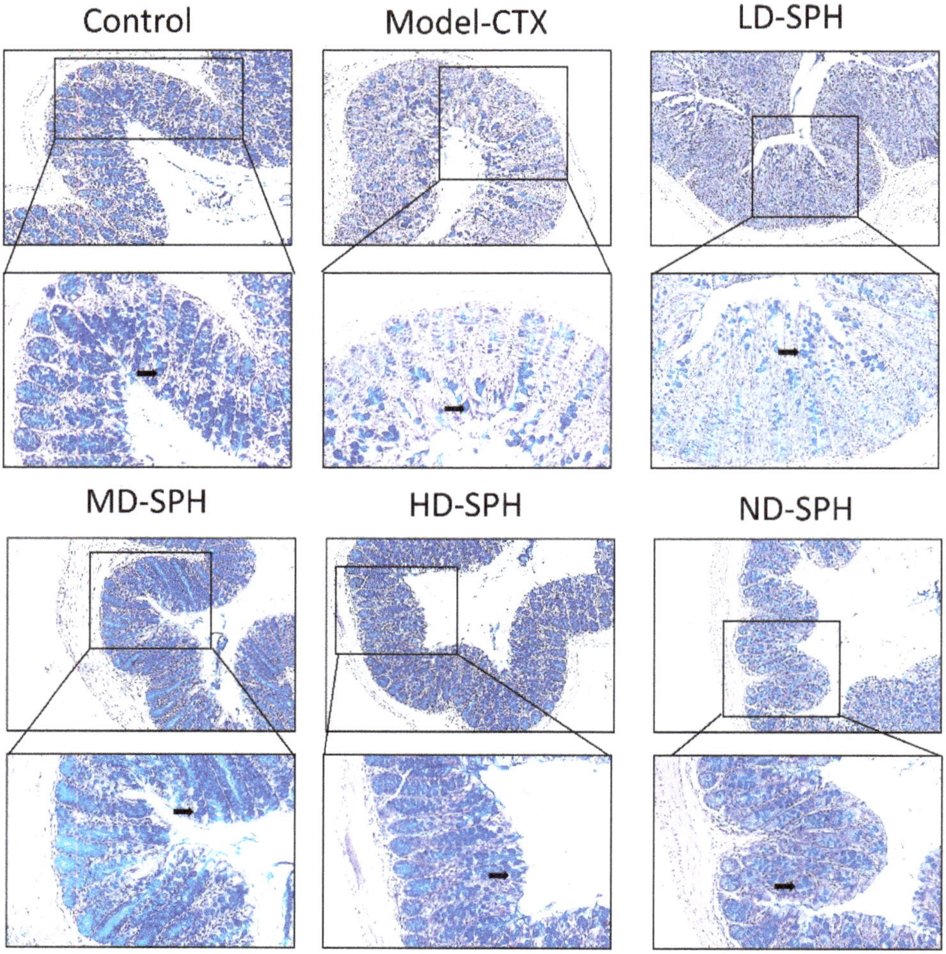

(**b**)

Figure 7. SPH treatment replenished goblet cells and enhanced mucin secretion in CTX-induced mice. (**a**). PAS staining (**b**). Images of colon sections are represented by AB-PAS staining. In each group, the number of goblet cells and mucin production were measured. Control mice revealed plenty of goblet cells with higher mucin expression (black arrows). In the CTX group, goblet cells were depleted completely. In SPH treated groups the number of goblet cells improved and the mucin expression is enhanced particularly at a higher dose. Magnification (upper 100×) (lower 200×).

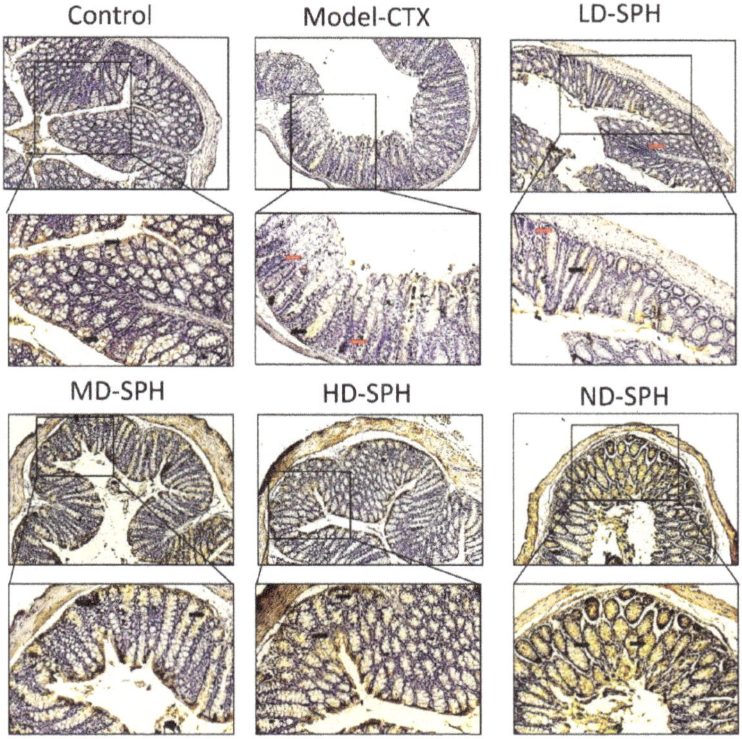

Figure 8. Mucin-2 immunohistochemistry staining in the colon of various groups Mucin expression (demonstrated by black arrows) and inflammatory cells are shown (demonstrated by red arrows), Original magnification 10×, 20×, scale bar: 100 μm.

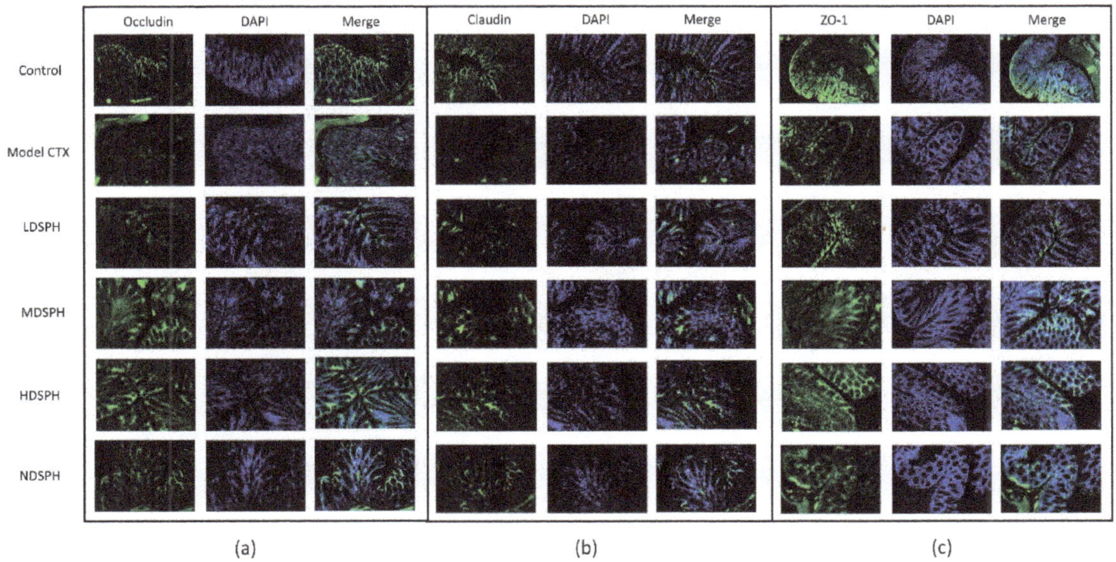

Figure 9. Immunofluorescent staining of tight junction proteins in the colonic epithelium (**a**) Occludin (**b**) Claudin (**c**) ZO-1. Original magnification 20×, scale bar: 100 μm.

2.6. SPH Modulates the Gut Microbiota Composition and Metabolic Functions Profile

The pyrosequencing examination of 16S rRNA was used to analyze after splicing and filtering dataset of good coverage clean 3,789,943 tags was obtained from fecal samples of all groups. The data was statistically analyzed after operational taxonomic units (OTUs) were clustered with a 97 percent similarity level. The taxonomic abundance of the intestinal microbiota was investigated at the phylum and genus level for the specific changes induced by CTX and SPH in all groups. *Bacteroidetes* and *Firmicutes* were the largest bacterial group at the phylum level (90%) as shown in Figure 10a. The next most prevalent phyla were *Proteobacteria* and *Actinobacteria.* In comparison to the control group, the CTX group demonstrated a rise in *Firmicutes* (50.90%), as well as a decrease in *Bacteroidetes* (45.70%). The CTX-induced dysbiosis in the gut microbiota was partially reversed after treatment with SPH shown In Table 1.

Bacteria at the taxonomic level of class and family were observed (Figure 10b,c). In comparison to the normal control group, the CTX group had an abundance of class; *Bacilli* (34.7%), *Coriobacteriia* (1.1%), and *Erysipelotrichi* (0.30%), while in the normal group, the abundance of class; *Bacteroides* (55.4%) and *Clostridia* (18.4%) are greater than CTX group. The abundance of fecal microbiota at family level in normal group *S24-7*(34.1%), *Prevotellaceae* (8.6%), *Lachnospiraceae* (4.9%), *Rikenellaceae* (3.5%), and in the CTX group, the greater level of *Bacteroidaceae* (7.5%), *Paraprevotellaceae* (5.7%), *Ruminococcaceae* (3.2%), *Lactobacillaceae* (34.7%), on the other hand, the CTX-induced changes were partially restored by SPH treatment. Next, we evaluated the relative abundance of bacterial genera through Heatmap; (Figure 10e) *Lactobacillus* (CTX 27.04 % vs. Control 33.83%), *Bacteroides* (CTX 5.65% vs. Control 1.6%), *Prevotella* (CTX 8.20% vs. Control 3.40%), after HD.SPH, MD.SPH, and LD.SPH in all treatment groups effectively restored the proportion of all affected phyla, demonstrating the clinical importance of SPH in the restoration of gut dysbiosis.

The community richness and diversity of each treatment group were described through a rank abundance and a rarefaction curve. The species abundance and evenness in, respectively, all treatment groups were explicitly represented by the rank abundance curve (Figure 10d), where the width of the curve and the horizontal direction represented the species richness and abundance, separately. The species richness is high in the normal and all treatment, groups compared with the CTX group. Alpha diversity reflects the richness and variation of the intestinal bacteria. The Chao1 and Ace indices are frequently used to evaluate the percentage of bacterial richness in various groups, whereas the Simpson and Shannon indices are used to assess the diversity of intestinal flora. There is no significant difference in bacterial richness and alpha diversity in separate groups analyzed with Chao1 and Ace indices, and the Shannon and Simpson indices. In Figure 10f, the rarefaction curve shows species diversity and abundance that reflects the analysis of several sequences. Shannon, Simpson, and the observed species show higher diversity and richness in the control and all treatment groups as compared to the CTX group.

The beta-diversity analysis revealed the dissimilarity or similarity of samples in species compositions shown in Figure 10g. The OTU abundance data and the Bray–Curtis inter-sample distance matrix were used for principal coordinates analysis (PCoA), and perform principal component analysis (PCA), shows apparent intestinal microbiota clustering for each group. Several parameters, including Pearson, unweighted UniFrac, weighted UniFrac, Bray–Curtis, and permutational multivariate analysis of variance (PERMANOVA), are used to measure distance and compare differences between groups in beta diversity. According to distance matrices, the intestinal microbiota of CTX-treated mice differs in species composition and clusters far from the NC group, whereas the microbiota of treatment groups is much similar, and clusters close to the NC group.

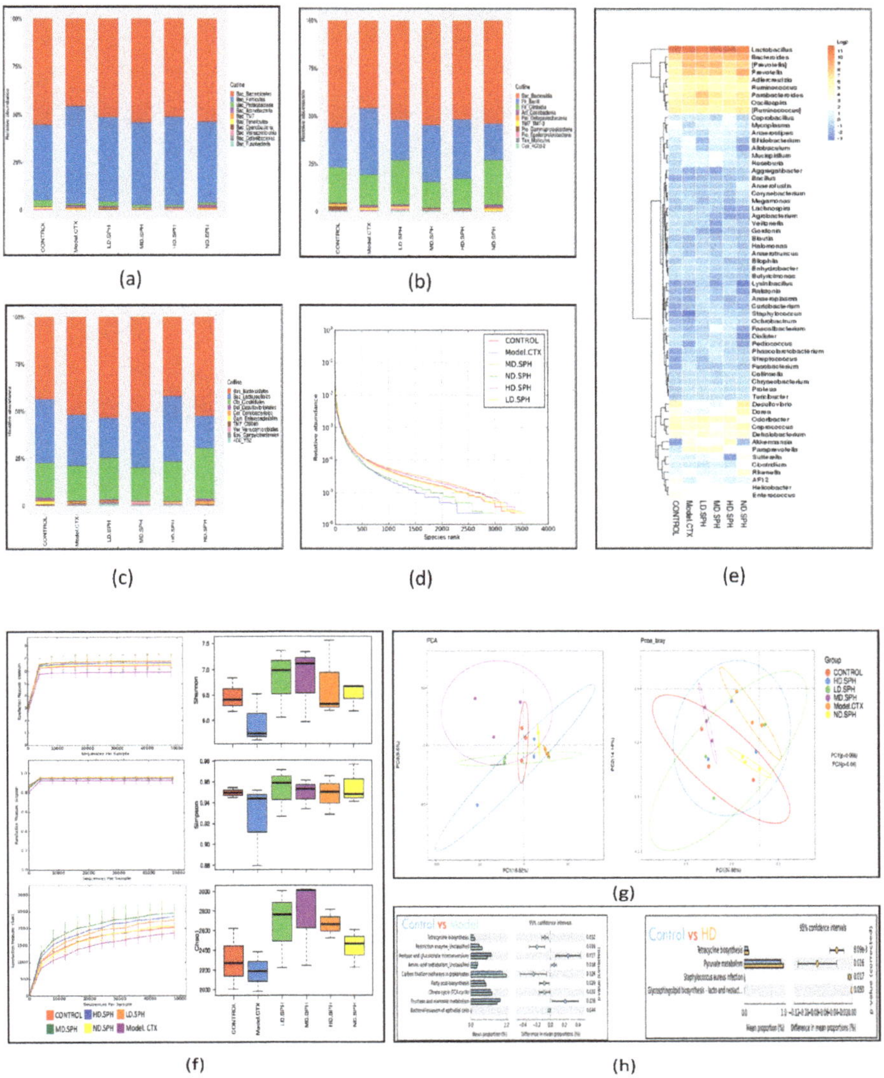

Figure 10. SPH's impacts on the gastrointestinal microbiota at the molecular basis. (**a**–**c**) Individual samples at the taxonomic levels of phylum, class, and family. (**d**) A rank abundance curve depicts the richness and evenness of the species. The x- and y-axes depict the number of operational taxonomic units based on the distribution, abundance, and abundance of each species (OTUs). (**e**) Hierarchical clustering of gut microbiome via heat map analysis of the highly characterized bacterial taxa genus level. (**f**) Rarefaction curves depicting species diversity and abundance. For each treatment group, the Shannon, Simpson, choa1 alpha coefficient depicts the microbial richness and diversity. (**g**) The proportion of the mice stool specimens predicated on the evolutionary appearance of their intestinal bacteria is shown in a principal coordinate analysis (PCoA) plot with Bray–Curtis dissimilarity. Every specimen is depicted by a dot in the graph, with various colors representing a diverse experimental group (n = 3). The deviation between the observations is represented by the range among each other; the nearer the specimens are, the more differently compared they are. (**h**) Analysis of functional genes associated pathways in the control, Model-CTX, and treated groups.

Table 1. Different treatment groups' percentages of bacterial phyla.

	Control	HD.SPH	LD.SPH	MD.SPH	Model. CTX	ND.SPH
p__Bacteroidetes	55.83%	51.47%	51.83%	54.47%	45.95%	54.12%
p__Firmicutes	38.95%	46.06%	43.78%	42.98%	50.72%	42.23%
p__Actinobacteria	0.21%	0.76%	0.79%	0.34%	1.13%	1.54%
p__Proteobacteria	3.22%	0.64%	2.21%	1.10%	1.03%	1.05%
p__TM7	0.26%	0.26%	0.41%	0.12%	0.80%	0.71%
p__Verrucomicrobia	0.61%	0.20%	0.09%	0.03%	0.02%	0.20%
p__Tenericutes	0.65%	0.19%	0.12%	0.34%	0.22%	0.10%
p__Cyanobacteria	0.04%	0.13%	0.49%	0.35%	0.05%	0.02%
p__Deferribacteres	0.11%	0.01%	0.05%	0.08%	0.04%	0.00%
p__Fusobacteria	0.00%	0.00%	0.00%	0.00%	0.00%	0.00%

PICRUSt 1.1.0 predicts microbial ecological functions based on 16S sequences against the spices in the Greengenes database using ortholog, Cog, and KEGG pathways. Statistical Analysis of Metagenomic Profile (STAMP) (version 2.1.3) was used to further assess outcomes. In CTX treated mice and control group 6 KEEG pathways show a significant variation detected in the pairwise abundance comparisons (Figure 10h). The most enriched metabolic pathways among these were replication, repair and recombination proteins, and restriction enzymes. Bacterial invasion of epithelial cells, pyruvate metabolism, ribosome biogenesis, tetracycline biosynthesis, carbon fixation pathways in prokaryotes, TCA cycle, fatty acid biosynthesis, energy production and depleted (Carbohydrate transport and metabolism. Amino acid metabolism, pentose and glucuronate interconversions, fructose and mannose metabolism, pentose phosphate pathway, RNA processing, modification, etc.) alleyways related to those of the NC group. KEGG pathways were altered in different groups, proving that SPH therapy can enhance immunity by balancing the gut microbiota's metabolism.

3. Discussion

The intestine is the largest digestive, absorbing, and immune organ, and gut intestinal microbiota and immunity continuously interact to enhance immune intestinal integrity and homeostasis [14,28]. Gut immune homeostasis is composed of epithelium, lamina propria, and Peyer's patches, which are the first barrier against damages [14,15]. The imbalance in intestinal/gut homeostasis leads to dysbiosis of gut microbial communities, and hence, will trigger a variety of diseases and abnormal immune responses [29]. Cyclophosphamide is a potent immunosuppressive agent, and a widely used anticancer drug, which may cause complications by the disruption of the mucosal barrier, and immune function that reduces the intestinal tight junctions, adherent junctions and enhances the potentially pathogenic bacteria [30,31]. Therefore, in this study, the effect of SPH on CTX induced mice were investigated to evaluate the gut intestinal integrity, gut microbiota, and immunomodulation. Factors; enzyme, PH, time, temperature, and solid–liquid ratio affected peptide extraction concentration from marine recourses. The aforementioned factors were assessed for improved product quality [32,33]. The enzymatic hydrolysis was affected by chymotrypsin (1%) at 50 °C. The SPH extracts isolated from shrimps have identified five low molecular weight peptides with MADIL-TOF-MS. Previous studies have shown that peptides fractions < 6 kDa have better immunomodulatory activity [34].

The thymus and spleen are the key component of the immune organ in the immune system and have an imperative role in nonspecific immunity [35]. Immune cell proliferation, differentiation, and activation lead to an increase in the weight of the immune organs [36–39], and vice versa represents the decreased activity of immune functions [40–43]. After the administration of CTX, the body weight and immune organ index of mice was significantly decreased and at post-treatment, with SPH the mice's body weight, spleen, and thymus were significantly increased ($p < 0.05$) in all SPH treatment groups as compared with model group. Zhang and Wang reported similarly and stated that peptide hydrolysate increases body weight and organ index [40,44]. The spleen and thymus, which are crucial organs in the immunological response, have been reported to be damaged when CTX is administered [41]. The micromorphological observation shows that spleens displayed a disrupted structure in the CTX Model group, with an unclear margin between the white and red pulp, lymphocyte count decreased, and reticular cells increased. Thymus shows uncertain margin, low thymocyte. These changes are due to atrophy of the immune organ and as an indication of immunosuppression. After treatment with SPH, the found changes were deprived and the number of lymphocytes, thymocytes increased, the pulps' margin was clear, and spleen and thymus tissues were regular.

The second line of defense of the gut intestinal barrier is epithelial cells, which play a direct role in the gut's immunological maintenance. Epithelial cells play an important role not only in the response of pathogens but also in the transmission of the signal to the intestinal immune system through the release of cytokines and inflammatory mediators [45]. Cytokines are small soluble glycoproteins derived from a variety of cells that stimulate differentiation, cell division, and multiplication through plasma receptor molecules in interacting cells [42]. Cytokines are important mediators and regulators of the immune response, and their secretion level may reflect the body's immune function [15]. In a CTX-induced immunosuppressed model, a peptide from yak collagen hydrolysates significantly increased serum levels of cytokines (IL-2, TNF-α, IFN-γ), immunoglobin (IgA, IgG), and improved humoral and cellular immune responses [43]. CTX-treated mice showed a substantial reduction in serum cytokine, similar to prior observations [19]. After the administration of SPH, cytokines expression TNF- α, IFN-γ, IL-1β, IL-6, and immunoglobulin IgA, IgM was found to be elevated in our studies. In previous studies, oyster peptide stimulated cytokines such as IL-2, IFN-γ, IL-4, and IL-10 [15], and Nibea japonica peptide significantly increased cytokine secretion such as IL-2, IFN-γ, and TNF-α [46].

Intestinal epithelial cells keep normal intestinal permeability, intact structure, and protect against pathogens or hazardous chemicals [47]. Goblet cells and tight junction protein play a key role to maintain intact intestinal structural, intestinal barrier function, regulation of gut permeability, and maintaining the epithelial cell barrier. It produces mucus, which effectively helps to protect invading pathogens and hence, maintenance of gut health [48]. Moreover, goblet cells produce multiple mucins that play an important role to maintain the mucosal barrier and protect against microbes [48,49]. Claudin and occludin are transmembrane proteins that interact with the extracellular environment and connect adjacent cells, ZO-1 connects claudin and occludin which may help to keep tight junctions intact [50,51]. In the CTX model group staining results showed intestinal and mucosal barrier disruption, and a decrease in the goblet cells and tight junction proteins. The SPH restores the damaged mucosal integrity caused by CTX through restore goblet cell populations and increasing the production of tight junction proteins. These results were consistent with other studies treated with cyclophosphamide [15,28].

Nutraceuticals not only contribute to a better gut structure, but also improve the host's functional activities, such as absorption and immune system [52]. The feces contain the most microbiota along with the gastrointestinal system contents, would easy to collect, is far less disruptive. It presents alterations in gut microbiota associated with health and disease [31]. CTX has shown arbitrarily dysregulation of immune cells and intestinal mucosa cells, resulting in enteritis by increasing permeability, reducing the intestinal barrier immune system, and changing the microbial populations of the small intestine [21,53,54]. Intestinal flora (microbiota) composition was compared between the normal and CTX treatment group at the phylum show the abundance of *Firmicutes*, and less abundant *Bacteroidetes* in the CTX group, these results were consistent with previous studies [18,21,54,55]. In addition, immune function dysbiosis was observed in immunosuppressed mice with an abundance of *Bacteroides* [56]. The fecal microbiota of chemotherapeutic induced mice,

Bacilli, Erysipelotrichi, TM7-3, Coriobacteriia were increased while *Bacteroides, Clostridia, Deltaproteobacteria* were decreased, Xu. X and Gu. S. reported similarly [31,57]. The healthy gut microbiota of mice has a well-balanced composition of microorganisms of several groups patterns were connected to low-grade intestinal inflammation irrespectively of mouse strain [58,59]. These data demonstrated associations between CTX-induced immune responses and variations in bacterial abundance in diverse bacterial groupings.

The Chao1 and Ace indices metrics were assessed to compare the number of OTU in a sample; a high value provides higher biodiversity of beneficial gut flora. Chao 1 and Ace indices were the most common indicators to predict the amount of group abundance, whereas Shannon and Simpson's indexes were the most common indicators of group biodiversity [60]. Alpha diversity reduces microbial richness and diversity after administration of CTX and after the administration of SPH, the richness and diversity of the spices reversed and enhanced in the treatment group shown in Figure 10f. Beta diversity indices indicate the dissimilarities among different treatments which exhibit that the CTX group where deviated as compared to the control group, whereas the control group and SPH treated groups were closely clustered shows more similarities between groups. Previous studies also show that the richness and diversity of microbiota decrease after administration of CTX [15,31]. In our study, metagenomic functional analysis prediction revealed a difference between the control and CTX groups. The metabolic and cellular pathways are linked to the higher and lower abundance of certain genes' carbohydrate transport and metabolism, amino acid metabolism, fructose and mannose metabolism, pentose and glucuronate interconversions, RNA processing, and modification. Pentose phosphate pathways all have low abundance in the CTX group.

The absence of carbohydrate metabolism, as well as RNA processing and modification, has an effect on organism physiological processes, the host immune system, and nutrient absorption. Changes in carbohydrate metabolism have previously been connected to the composition and metabolism of gut-associated intestinal bacteria [53,61]. In our research, we observed that SPH may improve carbohydrate metabolism induced by the CTX. Ultimately, we revealed that SPH improved nutrient absorption and energy utilization while also maintaining gut integrity, intestinal flora composition, preventing pathogens and their metabolites, and remodeling gut metabolome roles in immunocompromised mice, implying that SPH may play an immunologic protective role.

4. Materials and Methods

The shrimps (*Penaeus chinensis*) were purchased from the supermarket (Lvshunkou. Loaning, Dalian, China). RIPA lysis buffer was bought from Beyotime Biotechnology (Shanghai, China). Chymotrypsin was purchased from Sigma-Aldrich Trading Co., Ltd., (Shanghai, China). A protein quantification assay kit (BCA) was Purchased from Jiancheng Bioengineering Institute (Nanjing, China). ELISA assay kits for TNF-α, IFN-γ, IL-1β, IL-6, IgA, IgM, Rabbit antibodies were supplied by Jiangsu Mmmiology biological Co., Ltd., (Jiangsu, China); others all reagents were of analytical grade.

4.1. *Preparation of SPH from Shrimp*

SPH was prepared using the hydrolysis process approach with a few tweaks, as discussed earlier. Whole shrimps were mixed in a grinder and then washed in the double volume of distilled water at 98 °C for one hour. After wash, the smashed shrimp residues were filtered through meshes (150 µm) and combined with two volumes of distilled water before being digested at 50 °C for six hours with a 1 percent (*w*/*w*) chymotrypsin enzyme. The enzymes activity was then halted for 15 min by heating to 98 degrees Celsius. The digested lysates were centrifuged for 20 min at 14,000 rpm to extract the supernatant, which was used to make shrimp hydrolysate. The Bradford technique was used to measure the SPH concentrations.

4.2. Molecular Mass Distribution of SPH

Shrimp peptide hydrolysate was evaluated for molecular Mass distribution using Proteomics Analyzer for matrix-assisted laser desorption ionization-time of flight mass spectrometry (MALDI-TOF-MS) (Bruker, Germany).

4.3. Animals and Experimental Design

Animal Laboratory and Safety Research provided a total of 60 disease-free BALB/c mice (3 to 4 weeks old) with a bodyweight of 18 ± 2 g. The experimental and Animal Ethics Committee of Dalian Medical University approved the animal studies, which followed national and institutional guidelines for experimental animal handling. After one week of acclimating to their new surroundings, the mice were randomly divided into six groups (n = 10) based on their body weight. Normal Control, (NC), CTX model group, High dose (HD.SPH 400 mg/kg), Medium dose (MD.SPH 200 mg/kg), Low dose (LD.SPH 100 mg/kg), and Normal Control + Dose (ND.SPH 200 mg/kg). The four groups of mice were administered with 80 mg/kg/day of cyclophosphamide (CTX) intraperitoneally for 5 days according to body weight, with an equal volume of PBS administered to the NC and ND groups. From the day 6 to 30, the mice in the four groups were given 100, 200, and 400 mg/kg/day of SPH by gavage. Meanwhile, the NC and Model groups are given the same amount of distilled water orally. The procedure of animal experiments is shown in Figure S3.

4.4. Determination of Body Weight, Food/Water Intake, and Immune Organ Index

During the administration period, body weight was recorded every day, and food and water intake were measured and recorded every third day. After the mice were forfeited by cervical displacement, the immune organs, i.e., spleen and thymus were collected and weighed immediately, and the immune organ index was calculated using a formula:

$$\text{Spleen or thymus index (mg/g)} = \text{weight of organ (mg)/weight of mouse (g)}.$$

4.5. Measurement of Serum Antibodies, Serum Cytokines Level

After the mice were sacrificed, the whole blood samples were collected and centrifuged at 4000× g for 5 min. The serum was transferred into a 1.5 mL tube and stored at −20 °C. Serum IgA, IgM, and cytokine concentrations (IFN-γ, IL-1β, TNF-α, IL-6) were analyzed through the ELISA method, according to the manufacturer's instructions.

4.6. Determination of Intestine mRNA

Muc-2, ZO-1, occludin, and claudin-1 mRNA expression levels were measured. Total RNA was extracted from colonic tissue using Vazyme Total RNA Extraction Reagent (Vazyme biotech Co., Ltd.) according to the manufacturer's instructions. The total RNA was stored at −80 °C. The complementary DNA was then transcribed using the commercial kit HiScript II Q RT SuperMix (Vazyme biotech Co., Ltd.). The ChamQ SYBR qPCR MasterMix kit was used to perform quantitative PCR in Bioer light gene 9600 analyzers (Hi-tech (Binjiang) District, Hangzhou, 310053, China). Table S1 shows the primer sequences. The real-time PCR was run at 95 °C for 5 min, then 40 cycles of 95 °C for 20 s, primer annealing temperatures at 60 °C for 30 s, and extension at 72 °C for 30 s. Each sample was tested three times, and the instrument software gene 9660 was used to calculate and analyze relative expression, as well as GraphPad prism 8 to analyze differences between groups.

4.7. Histopathological Examination of Thymus, Spleen, and Colon

Colon, spleen, and thymus were collected, fixed with 10% formalin, and processed. Thin sections of 5 μ were made by microtome, stained with hematoxylin and eosin staining. Microscopic examination for histological changes was observed under a microscope.

Immunohistochemistry was used to examine the levels of expression of Mucin-2 in the colon tissue. Then, 5 μm paraffin-embedded spleen tissue cut, placed into positive charge slide, and deparaffinized in xylene. Rehydrate in a series of ethanol, follow the protocol of immunohistochemical staining kits SP-9001 (Zhongshan Goldenbridge Biotechnology, Beijing, China) according to manufacturer instruction. Results were evaluated semiquantitative method each slide was observed for immunolabeled cells in five fields for three-time randomly.

Immunofluorescent staining was used to examine the levels of expression of occludin, ZO-1, and claudin in the colon. Then, 5 μm paraffin-embedded spleen tissue was cut, placed into positive charge slide, deparaffinized in xylene, and rehydrated in series of ethanol. Tissue sections were treated for 30 min in citrate buffer for antigen retrieval at 100 watts on microwave and cool down for 1 h. Then, tissue slides were added to Blocking with 3% BSA solution for 1 h. Tissues were incubated overnight at 4 °C against occludin, claudin, or ZO-1 (1:200) antibodies. After washing, tissue sections were incubated with Alexa 488-conjugated secondary antibodies for 60 min, and nucleus staining DAPI was used. Images were taken using a confocal scanning microscope.

4.8. Microbiota 16S rRNA Pyrosequencing

PowerMax (stool/soil) DNA isolation kit (MoBio Laboratories, Carlsbad, CA, USA) was used for total fecal microbial genomic DNA was extracted for all the samples and stored at −20 °C. DNA quality and quantity were measured using nanodrop. Microbial 16S rRNA gene V4 region PCR amplification was done by using forward primer 515F (5′-GTGCCAGCMGCCGCGGTAA-3′) and the reverse primer 806R (5′-GGACTACHVGGGTW TCTAAT-3′), and sequenced with Illumina NovaSeq6000 platform at GUHE Info technology Co., Ltd. (Hangzhou, China). QIIME software version 1.9 pipeline was used for sequence data processing, as defined before [54]. Sequencing low-quality through the following criteria [55,62] paired-end reads and operational taxonomic unit (OTU) clustering, detection of chimera's clusters, and including dereplication using default parameters were conducted by Vsearch V2.4.4. The taxonomic unit was assigned for each dataset using Greengenes database.

4.9. Evolutionary Computation and Statistical Analysis

The QIIME and R programs(v3.2.0) were being used to analyze sequence data. OUT level alpha diversity indices and Beta diversity analysis were calculated using the QIIME. The data were further analyzed for ecologically relevant function, prokaryotic clades, and high-level phenotype using Statistical Analysis of Metagenomic Profiles (STAMP) software package v2.1.3 [63], FAPROTAX [64], and BugBase tool [65].

GraphPad Prism software was used for statistical analysis (v6.04). *p*-values of less than 0.05 were deemed significant. The Kruskal–Wallis test from the R stats package was used to determine whether there was a statistically significant difference among the groups based on 16S rRNA data. Mann–Whitney test was used to compare OTU counts and phenotype.

5. Conclusions

In this study, SPH immune modulates the effects of CTX on the gut intestine of immunosuppressed. SPH significantly increased immune organs index, reinstated the goblet cells, intestinal mucosa integrity, enhanced the serum levels of cytokines (IL-2, IFN-γ, IL1β, TNF-α, IL-6) and IgA, IgM, and increased the mRNA of tight-junction associated proteins (occludin, ZO-1, claudin-1, and Muc-2). SPH would also improve dysbiosis and modified the intestinal gut microbiota ecology by reducing pathogenic bacteria at various taxonomic levels. Firmicutes/Bacteroidetes proportion was reduced in CTX-induced mice. SPH can be employed as a prebiotic source with a modulatory influence on gut microbial ecology as a possible health-promoting regulator of the gut microbiota.

Supplementary Materials: The following supporting information can be downloaded online, Figure S1: Effect of enzyme different chymotrypsin concentration on SPH concentration; Figure S2: Tris-tricine gel electrophoresis; Figure S3. The animal experimental procedure was used in this research study; Table S1: PCR primers used for qPCR.

Author Contributions: Conceptualization, A.I.K., Y.X. and L.W.; methodology, A.I.K., A.U.R., N.Z.S., N.A.F. and Q.A.; software, A.I.K. and M.N.R.; formal analysis, A.I.K., N.A.F., A.U.R. and M.N.R.; investigation, A.I.K., A.U.R., Q.A., N.A.F. and N.Z.S.; resources, Y.X., data curation, A.I.K.; writing—original draft preparation, A.I.K.; writing—review and editing, Y.X., L.W., A.I.K., A.U.R. and Q.A.; visualization, Y.X.; supervision, Y.X.; project administration, Y.X.; funding acquisition, Y.X. and L.W. All authors have read and agreed to the published version of the manuscript.

Funding: This research was supported by the National Nature Science Foundation of China (Grant No. 31600614, 82072953); Top young talents of Liaoning Provincial Government (Grant No. XLYC1907009); Dalian outstanding young scientific and technological talents (Grant No. 2021RJ12).

Institutional Review Board Statement: The animal investigations were permitted by the experimental and Animal Ethics Committee of Dalian Medical University.

Informed Consent Statement: Not applicable.

Data Availability Statement: The original data for this work are available upon email request to the corresponding author.

Acknowledgments: We would like to thank Zhang Xiao Xiao and Wang Zexu for their technical assistance.

Conflicts of Interest: The authors declare no conflict of interest.

References

1. Frémont, M.; Coomans, D.; Massart, S.; De Meirleir, K. High-throughput 16S rRNA gene sequencing reveals alterations of intestinal microbiota in myalgic encephalomyelitis/chronic fatigue syndrome patients. *Anaerobe* **2013**, *22*, 50–56. [CrossRef] [PubMed]
2. Delcenserie, V.; Martel, D.; Lamoureux, M.; Amiot, J.; Boutin, Y.; Roy, D. Immunomodulatory effects of probiotics in the intestinal tract. *Curr. Issues Mol. Biol.* **2008**, *10*, 37–54. [PubMed]
3. De Santis, S.; Cavalcanti, E.; Mastronardi, M.; Jirillo, E.; Chieppa, M. Nutritional keys for intestinal barrier modulation. *Front. Immunol.* **2015**, *6*, 612. [CrossRef] [PubMed]
4. Chung, W.S.F.; Walker, A.W.; Louis, P.; Parkhill, J.; Vermeiren, J.; Bosscher, D.; Duncan, S.H.; Flint, H.J. Modulation of the human gut microbiota by dietary fibres occurs at the species level. *BMC Biol.* **2016**, *14*, 1–13. [CrossRef]
5. Sommer, F.; Bäckhed, F. The gut microbiota engages different signaling pathways to induce Duox2 expression in the ileum and colon epithelium. *Mucosal Immunol.* **2015**, *8*, 372–379. [CrossRef]
6. Viggiano, D.; Ianiro, G.; Vanella, G.; Bibbò, S.; Bruno, G.; Simeone, G.; Mele, G. Gut barrier in health and disease: Focus on childhood. *Eur. Rev. Med. Pharmacol. Sci.* **2015**, *19*, 1077–1085.
7. Zhang, L.; Cao, W.; Gao, Y.; Yang, R.; Zhang, X.; Xu, J.; Tang, Q. Astaxanthin (ATX) enhances the intestinal mucosal functions in immunodeficient mice. *Food Funct.* **2020**, *11*, 3371–3381. [CrossRef]
8. Kurashima, Y.; Goto, Y.; Kiyono, H. Mucosal innate immune cells regulate both gut homeostasis and intestinal inflammation. *Eur. J. Immunol.* **2013**, *43*, 3108–3115. [CrossRef]
9. Hansson, G.C. Role of mucus layers in gut infection and inflammation. *Curr. Opin. Microbiol.* **2012**, *15*, 57–62. [CrossRef]
10. Sommer, F.; Bäckhed, F. The gut microbiota—masters of host development and physiology. *Nat. Rev. Microbiol.* **2013**, *11*, 227–238. [CrossRef]
11. Kelly, C.J.; Zheng, L.; Campbell, E.L.; Saeedi, B.; Scholz, C.C.; Bayless, A.J.; Wilson, K.E.; Glover, L.E.; Kominsky, D.J.; Magnuson, A. Crosstalk between microbiota-derived short-chain fatty acids and intestinal epithelial HIF augments tissue barrier function. *Cell Host Microbe* **2015**, *17*, 662–671. [CrossRef] [PubMed]
12. Gamallat, Y.; Ren, X.; Meyiah, A.; Li, M.; Ren, X.; Jamalat, Y.; Song, S.; Xie, L.; Ahmad, B.; Shopit, A. The immune-modulation and gut microbiome structure modification associated with long-term dietary supplementation of Lactobacillus rhamnosus using 16S rRNA sequencing analysis. *J. Funct. Foods* **2019**, *53*, 227–236. [CrossRef]
13. Dwivedy, A.; Aich, P. Importance of innate mucosal immunity and the promises it holds. *Int. J. Gen. Med.* **2011**, *4*, 299. [PubMed]
14. Shi, N.; Li, N.; Duan, X.; Niu, H. Interaction between the gut microbiome and mucosal immune system. *Mil. Med. Res.* **2017**, *4*, 1–7. [CrossRef]
15. Xiang, X.-W.; Zheng, H.-Z.; Wang, R.; Chen, H.; Xiao, J.-X.; Zheng, B.; Liu, S.-L.; Ding, Y.-T. Ameliorative Effects of Peptides Derived from Oyster (Crassostrea gigas) on Immunomodulatory Function and Gut Microbiota Structure in Cyclophosphamide-Treated Mice. *Mar. Drugs* **2021**, *19*, 456. [CrossRef]
16. Yoo, H.J.; You, D.-J.; Lee, K.-W. Characterization and immunomodulatory effects of high molecular weight fucoidan fraction from the sporophyll of Undaria pinnatifida in cyclophosphamide-induced immunosuppressed mice. *Mar. Drugs* **2019**, *17*, 447. [CrossRef]
17. Shams ul Hassan, S.; Ishaq, M.; Zhang, W.-D.; Jin, H.-Z. An Overview of the Mechanisms of Marine Fungi-Derived Anti-Inflammatory and Anti-Tumor Agents and their Novel Role in Drug Targeting. *Curr. Pharm. Des.* **2021**, *27*, 2605–2614. [CrossRef]

18. Prendergast, G.C.; Jaffee, E.M. Cancer immunologists and cancer biologists: Why we didn't talk then but need to now. *Cancer Res.* **2007**, *67*, 3500–3504. [CrossRef]
19. Hartmann, R.; Meisel, H. Food-derived peptides with biological activity: From research to food applications. *Curr. Opin. Biotechnol.* **2007**, *18*, 163–169. [CrossRef]
20. Kiewiet, M.B.; Faas, M.M.; De Vos, P. Immunomodulatory protein hydrolysates and their application. *Nutrients* **2018**, *10*, 904. [CrossRef]
21. Ngo, D.-H.; Vo, T.-S.; Ngo, D.-N.; Wijesekara, I.; Kim, S.-K. Biological activities and potential health benefits of bioactive peptides derived from marine organisms. *Int. J. Biol. Macromol.* **2012**, *51*, 378–383. [CrossRef] [PubMed]
22. Chen, C.; Su, X.; Hu, Z. Immune promotive effect of bioactive peptides may be mediated by regulating the expression of SOCS1/miR-155. *Exp. Ther. Med.* **2019**, *18*, 1850–1862. [CrossRef] [PubMed]
23. Shah, S.A.A.; Bungau, S.; Si, Y.; Xu, H.; Rahman, M.; Behl, T.; Gitea, D.; Pavel, F.-M.; Corb Aron, R.A.; Pasca, B. Chemically diverse and biologically active secondary metabolites from marine Phylum chlorophyta. *Mar. Drugs* **2020**, *18*, 493. [CrossRef] [PubMed]
24. Mallet, J.-F.; Duarte, J.; Vinderola, G.; Anguenot, R.; Beaulieu, M.; Matar, C. The immunopotentiating effects of shark-derived protein hydrolysate. *Nutrition* **2014**, *30*, 706–712. [CrossRef] [PubMed]
25. Shimizu, M. Food-derived peptides and intestinal functions. *Biofactors* **2004**, *21*, 43–47. [CrossRef] [PubMed]
26. Korhonen, H.; Pihlanto, A. Bioactive peptides: Production and functionality. *Int. Dairy J.* **2006**, *16*, 945–960. [CrossRef]
27. Anjum, K.; Abbas, S.Q.; Akhter, N.; Shagufta, B.I.; Shah, S.A.A.; Hassan, S.S.U. Emerging biopharmaceuticals from bioactive peptides derived from marine organisms. *Chem. Biol. Drug Des.* **2017**, *90*, 12–30. [CrossRef]
28. Ying, M.; Zheng, B.; Yu, Q.; Hou, K.; Wang, H.; Zhao, M.; Chen, Y.; Xie, J.; Nie, S.; Xie, M. Ganoderma atrum polysaccharide ameliorates intestinal mucosal dysfunction associated with autophagy in immunosuppressed mice. *Food Chem. Toxicol.* **2020**, *138*, 111244. [CrossRef]
29. Berg, D.; Clemente, J.C.; Colombel, J.-F. Can inflammatory bowel disease be permanently treated with short-term interventions on the microbiome? *Expert Rev. Gastroenterol. Hepatol.* **2015**, *9*, 781–795. [CrossRef]
30. Brodsky, R.A. High-dose cyclophosphamide for autoimmunity and alloimmunity. *Immunol. Res.* **2010**, *47*, 179–184. [CrossRef]
31. Xu, X.; Zhang, X. Effects of cyclophosphamide on immune system and gut microbiota in mice. *Microbiol. Res.* **2015**, *171*, 97–106. [CrossRef] [PubMed]
32. Lee, J.K.; Jeon, J.-K.; Byun, H.-G. Antihypertensive effect of novel angiotensin I converting enzyme inhibitory peptide from chum salmon (Oncorhynchus keta) skin in spontaneously hypertensive rats. *J. Funct. Foods* **2014**, *7*, 381–389. [CrossRef]
33. Ahn, G.; Hwang, I.; Park, E.; Kim, J.; Jeon, Y.-J.; Lee, J.; Park, J.W.; Jee, Y.J.M.B. Immunomodulatory effects of an enzymatic extract from Ecklonia cava on murine splenocytes. *Mar. Biotechnol.* **2008**, *10*, 278–289. [CrossRef] [PubMed]
34. Feng, X.-L.; Liu, Q.-T.; Cao, R.-B.; Zhou, B.; Zhang, Y.-P.; Liu, K.; Liu, X.-D.; Wei, J.-C.; Li, X.-F.; Chen, P.-Y. Characterization and immunomodulatory function comparison of various bursal-derived peptides isolated from the humoral central immune organ. *Peptides* **2012**, *33*, 258–264. [CrossRef] [PubMed]
35. Chandrashekar, P.M.; Venkatesh, Y.P.J.I. Fructans from aged garlic extract produce a delayed immunoadjuvant response to ovalbumin antigen in BALB/c mice. *Immunopharmacol. Immunotoxicol.* **2012**, *34*, 174–180. [CrossRef] [PubMed]
36. Kraus, M.D. Splenic histology and histopathology: An update. *Semin. Diagn. Pathol.* **2003**, 84–93. [CrossRef]
37. Haque, R.; Bin-Hafeez, B.; Ahmad, I.; Parvez, S.; Pandey, S.; Raisuddin, S. Protective effects of Emblica officinalis Gaertn. in cyclophosphamide-treated mice. *Hum. Exp. Toxicol.* **2001**, *20*, 643–650. [CrossRef]
38. Zhou, Y.; Chen, X.; Yi, R.; Li, G.; Sun, P.; Qian, Y.; Zhao, X. Immunomodulatory effect of tremella polysaccharides against cyclophosphamide-induced immunosuppression in mice. *Molecules* **2018**, *23*, 239. [CrossRef]
39. Liu, L.-Q.; Li, H.-S.; Nie, S.-P.; Shen, M.-Y.; Hu, J.-L.; Xie, M.-Y. Tea polysaccharide prevents colitis-associated carcinogenesis in mice by inhibiting the proliferation and invasion of tumor cells. *Int. J. Mol. Sci.* **2018**, *19*, 506. [CrossRef]
40. Wang, Y.-K.; He, H.-L.; Wang, G.-F.; Wu, H.; Zhou, B.-C.; Chen, X.-L.; Zhang, Y.-Z. Oyster (Crassostrea gigas) hydrolysates produced on a plant scale have antitumor activity and immunostimulating effects in BALB/c mice. *Mar. Drugs* **2010**, *8*, 255–268. [CrossRef]
41. Li, W.-J.; Li, L.; Zhen, W.-Y.; Wang, L.-F.; Pan, M.; Lv, J.-Q.; Wang, F.; Yao, Y.-F.; Nie, S.-P.; Xie, M.-Y. Ganoderma atrum polysaccharide ameliorates ROS generation and apoptosis in spleen and thymus of immunosuppressed mice. *Food Chem. Toxicol.* **2017**, *99*, 199–208. [CrossRef]
42. Yu, J.; Cong, L.; Wang, C.; Li, H.; Zhang, C.; Guan, X.; Liu, P.; Xie, Y.; Chen, J.; Sun, J. Immunomodulatory effect of Schisandra polysaccharides in cyclophosphamide-induced immunocompromised mice. *Exp. Ther. Med.* **2018**, *15*, 4755–4762. [CrossRef] [PubMed]
43. Gao, S.; Hong, H.; Zhang, C.; Wang, K.; Zhang, B.; Han, Q.-A.; Liu, H.; Luo, Y. Immunomodulatory effects of collagen hydrolysates from yak (Bos grunniens) bone on cyclophosphamide-induced immunosuppression in BALB/c mice. *J. Funct. Foods* **2019**, *60*, 103420. [CrossRef]
44. Zhang, Z.; Hu, X.; Lin, L.; Ding, G.; Yu, F. Immunomodulatory activity of low molecular-weight peptides from Nibea japonica in RAW264. 7 cells via NF-κB pathway. *Mar. Drugs* **2019**, *17*, 404. [CrossRef]
45. Kuhn, K.A.; Pedraza, I.; Demoruelle, M.K. Mucosal immune responses to microbiota in the development of autoimmune disease. *Rheum. Dis. Clin.* **2014**, *40*, 711–725. [CrossRef] [PubMed]

46. Yu, F.; He, K.; Dong, X.; Zhang, Z.; Wang, F.; Tang, Y.; Chen, Y.; Ding, G. Immunomodulatory activity of low molecular-weight peptides from Nibea japonica skin in cyclophosphamide-induced immunosuppressed mice. *J. Funct. Foods* **2020**, *68*, 103888. [CrossRef]
47. Lu, J.T.; Xu, A.T.; Shen, J.; Ran, Z.H. Crosstalk between intestinal epithelial cell and adaptive immune cell in intestinal mucosal immunity. *J. Gastroenterol. Hepatol.* **2017**, *32*, 975–980. [CrossRef] [PubMed]
48. Maidana, L.; Gerez, J.R.; El Khoury, R.; Pinho, F.; Puel, O.; Oswald, I.P.; Bracarense, A.P.F. Effects of patulin and ascladiol on porcine intestinal mucosa: An ex vivo approach. *Food Chem. Toxicol.* **2016**, *98*, 189–194. [CrossRef]
49. Okumura, R.; Takeda, K. Roles of intestinal epithelial cells in the maintenance of gut homeostasis. *Exp. Mol. Med.* **2017**, *49*, e338. [CrossRef] [PubMed]
50. Buckley, A.; Turner, J.R. Cell biology of tight junction barrier regulation and mucosal disease. *Cold Spring Harb. Perspect. Biol.* **2018**, *10*, a029314. [CrossRef]
51. Förster, C. Tight junctions and the modulation of barrier function in disease. *Histochem. Cell Biol.* **2008**, *130*, 55–70. [CrossRef] [PubMed]
52. Garagnani, P.; Pirazzini, C.; Franceschi, C. Colorectal cancer microenvironment: Among nutrition, gut microbiota, inflammation and epigenetics. *Curr. Pharm. Des.* **2013**, *19*, 765–778. [CrossRef] [PubMed]
53. Chen, Y.; Duan, J.-A.; Guo, J.; Shang, E.; Tang, Y.; Qian, Y.; Tao, W.; Liu, P. Yuanhuapine-induced intestinal and hepatotoxicity were correlated with disturbance of amino acids, lipids, carbohydrate metabolism and gut microflora function: A rat urine metabonomic study. *J. Chromatogr. B* **2016**, *1026*, 183–192. [CrossRef] [PubMed]
54. Caporaso, J.G.; Kuczynski, J.; Stombaugh, J.; Bittinger, K.; Bushman, F.D.; Costello, E.K.; Fierer, N.; Peña, A.G.; Goodrich, J.K.; Gordon, J.I. QIIME allows analysis of high-throughput community sequencing data. *Nat. Methods* **2010**, *7*, 335–336. [CrossRef]
55. Gill, S.R.; Pop, M.; DeBoy, R.T.; Eckburg, P.B.; Turnbaugh, P.J.; Samuel, B.S.; Gordon, J.I.; Relman, D.A.; Fraser-Liggett, C.M.; Nelson, K.E. Metagenomic analysis of the human distal gut microbiome. *Science* **2006**, *312*, 1355–1359. [CrossRef]
56. Elinav, E.; Strowig, T.; Kau, A.L.; Henao-Mejia, J.; Thaiss, C.A.; Booth, C.J.; Peaper, D.R.; Bertin, J.; Eisenbarth, S.C.; Gordon, J.I. NLRP6 inflammasome regulates colonic microbial ecology and risk for colitis. *Cell* **2011**, *145*, 745–757. [CrossRef]
57. Zwielehner, J.; Lassl, C.; Hippe, B.; Pointner, A.; Switzeny, O.J.; Remely, M.; Kitzweger, E.; Ruckser, R.; Haslberger, A.G. Changes in human fecal microbiota due to chemotherapy analyzed by TaqMan-PCR, 454 sequencing and PCR-DGGE fingerprinting. *PLoS ONE* **2011**, *6*, e28654. [CrossRef]
58. Hildebrand, F.; Nguyen, T.L.A.; Brinkman, B.; Yunta, R.G.; Cauwe, B.; Vandenabeele, P.; Liston, A.; Raes, J. Inflammation-associated enterotypes, host genotype, cage and inter-individual effects drive gut microbiota variation in common laboratory mice. *Genome Biol.* **2013**, *14*, 1–15. [CrossRef]
59. Gu, S.; Chen, D.; Zhang, J.-N.; Lv, X.; Wang, K.; Duan, L.-P.; Nie, Y.; Wu, X.-L. Bacterial community mapping of the mouse gastrointestinal tract. *PLoS ONE* **2013**, *8*, e74957. [CrossRef]
60. Yang, Y.; Ye, H.; Zhao, C.; Ren, L.; Wang, C.; Georgiev, M.I.; Xiao, J.; Zhang, T. Value added immunoregulatory polysaccharides of Hericium erinaceus and their effect on the gut microbiota. *Carbohydr. Polym.* **2021**, *262*, 117668. [CrossRef]
61. Monk, J.M.; Lepp, D.; Wu, W.; Graf, D.; McGillis, L.H.; Hussain, A.; Carey, C.; Robinson, L.E.; Liu, R.; Tsao, R. Chickpea-supplemented diet alters the gut microbiome and enhances gut barrier integrity in C57Bl/6 male mice. *J. Funct. Foods* **2017**, *38*, 663–674. [CrossRef]
62. Chen, H.; Jiang, W. Application of high-throughput sequencing in understanding human oral microbiome related with health and disease. *Front. Microbiol.* **2014**, *5*, 508. [CrossRef]
63. Parks, D.H.; Tyson, G.W.; Hugenholtz, P.; Beiko, R.G. STAMP: Statistical analysis of taxonomic and functional profiles. *Bioinformatics* **2014**, *30*, 3123–3124. [CrossRef] [PubMed]
64. Louca, S.; Jacques, S.M.; Pires, A.P.; Leal, J.S.; Srivastava, D.S.; Parfrey, L.W.; Farjalla, V.F.; Doebeli, M. High taxonomic variability despite stable functional structure across microbial communities. *Nat. Ecol. Evol.* **2016**, *1*, 1–12. [CrossRef] [PubMed]
65. Thomas, A.M.; Jesus, E.C.; Lopes, A.; Aguiar Jr, S.; Begnami, M.D.; Rocha, R.M.; Carpinetti, P.A.; Camargo, A.A.; Hoffmann, C.; Freitas, H.C. Tissue-associated bacterial alterations in rectal carcinoma patients revealed by 16S rRNA community profiling. *Front. Cell. Infect. Microbiol.* **2016**, *6*, 179. [CrossRef] [PubMed]

molecules

MDPI

Article

An Extensive Pharmacological Evaluation of New Anti-Cancer Triterpenoid (Nummularic Acid) from *Ipomoea batatas* through In Vitro, In Silico, and In Vivo Studies

Muhammad Majid [1], Anam Farhan [2], Muhammad Imran Asad [3], Muhammad Rashid Khan [4], Syed Shams ul Hassan [5,6,*], Ihsan-ul Haq [3,*] and Simona Bungau [7,*]

1 Faculty of Pharmacy, Capital University of Science and Technology, Islamabad 44000, Pakistan; majidpharma808@gmail.com
2 Department of Biology, School of Science and Engineering, Lahore University of Management Sciences, Lahore 54810, Pakistan; anam.farhan764@gmail.com
3 Department of Pharmacy, Faculty of Biological Sciences, Quaid-i-Azam University, Islamabad 44000, Pakistan; miasad@bs.qau.edu.pk
4 Department of Biochemistry, Faculty of Biological Sciences, Quaid-i-Azam University, Islamabad 44000, Pakistan; mrkhanqau@yahoo.com
5 Shanghai Key Laboratory for Molecular Engineering of Chiral Drugs, School of Pharmacy, Shanghai Jiao Tong University, Shanghai 200240, China
6 Department of Natural Product Chemistry, School of Pharmacy, Shanghai Jiao Tong University, Shanghai 200240, China
7 Department of Pharmacy, Faculty of Medicine and Pharmacy, University of Oradea, 410028 Oradea, Romania
* Correspondence: shams1327@yahoo.com (S.S.u.H.); ihsn99@yahoo.com (I.-u.H.); simonabungau@gmail.com (S.B.)

Citation: Majid, M.; Farhan, A.; Asad, M.I.; Khan, M.R.; Hassan, S.S.u.; Haq, I.-u.; Bungau, S. An Extensive Pharmacological Evaluation of New Anti-Cancer Triterpenoid (Nummularic Acid) from *Ipomoea batatas* through In Vitro, In Silico, and In Vivo Studies. *Molecules* **2022**, *27*, 2474. https://doi.org/10.3390/molecules27082474

Academic Editor: Lillian Barros

Received: 3 March 2022
Accepted: 9 April 2022
Published: 12 April 2022

Abstract: Prostate cancer (PCa) is the most common cancer in men, accounting for approximately 10% of all new cases in the United States. Plant-derived bioactive compounds, such as pentacyclic triterpenoids (PTs), have the ability to inhibit PCa cell proliferation. We isolated and characterized nummularic acid (NA), a potent PT, as a major chemical constituent of *Ipomoea batatas*, a medicinal food plant used in ethnomedicine for centuries. In the current study, in vitro antiproliferative potential against PCa cells (DU145 and PC3) via 3-(4,5-dimethylthiazol-2-yl)-2,5-diphenyl-2H-tetrazolium bromide (MTT) assay; Western blot protein expression analysis; absorption, distribution, metabolism, excretion (ADME); pharmacokinetic prediction studies; and bisphenol A (BPA)-induced prostate inhibition in Sprague Dawley rats were conducted to gauge the anti-cancer ability of NA. Significant ($p < 0.05$ and $p < 0.01$) time- and dose-dependent reductions in proliferation of PCa cells, reduced migration, invasion, and increased apoptotic cell population were recorded after NA treatment (3–50 µM). After 72 h of treatment, NA displayed significant IC_{50} of 21.18 ± 3.43 µM against DU145 and 24.21 ± 3.38 µM against PC3 cells in comparison to the controls cabazitaxel (9.56 ± 1.45 µM and 12.78 ± 2.67 µM) and doxorubicin (10.98 ± 2.71 µM and 15.97 ± 2.77 µM). Further deep mechanistic studies reveal that NA treatment considerably increased the cleavage of caspases and downstream PARP, upregulated BAX and P53, and downregulated BCL-2 and NF-κB, inducing apoptosis in PCa cells. Pharmacokinetic and ADME characterization indicate that NA has a favorable physicochemical nature, with high gastrointestinal absorption, low blood–brain barrier permeability, no hepatotoxicity, and cytochrome inhibition. BPA-induced perturbations of prostate glands in Sprague Dawley rats show a potential increase (0.478 ± 0.28 g) in prostate weight compared to the control (0.385 ± 0.13 g). Multi-dose treatment with NA (10 mg/kg) significantly reduced the prostate size (0.409 ± 0.21 g) in comparison to the control. NA-treated groups exhibited substantial restoration of hematological and histological parameters, reinstatement of serum hormones, and suppression of inflammatory markers. This multifaceted analysis suggests that NA, as a novel small molecule with a strong pharmacokinetic and pharmacological profile, has the potential to induce apoptosis and death in PCa cells.

Keywords: prostate cancer; nummularic acid; docking; apoptosis; BAX; p53

1. Introduction

Prostate cancer (PCa) is one of the most common cancers worldwide and has a high mortality rate. Enlarged prostate, with reduced volume and intensity of the urinary stream as a result, is considered a leading cause of illness and death in PCa. It is the fourth most significant type of cancer on Earth, prevailing in 1.41 million people (WHO). Surveys suggest that by 2030, PCa will cross 1.7 million new cases and 499,000 new deaths [1,2]. The role of endocrine function in prostatic growth, physiology, and neoplasia is demonstrated by numerous observations, including those showing that the growth of animal and human prostates depends on androgen stimulation. Androgens play an essential role in the differentiation, development, and normal functioning of the prostate and thus likely have a role in developing prostate carcinogenesis [3]. The exact etiology of sporadic PCa cannot be explained well, but it involves environmental and genetic factors. Growing evidence indicates that environmental contaminants can mimic and alter the actions of endogenous hormones, some potentially disrupting endocrine function in humans; thus, they are referred to as endocrine disruptors (EDC). Exposure to ecological contaminants and xenobiotics such as bisphenol A (BPA), alcohol, smoking, carbon tetrachloride (CCl4), thioacetamide, etc., is considered responsible for oxidative stress propagation [4]. Excessive and prolonged exposure of the body to such factors leads to overexpression of reactive oxygen species (ROS), inducible forms of nitric oxide synthase (iNOS), cyclooxygenase-2 (COX-2), and 5-lipoxygenase (5-LOX). Raised levels of ROS and iNOS are directly correlated to the inflammation and tumorigenesis of vital organs, including the liver, kidney, and testes [5,6]. It is well understood that infiltration of mediators such as NO, interleukins (IL), and prostaglandins (PGs) at the site of inflammation may lead to chronic disorders, including cancer. Apoptosis inhibition, enhanced cell proliferation, immunity suppression, and augmented cancer cell invasion are key events followed by mediator infiltration [6,7].

Current therapeutic approaches include surgery, radiations, chemotherapy, hormonal therapy, cryosurgery, and other methods. These approaches are more or less effective either as monotherapy or in a multimodal approach. Due to limited treatment options, sustained, targeted, and well-tolerated therapeutic alternatives are needed to combat high-grade resistant and progressive diseases [8]. Dietary modifications may also contribute to decreasing prostate cancer risk. Thus, the development of nutraceuticals against prostate cancer is an area of divine concern. Medicinal plants have plentiful therapeutic agents that can be used against inflammation, algesia, pyrexia, carcinoma, and infectious diseases [9]. It has long been accepted that structures derived from natural products possess biochemical specificity, high chemical diversity, and other molecular characteristics that make them promising lead structures for drug development, differentiating them from synthetic and combinatorial compound libraries [10,11]. Triterpenoids (TPs) belong to isopentenyl pyrophosphate oligomers which are thoroughly distributed within kingdom Plantae as cyclic (tetra/Penta) TPs, glycosidic triterpenes (saponins), phytosterols, and other related compounds with substantial therapeutic potential. The anti-inflammatory effects of PTs are primarily attributed to their capacity to inhibit abnormalities induced by 5-LOX, iNOS, COX-2, and nuclear factor-κB (NF-κB). PTs such as oleanolic acid (OA) and glycyrrhizic acid (GA) are extensively used to treat liver disorders, as well as betulinic acid for prostatitis, corosolic and gymnemic acids for diabetic complications, and Asiatic acid in wound healing. Thus, the whole class of PTs is a worthy treasure of novel multi-targeted bioactive components with commendable therapeutic potential [12].

Ipomoea batatas (L.) Lam, commonly known as sweet potato, shaker-kandi, and cultivar, is a perennial crop extensively cultivated in China, Nigeria, India, the USA, Uganda, Tanzania, and Vietnam [13]. In Pakistan, it is called "Shakar-Kandi" and is cultivated in Khyber Pakhtunkhwa, lower and central Punjab, and Kashmir and used as an aphrodisiac, anti-inflammatory, energizing, antimicrobial, purgative, laxative, and antifungal agent; for ulcers of the mouth and throat; burns; bug bites; GIT problems [14]; and against anemia, hypertension, diabetes, and enlarged prostates [15]. Nummularic acid, (19S) 3β-hydroxy-urs-12-en-29β-oic acid, is a PT of ursane skeleton isolated from the ethyl acetate extract of its aerial part. The chemical formula of NA is $C_{30}H_{48}O_3$, and it was isolated for the first time in *I. batatas*. It was earlier isolated in 2007 from *E. nummularius* [16]. In vitro, in silico, and in vivo assays have been conducted to refine the search for a new drug that presents better pharmacokinetic and chemotherapeutic properties with economic viability for use in the treatment of prostate cancer.

2. Results

2.1. Structure Elucidation of NA

The purified compound was obtained as a white amorphous powder with molecular formula $C_{30}H_{48}O_3$ and a molecular weight of 456.3 g/mol. The presence of hydroxyl (3421 cm^{-1}), carboxyl (1693 cm^{-1}), and an olefinic double bond (1652 cm^{-1}) in FTIR and the presence of 30 C signals in ^{13}C NMR spectrum confirm its triterpenoidal structure. In ^{13}C NMR spectrum (Figures S1 and S2), COOH at δ 181.06 (s), OH at δ 78.57 (d), and signals for olefinic carbons at δ145.32 (s) and 121.75 (d) were detected. HMQC spectra showed other carbon resonances, namely, nine methylene (CH_2), five quaternary (C), five methane (CH), and seven methyl carbons (CH_3). The de-shielded olefinic methine carbon resonance at δ 121.75 and shielded olefinic quaternary carbon resonance at δ 145.32 recommended the position of olefinic unsaturation at the C-12 location of the ursane skeleton. Six methyl singlets [δ 0.92, δ 0.95, δ 1.02, δ 1.05, δ 1.22, and δ 1.24], one methyl doublet [δ 1.03 (J = 6.3 Hz)], one carbinol methine [δ 3.45 (dd, J = 10.3 and 5.5 Hz)], and one olefinic proton [δ 5.51 (t, J = 3.5 Hz)] typical of hydroxyursane skeleton were prominent in ^1H NMR spectra. The HMBC correlation of H_3 (δ 3.45) with C_2 (δ 27.25), C_4 (δ 38.55), C_{23} (δ 28.73), and C_{24} (δ 16.65) suggested OH at C_3. The coupling constants of H_3 proposed β and equatorial positioning of OH at C_3. The HMBC interface of H_{18} (δ 2.63) with C13 (δ 138.41), C_{17} (δ 31.23), C_{19} (δ 46.13), and C_{29} (δ 181.18) and of H_{19} (δ 1.85) with C_{18} (δ 55.53), C_{20} (δ 38.56), C_{21} (δ 30.22), C_{29} (δ 181.06), and C_{30} (δ 21.58) collectively showed the location of a carboxyl group at C_{19} locus. From the NMR data, the compound was identified as a known compound 3β-Hydroxyurs-12-en-29-oic acid [16], and its trivial name is Nummularic acid (Figures 1 and S1–S4, Table S1).

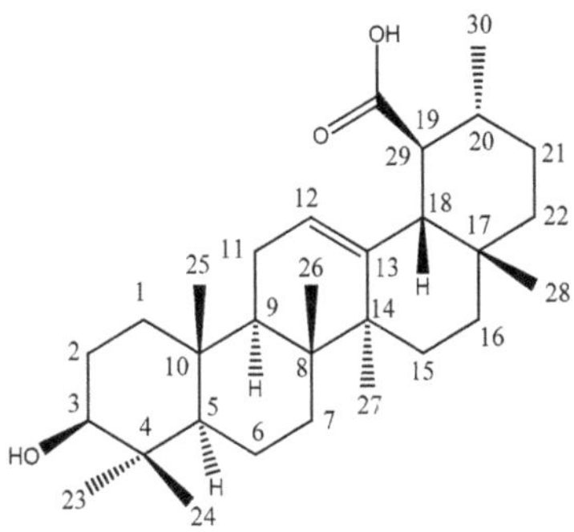

Figure 1. Chemical structure of NA.

2.2. Molecular Docking

The affinity among the protein targets and the ligands was investigated using molecular docking. AutoDock Vina program was used for the docking analysis through the PyRx user interface. E-value (kcal/mol) was used to assess the protein affinity and best-docked pose complex. It provided a prediction of binding free energy and binding constant for docked ligands. In the current study, NA has the most stable complexes, showing −8.5 kcal/mol binding energy (E-value) with BAX (PDB-ID: 2K7W), −8.2 kcal/mol E-value with BCL-2 (PDB-ID 1K3K), −1.9 kcal/mol E-value with NF-κB (PDB-ID 1NFK), and −8.4 kcal/mol E-value with P53 (PDB-ID 1AIE). In the current study, NA binding affinities with ligands for target proteins and amino acid residues involved in the binding pocket interactions were comparable to selective standard inhibitors, i.e., doxorubicin, lapatinib, and vincristine. This study strongly suggests that NA has significant potential to be developed as a selective inhibitor of extrinsic and intrinsic apoptotic proteins. All 2D and 3D images of binding interactions of NA with target proteins are represented as Supplementary materials (Figures S5–S16).

2.3. *Antiproliferative Potential of NA*

Significant reductions in cell viability were observed upon exposure to NA for 24, 48, and 72 h. NA showed significant time-dependent inhibition against DU145 after 48 h and 72 h exposure time, with IC_{50} of 41.23 ± 4.84 μM and 21.18 ± 3.43 μM. In contrast, IC_{50} of NA was recorded 46.18 ± 4.80 μM and 24.21 ± 3.38 μM at 48 h and 72 h exposure time, respectively, against PC3. Standard drugs (Cabazitaxel and Doxorubicin) exhibited maximum cytotoxicity against DU145 (9.56 ± 1.45 μM and 10.98 ± 2.71 μM, respectively) and PC3 (12.78 ± 2.67 μM and 15.97 ± 2.77 μM, respectively) after maximum exposure time (72 h). All results are displayed in Figure 2. After assessing antiproliferative potential, NA was tested for cell migration, inhibiting DU145 using an established in vitro scratch test for 24 h. NA (20 μM) tended to cause a considerable reduction in cell migration at 12 and 24 h treatment of DU145. This was calculated by dividing the scratch area at each observation period by the control area at 0 h. Compared to the control (16.22 ± 4.4%), the scratch area after 24 h of NA treatment was 65.3 ± 3.7% (Figure 3).

In successive experiments, the inhibitory effect of NA on the multiplication of prostate cancer cells was assessed by clonogenic assay. DU145 cells were treated for 48 h and permitted to grow until untreated cells formed adequate colonies. Visual observation of cells demonstrated that cells could not develop sufficient colonies in the presence of NA. In contrast, there was an unchecked proliferation of untreated prostate cancer cells and those treated with 1% DMSO as a negative control (Figure 4). As determined using ImageJ software, the colony number was significantly reduced with compound treatment. NA (20 μM) followed by 10 μM exhibited gray values of 12.54 ± 4.5% and 36.87 ± 3.7%, respectively, versus DMSO (84.3 ± 4.4%).

2.4. *NA Induces Apoptosis*

To uncover the probable mechanism of cancer cell antiproliferation in MTT assays, cleavage of PARP and caspase 3 in treated cells, as well as expression analysis of apoptotic (BAX and P53) and anti-apoptotic (NF-κB and BCL-2) proteins were analyzed through Western blotting (Figure 5). The loading control was set as GAPDH. PCa cells DU145 were treated with NA at two concentrations (5 μM and 10 μM) over 48 h. Protein (40–60 μg) was separated from protein lysate and stained with monoclonal antibodies. Cleavage of PARP by caspase 3 is the hallmark of apoptosis. Downregulation of NF-κB and BCL-2 and upregulation of P53 and BAX expression indicate that NA induces apoptosis to variable degrees. Figure 5 shows the fold change (extent of increase or decrease) in protein expression after treatment of DU145 cells with NA for 48 h.

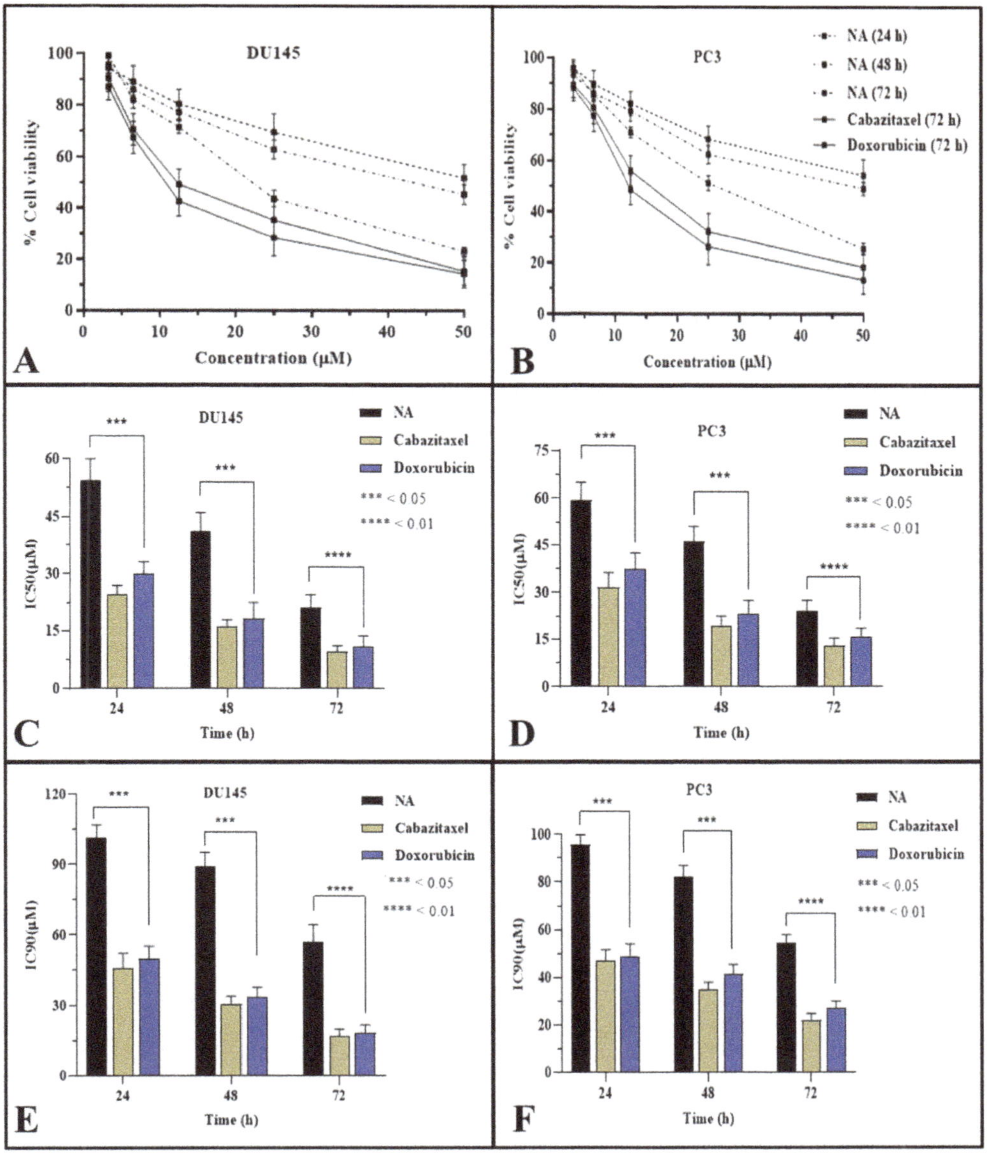

Figure 2. Effect of NA on the viability of prostate cancer cells and IC_{50} values. Note: MTT assay was used to determine the viability of cancer cells. (**A**) 24, 48, and 72 h treatment of DU145 cells and (**B**) 24, 48, and 72 h treatment of PC3 cells. (**C**) IC_{50} values of NA against DU145. (**D**) IC_{50} values of NA against PC3. (**E**) IC_{90} values of NA against DU145. (**F**) IC_{90} values of NA against PC3. Data are mean ± SEM of % cell viability (n = 3) at $p < 0.05$ and $p < 0.01$.

A

B

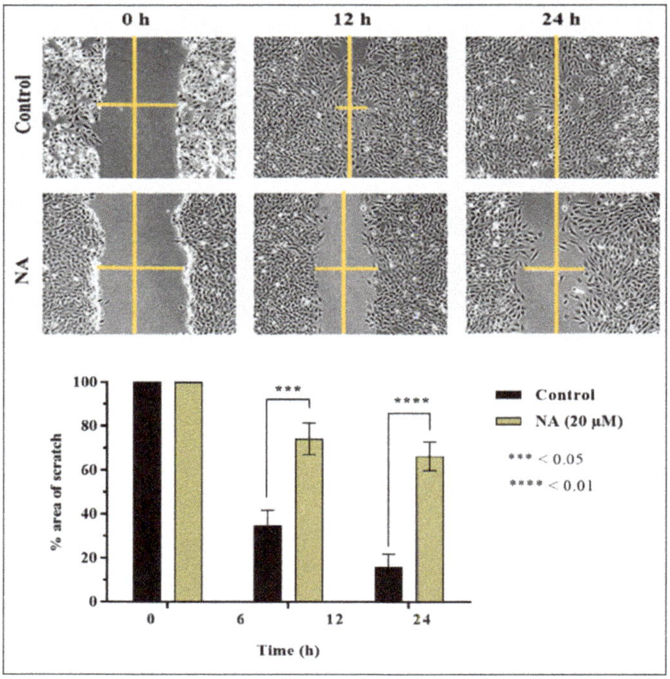

Figure 3. In vitro scratch assay on NA-treated prostate cancer cells. Note: DU145 cells were plated in 6-well plates and scratched at full confluency. (**A**) Migration of cells to heal the area of scratch was observed at 0 h, 12 h, and 24 h after treatment. (**B**) Reduction in area of scratch was photographed using Olympus CKX41 microscope and measured using ImageJ software. Data are mean ± SEM percent area of scratch in triplicate with a marked difference at $p < 0.05$ and $p < 0.01$.

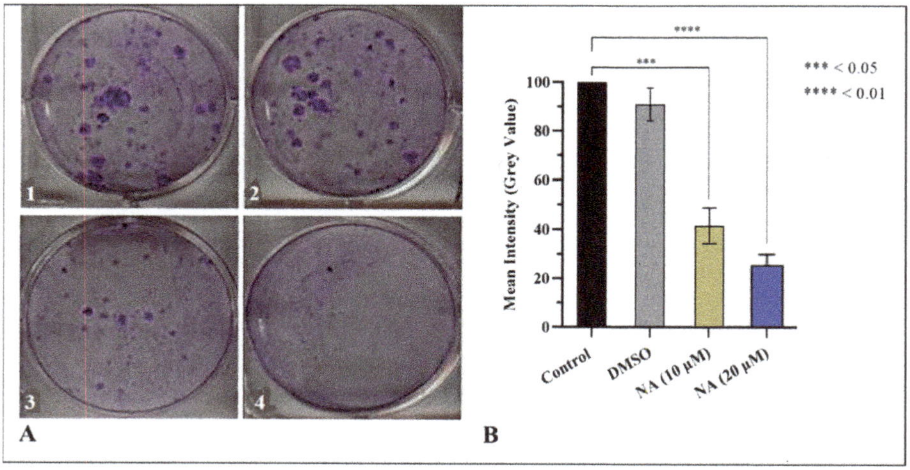

Figure 4. Inhibition of colony formation by NA in prostate cancer cell lines. Note: DU145 cells were treated with 10 µM and 20 µM concentrations of NA for 48 h and then allowed to form colonies for 5–7 days. (**A**) Pictures of colonies for DU145 cell lines were taken using Olympus CKX41 microscope: (1) Control, (2) 1% DMSO, (3) NA (10 µM), and (4) NA (20 µM). (**B**) Densitometry analysis of colonies was performed in control and treated wells. Data are mean ± SEM (n = 3) of colony intensities measured using ImageJ software.

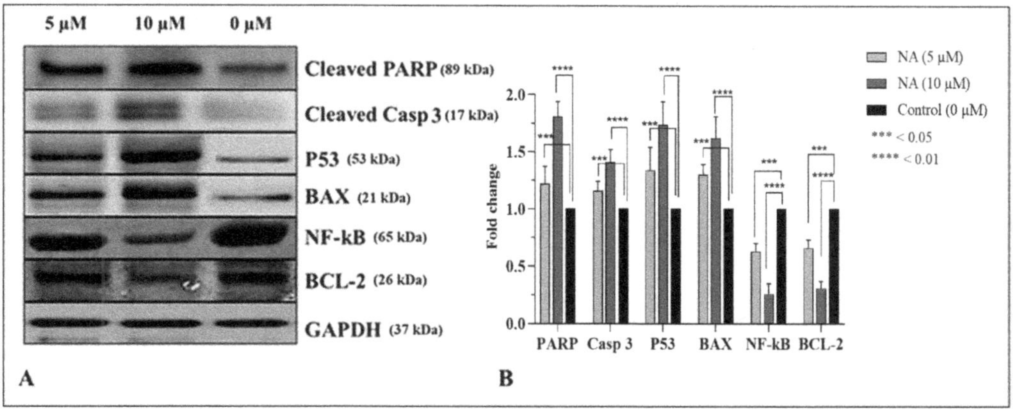

Figure 5. Western blot analysis of proteins associated with NA induced apoptosis. Note: Prostate cells and DU145 cells were treated with NA at concentrations of 5 µM and 10 µM for 48 h. (**A**) Data show increased cleavage of caspase 3 and downstream PARP, increased expression and upregulation of pro-apoptotic P53 and BAX, and decreased expression downregulation of anti-apoptotic NF-κB and BCL2. GAPDH was used as a loading control. (**B**) Fold change in cleaved PARP, cleaved caspase 3, P-53, BAX, BCL-2, and NF-κB expression after treating for 48 h with different concentrations of NA compared to control.

2.5. Pharmacokinetic and ADME Properties

Absorption, distribution, metabolism, excretion (ADME), and pharmacokinetic prediction studies were conducted for the compound NA.

The physico-chemical characteristics of NA are discussed in Table 1. According to Table 1, the lipophilicity, insolubility, size, insaturation, polarity, and flexibility of NA were studied and classified into six sections with appropriate ranges for oral bioavailability (Figure 6a). The oral bioavailability graph of the NA is shown in Figure 6a, which is based on the six sections stated in the physicochemical characteristics section. The results of the compound NA were within these limits, demonstrating that NA has a favorable physiochemical profile, which is one of the factors that must be monitored in pharmaceutical and clinical studies.

Table 1. Predicted physicochemical parameters and lipophilicity properties of NA.

Properties	Parameters	NA
Physicochemical properties	MW [a] (g/mol)	456.70
	Rotatable bonds	1
	HBA [b]	3
	HBD [c]	2
	Fraction Csp3	0.90
	TPSA [d]	57.53
Lipophilicity Log $P_{o/w}$	iLOGP	3.90
	XLOGP3	7.75
	MLOGP	7.09
	Consensus	6.00

[a] Molecular weight, [b] H-bond acceptor, [c] H-bond donor, [d] topological polar surface area.

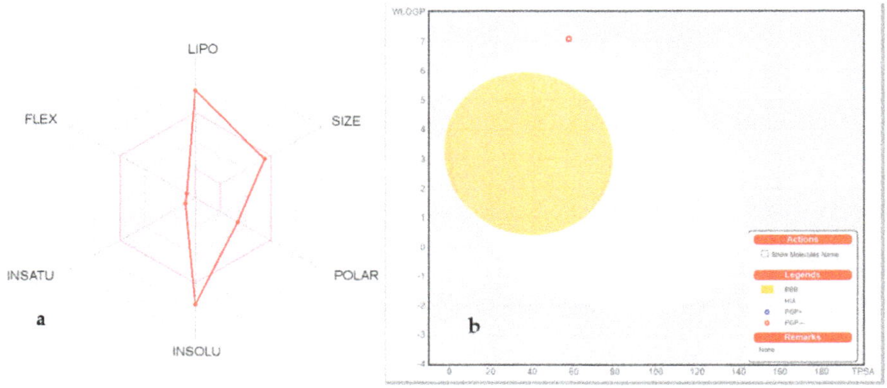

Figure 6. (**a**) Bioavailability radar chart for NA. The pink zone represents the physicochemical space for oral bioavailability, and the red line represents the oral bioavailability properties. (**b**) Predicted BOILED-Egg plot from *swiss ADME* online web tool for NA.

NA has a good gastrointestinal absorption (HIA) and no BBB permeability, as seen in Table 1. The BOILED-Egg graph, shown in Figure 6b, predicts NA absorption in the GI tract (HIA) and BBB penetration. The absorption region for the HIA is white, while the BBB penetration region is yellow. Furthermore, the compound NA has a lower Log Kp (Table 2) for skin permeation. Table 2 shows that the biomolecule NA does not inhibit any cytochrome isoform and is rapidly metabolized; hence, it cannot cause any drug–drug interactions with any cytochrome, and no hepatotoxicity is expected. Drug clearance is determined as the combination of hepatic and renal clearances in the frequency of excretion, and it is vital for determining dosing rates to achieve steady-state concentrations. The compound NA has an inadequate clearance value. Overall pharmacokinetic and ADME characterization indicate that NA has a favorable physicochemical nature with high gastrointestinal absorption, low BBB permeability, no hepatotoxicity, and cytochrome inhibition. Hence, NA is a potential biomolecule for anti-cancer pharmaceutical preparations with higher bioavailability and lesser toxicity.

Table 2. Predicted ADME parameters of NA.

Properties	Parameters	NA
Absorption	Water solubility	−4.315
	Caco permeability (cm/s)	1.327
	GI [a]	97.334
	Log K_p (skin permeation) cm/s	−2.706
	P-gp substrate	No
Distribution	BBB [b]	−0.235
	CNS permeation (Log PS)	−1.014
	V_D [c] (human)	−0.723
Metabolism	CYP1A2 inhibitor	No
	CYP2C19 inhibitor	No
	CYP2C9 inhibitor	No
	CYP2D6 inhibitor	No
	CYP3A4 inhibitor	No
Excretion	Total clearance (log mL/min/kg)	0.048
	Renal OCT2 substrate	No

[a] Gastrointestinal, [b] blood–brain barrier, [c] volume of distribution.

2.6. *Effect on Size, Weight, and ROW of Prostate*

Excised gonads on day 21 of experimentation were measured. The results (Table 3) depict that the weight of the prostate of BPA-intoxicated rats potentially ($p < 0.05$) increased, measuring 0.478 ± 0.28 g in comparison to the control (0.385 ± 0.13 g) with ROW (0.274 ± 0.16). Maximum protection was examined in the BPA + NA (10 mg/kg) group followed by the BPA + NA (5 mg/kg) group, showing significant ($p < 0.05$) decreases in prostate size (0.409 ± 0.21 and 0.432 ± 0.19 g, respectively). The test group receiving NA (10 mg/kg) showed no obvious increase in prostate size (0.393 ± 0.16 g), validating its safety in multidose treatment.

Table 3. Assessment of size, weight, and ROW of prostate.

Groups	Weight of Prostate (g)	Final Body Weight (g)	ROW
Control	0.385 ± 0.13 [d]	213 ± 8.0 [c]	0.180 ± 0.02 [c]
Vehicle (10% DMSO)	0.379 ± 0.22 [d]	208 ± 11 [b]	0.182 ± 0.03 [c]
BPA (50 mg/kg)	0.478 ± 0.28 [a]	174 ± 13.0 [a]	0.274 ± 0.16 [a]
NA (10 mg/kg)	0.393 ± 0.16 [b]	217 ± 8.0 [c]	0.181 ± 0.07 [c]
BPA + NA (10 mg/kg)	0.409 ± 0.21 [c]	216 ± 6.0 [c]	0.188 ± 0.08 [c]
BPA + NA (5 mg/kg)	0.432 ± 0.19 [b]	208 ± 10.0 [b]	0.207 ± 0.10 [b]

Note: ROW, relative organ weight; BPA, bisphenol A. Data values represent mean ± SD (n = 7). Means with dissimilar superscript ([a–d]) letters in the column are significantly ($p < 0.05$) different from one another.

2.7. *Effect on Hematology and Histology*

The blood profiling of the BPA-intoxicated rats treated with NA (10 mg/kg and 5 mg/kg) doses is listed in Table S3. BPA intoxication altered several hematological parameters, causing a high degree of toxicity. Severe depression in Hb levels, platelet count, and RBC count with raised WBC and ESR was noticed in BPA-treated rats compared to controls. Contrarily, test groups receiving NA (10 mg/kg) showed no substantial variations in hematological parameters compared to control, justifying its biosafety and use in in vivo systems. RBC count (5.62 ± 0.19 and 5.18 ± 0.13 × 10^6/μL), WBC (4.19 ± 0.11 and 4.76 ± 0.12 × 10^3/μL), platelet count (468.1 ± 9.81 and 437 ± 03.21 × 10^3/μL), Hb level (11.33 ± 0.47 and 10.31 ± 0.54 g/dL), and ESR (4.18 ± 0.53 and 5.09 ± 0.19 mm/h) of test groups BPA + NA (10 mg/kg) and BPA + NA (5 mg/kg), respectively, were significantly restored in comparison to control. BPA intoxication considerably altered the hematological profile of treated rats.

Histological investigations ascertain the gonadoprotective potential of NA and validate its anti-prostate carcinoma potential. Slides attest that NA at high (10 mg/kg) and low (5 mg/kg) doses is proactive against BPA-induced toxicity (Figure S17). The control group displayed normal morphology of testes with spermatocytes, spermatids, spermatogonia, Sertoli and Leydig cells, normal architecture of seminiferous tubules, normal developmental stages, and concentration of sperms in the seminiferous tubules. Test groups showed marked protection in the morphology of the seminiferous tubules and high density of germ cells, while BPA caused significant damage and abrasions to seminiferous tubules with low cellular density.

2.8. *Effect of NA on Hormonal and Biochemical Levels*

Levels of circulating hormones (testosterone, FSH, LH, and estradiol) were recorded, with no marked ($p < 0.05$) alteration in rats receiving NA (10 mg/kg) in comparison to the control and vehicle, hence validating their nontoxic behavior towards gonads and sex hormones. On the other hand, test groups BPA + NA (10 mg/kg) and BPA + NA (5 mg/kg) showed significant protection against BPA-induced gonadotoxicity as testosterone, FSH, LH, and estradiol were recorded as 4.02 ± 0.09 ng/mL, 10.12 ± 0.27 mIU/mL, 2.90 ± 0.13 mIU/mL, and 21.07 ± 1.17 pg/mL, respectively, and 4.37 ± 0.11 ng/mL, 11.09 ± 0.42 mIU/mL, 3.22 ± 0.10 mIU/mL, and 20.39 ± 1.12 pg/mL, respectively. Raised estradiol levels (26.19 ± 2.16 pg/mL) in rats receiving BPA indicate gonadotoxicity, which ultimately leads to prostatitis (Table 4). NA significantly maintained the physiological concentrations of serum estradiol in a dose-dependent manner, which suggests its potential against prostate carcinoma. The effect of NA on biochemical levels in testicular homogenates is tabulated compared to the control in Table 5. Maximum alterations in the serum concentrations of biochemical markers were recorded in the BPA-intoxicated group, indicating a high degree of gonadotoxicity. Test groups BPA + NA (10 mg/kg) and BPA + NA (5 mg/kg) showed significant ($p < 0.05$) restoration of these biochemicals.

Table 4. Appraisal of reforms to hormonal levels by NA.

Groups	Testosterone (ng/mL)	FSH (mIU/mL)	LH (mIU/mL)	Estradiol (pg/mL)
Control	4.41 ± 0.12^{c}	11.27 ± 0.52^{c}	3.14 ± 0.23^{c}	18.02 ± 0.94^{c}
Vehicle (10% DMSO)	4.33 ± 0.19^{c}	10.93 ± 0.75^{c}	3.29 ± 0.33^{c}	19.14 ± 1.17^{c}
BPA (50 mg/kg)	1.47 ± 0.16^{a}	5.71 ± 0.31^{a}	1.41 ± 0.14^{a}	26.19 ± 2.16^{a}
NA (10 mg/kg)	4.56 ± 0.10^{d}	12.07 ± 0.29^{d}	3.31 ± 0.15^{c}	19.21 ± 1.02^{c}
BPA + NA (10 mg/kg)	4.37 ± 0.11^{c}	11.09 ± 0.42^{c}	3.22 ± 0.10^{c}	20.39 ± 1.12^{c}
BPA + NA (5 mg/kg)	3.88 ± 0.12^{b}	8.81 ± 0.35^{b}	2.75 ± 0.09^{b}	23.18 ± 1.31^{b}

Note: FSH, follicle-stimulating hormone; LH, luteinizing hormone; BPA, bisphenol A. All the data are represented as mean ± SD ($n = 7$). Means with different superscript letters ($^{a-d}$) in a column specify significant difference at $p < 0.05$.

Table 5. Effect of NA on biochemical levels.

Groups	CAT (U/min)	POD (U/min)	SOD (U/min)	GSH (μM/mg Protein)	Nitrite (μM/mg Protein)
Control	3.77 ± 0.12^{d}	9.14 ± 0.63^{d}	18.02 ± 1.54^{d}	24.81 ± 3.14^{d}	56.78 ± 2.34^{d}
Vehicle (10% DMSO)	3.83 ± 0.09^{d}	9.29 ± 0.39^{d}	18.14 ± 1.37^{d}	23.68 ± 2.91^{cd}	57.23 ± 4.58^{d}
BPA (50 mg/kg)	1.71 ± 0.07^{a}	3.41 ± 0.14^{a}	8.19 ± 1.56^{a}	12.71 ± 2.84^{a}	83.54 ± 3.12^{a}
NA (10 mg/kg)	3.66 ± 0.11^{cd}	9.60 ± 0.38^{e}	17.84 ± 1.21^{c}	22.90 ± 2.10^{c}	58.61 ± 2.59^{d}
BPA + NA (10 mg/kg)	3.44 ± 0.32^{c}	8.85 ± 0.40^{c}	17.69 ± 1.3^{c}	22.45 ± 5.12^{c}	62.87 ± 3.15^{c}
BPA + NA (5 mg/kg)	3.11 ± 0.09^{b}	7.89 ± 0.40^{b}	15.96 ± 0.86^{b}	20.05 ± 3.45^{b}	66.56 ± 1.81^{b}

Note: All the data are represented as mean ± SD ($n = 7$), Means with different superscript letters ($^{a-e}$) in a column specify significant difference at ($p < 0.05$). CAT, catalase; POD, peroxidase; SOD, superoxide dismutase; GSH, reduced glutathione.

3. Discussion

The significant findings of the current study are (i) the characterization of Nummularic acid, specifically that (ii) NA reduces cell proliferation, migration, and invasion of prostate cancer cells; (iii) induces apoptosis at very low concentrations; (iv) has excellent pharmacokinetic and ADME properties; and (v) reduce prostatitis in vivo.

Compounds of natural origin with fewer side effects and admirable therapeutic efficacy have gained a repute in anti-cancer drug development. In recent years, various potent plant-based triterpenoids have been discovered, which have shown great promise as chemopreventive and therapeutic agents. They were principally inhibiting key signaling molecules, inflammatory mediators, tumor cell proliferation, invasion, metastasis, and angiogenesis in various in vitro and in vivo models of cancer. NA is obtained as needle-shaped white crystals with molecular formula $C_{30}H_{48}O_3$ and a molecular weight of 456.3 g/mol. The presence of hydroxyl (3421 cm^{-1}), carboxyl (1693 cm^{-1}), and olefinic double bond (1652 cm^{-1}) in FTIR and the presence of 30 C signals in ^{13}C NMR spectrum confirm the triterpenoid structure of its ursane skeleton. Ursane triterpenoids have been reported to suppress the proliferation of various tumor cells; induce apoptosis; and inhibit tumor promotion, metastasis, and angiogenesis in animal cancer models. The presence of the OH group at C-3, COOH at C-29, and dimethyl substitution at C-17 and C-20 of NA plays a vital role in enhancing their cytotoxic effect [17]. In the present study, NA majorly exhibited (i) significant inhibition of prostate cancer cells proliferation; (ii) induction of apoptosis by altering pro- and anti-apoptotic proteins expression; (iii) significant anti-clonogenic ability, which reconfirmed the antiproliferation efficacy of compounds; and (iv) significant cessation of migration capacity assessed via wound scratch assay.

Apoptosis is a complex pathway to thoroughly understand, but is generally divided into extrinsic (FAS/FasL ligand) and intrinsic (mitochondrial) pathways. Caspase activation by intrinsic or extrinsic pathways is mediated by the activation of caspase 8 and caspase 9, respectively. These can activate and cleave downstream protein caspase 3 that can elicit morphological hallmarks of apoptosis, including DNA fragmentation by proteolytic cleavage of PARP. NF-κB is a protein complex that monitors the transcription of specific genes, cytokine production, and cell survival. NF-κB regulates anti-apoptotic genes, particularly TRAF1/TRAF2, and blocks the caspase family of enzymes involved in various apoptotic pathways. Thus, downregulation of NF-κB will trigger caspases and ultimately upregulate apoptosis [8]. Transcription factor P53 suppresses the tumor by inducing the expression of another apoptotic protein BAX. Hence, the upregulation of both P53 and BAX enhances the programmed death of the cell. While BCL-2 is an anti-apoptotic protein, it inhibits mitochondrial apoptosis by blocking the release of cytochrome c and ultimately inhibiting the activation of caspase 3 [18,19]. It was observed in the current study that after binding at extrinsic or intrinsic receptors, ligand (NA) activated the downstream caspase cascade to induce apoptosis. Increased expression of cleaved caspase 3 as well as PARP, significant upregulation of pro-apoptotic proteins (P53 and BAX), and downregulation of anti-apoptotic (BCL-2 and NF-κB) indicate that the compound induced apoptosis in PCa cells. Previously, a team of researchers reported that NA isolated from *F. xanthoxyloides* induces apoptosis in prostate cancer (DU145 and C4-2) cells by activating AMP-kinases, altering the metabolic rate, and triggering an immediate energy crisis that causes the ultimate death of the cells [20]. Earlier, Rengarajan et al., 2014 [21], reported that D-pinitol instigates apoptosis in breast cancer (MCF-7) cells by expressing BAX and P53 while downregulating BCL-2 and NF-κB levels. Ursolic acid (UA) has been repeatedly reported to induce apoptosis in various cancer cells by activating caspase cascade and upregulating pro-apoptotic (BAX and P53) while downregulating anti-apoptotic (BCL-2 and NF-κB) proteins. UA has the ability to trigger the activation of protein kinase C (PKC) which in turn is involved in the apoptosis of many cancer cells, including PCa cells [22].

Pharmacokinetic and ADME profiles are key steps in the biotransformation of any biomolecule into a medication [23]. According to the ADME profile, NA possesses moderate absorption and distribution, and high GIT solubility with less BBB permeability. Therefore, NA cannot create any serious adverse effects related to CNS. To minimize deleterious effects on the CNS, substances that are inert to the CNS should not cross the BBB. [24]. Furthermore, NA has demonstrated that it is not a P-glycoprotein (P-gp) substrate and thus not sensitive to the P-gp efflux mechanism, which is used as a drug resistance mechanism by many cancer cell lines. CYP enzymes are critical for drug excretion, and its isoforms metabolize almost 75% of commercially available drugs. Inhibition of any of these isoforms results in pharmacokinetically significant drug–drug interactions [25]. NA has not inhibited any CYP enzymes, thus not creating drug–drug interactions for those CYP enzyme-targeted drugs. As the liver serves as the metabolic factory for a large number of medications, one of the primary disadvantages of many pharmaceuticals is that they cause hepatotoxicity [26]. NA has not shown any hepatotoxicity with a 70% confidence value. Drug clearance is determined as the combination of hepatic and renal clearances in the frequency of excretion, and it is vital for dose calculations to achieve steady-state concentrations. The compound NA has an inadequate clearance value. Organic Cation Transporter 2 (OCT2) substrates may impact adverse interactions with OCT2 inhibitors in combination. The compound NA has been predicted as a non-substrate of OCT2.

BPA is a weak agonist to the androgen receptor (AR); thus, it can disrupt the estrogen-triggered pathways by forming a transcriptional complex that can bind the estrogen-responsive element (ERE). Intracellular enzyme 5α-reductase (5αR) plays a vital role in converting testosterone to 5α-dihydrotestosterone (DHT) inside the stroma and basal cells. DHT has a 10× greater affinity with androgen receptors than testosterone and plays a vital role in prostate enlargement. So, BPA could compete with DHT to bind to the androgen receptor and initiate anti-androgenic activity in cell systems by forming an AR/BPA complex that prevents endogenous androgens from regulating androgen-dependent gene transcription. Continuous exposure to xenobiotics such as BPA causes oxidative stress, leading to genitourinary abnormalities, sperm deformity, epigenetic variations, enlarged prostate mass, and reduced epididymal weight [6].

Currently, multidose intoxication of BPA significantly ($p < 0.05$) increases prostate mass (0.478 ± 0.28 g) of male Sprague Dawley rats and reduced epididymal weight to 0.334 ± 0.17 g. Triterpenoids with oleanane, ursane, and lupine skeleton exhibit significant anti-oxidant potential and have a greater tendency to mitigate ROS- and iNOS-induced pathological conditions in the body. The presence of the OH group at position C-3 and COOH at C-29 turn NA into more anti-oxidant molecules as OH and COOH have a commendable H^+ donating feature [12]. Oral treatment of rats with NA doses significantly reduced prostate size and restored epididymal mass compared to the control and BPA-intoxicated rats, indicating the protective aptitude of the compound. As previously discussed in detail [20,27], PTs (especially oleananes and ursanes) in *S. cumini* have shown promising anti-oxidant and gonadoprotective potential. Similarly, Olasantan et al. [28] have also reported gonadoprotective ability of *A. floribunda* triterpenoids.

Hematology is the most important predictive tool for determining the existence and intensity of any type of inflammation. Persistent oxidative stress damages vital organs such as the testes, causing inflammation that can be identified with hematological tests. It is evident in the literature that raised ESR, higher WBC count, lower platelet numbers, and elevated levels of circulatory NO indicate bodily inflammations and uremic toxicity [9]. All these disquiets to hematological indices have been observed in BPA-intoxicated rats, indicating oxidative stress. Co-treatment of NA in both doses significantly reserved the hematological parameters of the test groups. This is the first report of the biological effectiveness of NA in restoring hematological parameters related to oxidative stress. Continuous oxidative stress generated by ROS leads to disturbing levels of gonadotropins in serum. In the present study, BPA intoxication is responsible for long-term generation of free radicals, which causes unsettled levels of gonadotropins leading to prostatitis. Test groups receiving NA (10 and 5 mg/kg) resisted significant ROS-induced alterations, and considerable serum concentrations of gonadotropic hormones have been observed. Previously, a team of researchers [20] discussed that PTs in *S. cumini* have shown promising anti-oxidant and gonadoprotective potential. Testicular injuries due to BPA are associated with a depleted amount of endogenous anti-oxidant enzymes and raised nitrite production. Raised nitrite levels in BPA-intoxicated rats indicate injuries to the vascular endothelium or the activation of neutrophils in damaged testicular tissue, which causes the synthesis of NO. No significant changes in CAT, SOD, and POD were noticed in groups treated with NA (5 and 10 mg/kg) in comparison to the control and vehicle. ROS is the ultimate product of sustained bodily exposure with endocrine-disrupting chemicals (EDC) such as BPA that cause oxidative stress. As a result, hypomethylation, a mutation in genetic makeup, and testicular disruption occur, which cause immature and de-morphed spermatogenesis [29]. BPA-intoxicated rats showed apparent deleterious effects during the current study as a cross-section of seminiferous tubules indicated destroyed Leydig cells, damaged seminiferous tubules, and desorbed spermatids. The substantial anti-oxidant, antiproliferative, and anti-inflammatory potential contribute to the increasing body of evidence demonstrating the chemopreventive aptitude of NA. These findings strongly suggest the potential of NA for the prevention of the multifocal development of PCa as well as to prolong survival in the growing population of PCa survivors of primary therapy. Therefore, clinical trials with well-characterized and standardized NA formulations, as primary or adjuvant therapy, in men with PCa are suggested.

4. Material and Methods

4.1. Chemicals and Reagents

Nummularic acid was isolated from ethyl acetate extract of the aerial part of *I. batatas*, and its stock solution was prepared in dimethyl sulfoxide (DMSO) and stored at −20 °C.

Doxorubicin and Cabazitaxel (Sigma, Ronkonkoma, NY, USA) were used as positive control and their stock solutions were also prepared in DMSO. Primary antibodies Cleaved-PARP (Asp214) (E2T4K) Mouse mAb #32563, Cleaved Caspase-3 (Asp175) (5A1E) Rabbit mAb # 9664, BCl-2 (124) Mouse mAb #15071, BAX (2D2) Mouse mAb #89477, P53 (1C12) Mouse mAb #2524, and NF-κB p65 (D14E12) XP® Rabbit mAb #8242 were supplied by Cell Signaling Technology (Beverly, MA, USA). Anti-mouse and anti-rabbit secondary antibodies were purchased from GE healthcare (Pittsburgh, PA, USA). Prostate cancer cell lines DU145 (HTB-81) and PC3 (CRL-1435) were purchased from American Type Culture Collection (ATCC; Manassas, VA, USA). Dimethyl sulfoxide (DMSO), phosphate buffer, and Folin–Ciocalteu reagent were purchased from Riedel-de Haen (Seelze, Germany). Bisphenol A (BPA), trichloroacetic acid (TCA), and tryptone soy broth (TSB) were procured from Sigma-Aldrich (USA), and Tween-20 from Merck-Schuchardt (Savannah, GA, USA). We also used Medium 199, heat-inactivated Biowest FBS (South America), and RPMI-1640 culture media (Gibco BRL, Life Technologies, Inc., Carlsbad, CA, USA). Dulbecco's Modified Eagle Medium (DMEM), DMEM/F1 supplemented with L-glutamine and 2.438 g/l H_2CO_3 (Gibco life technologies, Carlsbad, CA, USA), 3-(4.5-Dimethylthiazol-2-yl)-2,5-diphenyltetrazolium bromide (MTT) powder, phosphate buffer saline (PBS), SRB (sulforhodamine B), and acetic acid (Merck Millipore, Burlington, MA, USA) were also purchased locally. Pre-coated silica gel 60 F_{254} TLC plates, normal phase silica gel 60 (63–200 µm particle size) and silica gel 60 (particle size = 40–60 µm), and chromatography columns were purchased from Merck (Hohenbrunn, Munich, Germany).

4.2. Isolation and Characterization of NA

Ethyl acetate fraction from the aerial part of *I. batatas* (IPA-EAf) was dissolved in ethyl acetate. The slurry was adsorbed on silica gel 60 (63–200 microns; 106 g) and loaded on a gravity column packed with 580 g silica gel 60 for chromatography. The amorphous powder as a purified compound was eluted with DCM and EA (1:0–0:1) using flash column, and purity was observed using the LC-PDA-ELSD method from a single peak obtained during analysis. The isolated compound was characterized using 1D and 2D NMR, FTIR, and mass spectroscopy.

4.3. Molecular Docking

The three-dimensional (3D) structures of BAX (Homo sapiens, PDB-ID 2K7W), BCL2 (Human gamma herpes virus 8, PDB-ID 1K3K) NF-κB (Musmusculus PDB-ID 1NFK), and P53 (Homo sapiens, PDB-ID 1AIE) were accessed from Protein Data Bank (PDB) (www.rcsb.org, accessed on 20 February 2022) with PDB IDs of 1M9K and 1O86, respectively. The Autodock Tools program was used to prepare target proteins to be docked. The proteins were energy-minimized, and Gasteiger charges were added and saved in PDBqt format. Discovery Studio 4.1 Client (2012) was used to generate the hydrophobicity and Ramachandran graphs. The protein construction and statistical percentage values of helices, β-sheets, coils, and turns were assessed by VADAR 1.8 [30].

4.4. Ligands Molecular Docking

The compounds were designed in Discovery Studio Client and saved in PDB format as ligands after energy minimization. Autodock Tools were used to prepare ligands in their most stable conformation. After the addition of the Kolman and Gasteiger charges, the ligands were saved in PDBqt format. Molecular docking analysis was used for all the synthesized ligands against BAX, BCL-2, NF-κB, and P53 using the PyRx virtual screening tool with the Auto Dock VINA Wizard approach [31]. The grid box center values for BAX (PDBID: 2K7W) (center X = 0.273 center Y = 4.614 center Z = 2.980) and size values were adjusted (X = 126, Y = 64, Z = 62). BCL2 (center X = 226.442 center Y = 12.598 center Z = 112.161) and size values were adjusted (X = 64, Y = 108, Z = 82). NF-κB (center X = −10.754 center Y = 12.598 center Z = 112.161) and size values were adjusted (X = 116, Y = 126, Z = 116). P53 (center X = 6.586 center Y = 22.334 center Z = −0.939) was adjusted for better conformational position in the active region of the target protein. Ligands were docked individually against nitric oxide synthase and angiotensin-converting enzyme with a default exhaustiveness value of 25. The predicted docked complexes were evaluated based on the lowest binding energy values (Kcal/mol). The 3D graphical depictions of all the docked complexes were accomplished by Discovery Studio (2.1.0) (Discovery Studio Visualizer Software, Version 4.0., 2012).

4.5. Structural Analysis of Target Proteins

BAX consisted of 70% helices (148 residues), 0% β-sheets, 30% coils (63 residues), 14% turns (30 residues), and a total of 212 amino acid residues R = 0.210 and resolution A° = 1.22. Unit cell dimensions for the lengths were observed to be a = 69.786, b = 91.573, and c = 156.096 with 90° angle for α, β, and γ. The Ramachandran plot confirmed that 96% amino acids were in the allowed regions for the phi (φ) and psi (ψ) angles. BCL2 consisted of 69% helices (101 residues), 0% β-sheets, 30% coils (45 residues), 35% turns (52 residues), and a total of 158 amino acid residues R = 0.150 and resolution A° = 1.19. Unit cell dimensions for the lengths were observed to be a = 69.786, b = 91.573, and c = 156.096 with 90° angles for α, β, and γ. The Ramachandran plot confirmed that 96% amino acids were in the allowed regions for the phi (φ) and psi (ψ) angles. NF-κB consisted of 90% helices (59 residues), 45% β-sheets (287 residues), 44% coils (278 residues), 7% turns (44 residues), and a total of 672 amino acid residues R = 0.340 and resolution A° = 2.30. Unit cell dimensions for the lengths were observed to be a = 84.2, b = 132.1, and c = 80.1 with 90° angles for $\alpha = 90$, $\beta = 93.1$, and $\gamma = 90$. The Ramachandran plot confirmed that 97% amino acids were in the allowed regions for the phi (φ) and psi (ψ) angles. P53 consisted of 64% helices (20 residues), 0% β-sheets, 35% coils (11 residues), 0% turns, and a total of 31 amino acid residues R = 0.252 and resolution A° = 1.50. Unit cell dimensions for the lengths were observed to be a = 69.786, b = 91.573, and c = 156.096 with 90° angle for α, β, and γ. The Ramachandran plot confirmed that 93% amino acids were in the allowed regions for the phi (φ) and psi (ψ) angles. The Ramachandran plots for target proteins are presented in the Supplementary material Table S2, (Figures S6, S8, and S10).

4.6. Cytotoxicity against Prostate Cancer Cell Line

To assess the cytotoxic ability of NA against prostate cancer cell lines DU145 (HTB-81) and PC3 (CRL-1435), MTT assay was carried out according to the pre-established protocol [32]. Briefly, PCa cells (DU145 and PC3) were treated with multiple concentrations of NA for 24, 48, and 72 h duration and the percentage of the viable cells was calculated by defining the cell viability without treatment as 100%. Analysis was carried out in triplicate and IC_{50} was determined after exposing cells with extracts for 24, 48, and 72 h.

4.7. Cell Migration Assay

After assessing cytotoxic potential, NA was tested for cell migration of DU145 using an established in vitro scratch test for 24 h [19,32]. In short, 100 k cells/well (DU145) were cultured in 6-well tissue culture plates. After a few days, when plates were fully confluent, with the help of a 10 µL pipette tip, a scratch line was applied in the middle of the confluent cell monolayer. Wells ($n = 3$) designated as the control were seeded with freshly prepared media without any extract or drug. Test group wells ($n = 3$) were supplied with freshly prepared media containing 20 µM of each extract. The whole event was photographed at intervals of 0, 12, and 24 h on an inverted microscope (Olympus CKX41, Tokyo, Japan). The distance between migration edge and wound edge was measured with Image Pro Plus software, and the % scratch area of test groups compared to controls was plotted.

4.8. Clonogenic Assay

The prostate cancer cell line (DU145) was plated on 6-well tissue culture plates with 1 million cells/well and placed at 37 °C in 5% CO_2 for 2 days. Cells were then treated with 10 µM and 20 µM concentrations of NA for 48 h. After treatment for 48 h, the cells were allowed to form colonies for 5–7 days and media in the test and control groups were regularly changed on alternate days. Following the methodology of [21], cells were then dyed using 0.5% crystal violet comprising equimolar methanol and water (1:1) and pictures were taken under an inverted microscope (Olympus CKX41, Tokyo, Japan). Densitometry analysis of colonies was performed in control and treated wells. Colony intensities were measured using ImageJ software, and the experiment was repeated in triplicate.

4.9. Western Blotting

Protein extraction and Western blot analysis were performed following the protocol described previously [23]. Cells (DU145) were cultured in a T75 flask (1×10^6/flask). After 48 h, cells were treated with NA (10 μM and 20 μM) for 24 h. After treatment, media were aspirated, cells were washed with cold PBS with pH maintained at 7.4, trypsinized and pelleted in 15 mL falcon tubes, and cold lysis buffer was added to the pellet. To perform Western blotting, 40–60 μg of protein was mixed with an equal volume of sample buffer and denatured for 10 min on a heat block at 95 °C. Samples after cooling at room temperature were centrifuged for a split second. Marker (5 μL) and denatured samples loading were carried out on 4–20% mini-protean TGX® stain-free gels (50 μL well) fixed in a mini-Protean® Tetra system vertical electrophoresis tank. The tank was filled with running buffer and proteins were resolved with Power Pac Firmware version 1.07 at 3 A and 100 V for 1.5 h. After separation, the protein transfer was conducted to a 0.2 μm nitrocellulose membrane on a Trans-Blot turboTM transfer pack at 2.5 A for 10 min using the Trans-Blot Turbo transfer system. After transferring the proteins, blots were washed with 1× wash buffer for 5 min, blocked for 30 min by blocking buffer, and probed with the appropriate primary monoclonal antibody (3 μL/3 mL blocking buffer) overnight at 4 °C. Later, 5 min washing of blots was carried out with 1× wash buffer, probed with specific secondary antibody (rabbit/mouse IgG) at room temperature for 2 h and developed with ECLTM Prime Western blotting detection reagent for 5 min. Protein bands were detected by chemiluminescence autoradiography using the ChemDocTM MP imaging system.

4.10. ADME Predictions

ADME (absorption, distribution, metabolism, and excretion) are the essential measurement tools for any compound before it is elected as a drug candidate. The online web tool swiss ADME (http://www.swissadme.ch/index.php, accessed on 20 February 2022) was used to obtain ADME properties of NA [23], and the online web tool pkCSM (http://biosig.unimelb.edu.au/pkcsm/prediction, accessed on 20 February 2022) was used to predict the pharmacokinetic scores.

4.11. Animals and Ethical Statements

Healthy Sprague Dawley rats were purchased from the National Institute of Health (NIH), Islamabad, Pakistan, and housed at Primate Facility of Faculty of Biological Sciences, Quaid-i-Azam University Islamabad, Pakistan. The approved guidelines of the ethical committee of Quaid-i-Azam University, Islamabad, Pakistan, for animal care and experiments (letter number # QAU-PHM-017/2016) were strictly followed. The experimental design was executed according to the guidelines of NIH, Islamabad. During the study, it was ensured that the test animals experienced minimum distress, discomfort, and pain. Euthanasia and blood sampling for hematological, biochemical, and serological studies were performed under anesthesia.

4.12. Experimental Design

Standardized conditions (12 h light/dark cycle, 25 ± 1 °C temperature) were provided to the experimental animals (*Rattus norvegicus*) weighing approximately 170–220 g/each. All test animals were properly fed and supplied with plentiful fresh water. To gauge the protective ability of NA, a detailed investigation was conducted ($n = 7$) against BPA-induced gonadotoxicity and prostatitis. A total of 21 days of experimentation were carried out as previously described by following published protocol [29].

Group I. (Control) group remained untreated.

Group II. (Vehicle) group received 10% DMSO in water, administered orally at 10 mL/kg body weight (BW) on alternate days.

Group III. (BPA) group received 50 mg/kg BW dissolved in 10% DMSO and injected intraperitoneally on alternate days.

Group IV. (NA) group received 10 mg/kg in 10% DMSO given orally on alternate days.

Group V. (BPA + NA) group received 10 mg/kg (50 mg/kg BPA injected intraperitoneally + 10 mg/kg NA administered).

Group VI. (BPA + NA) group received 5 mg/kg (50 mg/kg BPA injected intraperitoneally + 5 mg/kg NA administered).

On the final day of the study, rats were weighed and euthanized by cervical dislocation under chloroform anesthesia. The blood samples were collected under anesthesia via the abdominal aorta for hematological, biochemical, and serological investigations.

4.13. Size, Weight, and Relative Organ Weight (ROW) of Gonads

Rats from each experimental group were anesthetized adequately before euthanizing on the final day of the experiment. Gonads were excised, cleaned out of extra connective tissues, weighed, and measured for size. ROW was calculated as follows:

$$\text{ROW} = \left(\frac{\text{AOW}}{\text{BW}}\right) \times 100$$

where AOW denotes absolute organ weight (g) and BW denotes body weight (g) on the final day of the experiment

4.14. Hematological and Histological Parameters

The blood samples were collected under anesthesia via the abdominal aorta in specific tubes (BD vacutainer) for hematological, biochemical, and serological investigations. Serum was separated by centrifuging blood samples at 6000 rpm for 15 min at 4 °C that were either analyzed or stored at −20 °C. A Neubauer hemocytometer (Feinoptik, Niedersachsen, Germany) was used to count platelets, red blood cells (RBC), and white blood cells (WBCs). Sahli's hemoglobin meter was used to estimate hemoglobin (Hb) content. A modified Westergren method was followed to measure erythrocyte sedimentation rate (ESR) [33]. Afterward, the animals were dissected via ventral longitudinal abdominal incision. Gonads were identified and dissected to assess histopathology caused by BPA intoxication and the protective aptitude of NA [29].

4.15. Hormonal and Biochemical Assessment

To investigate the fate of sex hormones in BPA intoxication and the protective ability of NA, testosterone, FSH (follicle stimulating hormone), LH (luteinizing hormone), and estradiol were quantified according to a well-established method as previously described [34]. Quantification serum testosterone concentrations were estimated using Astra Biotech kit (Immunotech Company, Philadelphia, PA, USA). The sensitivity of the kit is 0.2–50 nmol/L. LH, FSH, and estradiol were purchased from Erba Fertikit, Germany. These hormones were measured via the immune enzymatic method using an ELISA reader. To monitor biochemical abundance, catalase (CAT), peroxidase (POD), superoxide dismutase (SOD), glutathione (GSH), and nitric oxide (NO) were quantified using pre-established protocols [35]. Activities of CAT, POD, and SOD were evaluated by monitoring the rate of H_2O_2 hydrolysis at 240 nm after every minute, and one unit of catalase activity was determined as an absorbance change of 0.01 units per minute. The results are expressed as unit per milligram protein (U/mg protein). GSH was measured at 405 nm by oxidizing the serum with DTNB as μM/mg protein and NO was measured by reducing Griess reagent at 540 nm. A sodium nitrite curve was used to quantify NO amount in serum as μM/mg protein.

4.16. Statistical Analysis

Data obtained in this study are presented as mean ± SD. One-way analysis of variance was performed to determine the variability among groups by Statistix 8.1. Graph Pad Prim 8.1 was used to construct different graphs. Tukey's multiple comparisons were used to calculate significant differences among groups at $p < 0.05$ and $p < 0.01$.

5. Conclusions and Future Perspectives

The present study validates the ethnomedicinal use of *I. batatas* in prostatitis and suggests that NA, as a small, potent molecule, has the ability to induce apoptosis and death in PCa cells. We created a simple, quick, and easy approach for multifaceted screening of compounds for antiproliferative and apoptotic potential. ADME and pharmacokinetic profile also reveal that NA can be a potential biomolecule for anti-cancer pharmaceutical preparations with higher bioavailability and less toxicity. Targeted and specified oncological formulations can also be used to evaluate the anti-cancer potential of the compound. Therefore, more detailed investigations are warranted to declare NA as a potential therapeutic agent in formulations for the prevention and treatment of prostate cancer and other types of cancers.

Supplementary Materials: The following supporting information can be downloaded at: https://www.mdpi.com/article/10.3390/molecules27082474/s1, Table S1: ^{1}H and ^{13}C data of NA dissolved in $CDCl_3$; Table S2: Binding affinities of ligands for target proteins. Amino acid residues are involved in the binding pocket interactions with NA; Table S3: Effect of NA on hematological parameters; Figure S1: ^{1}H spectrum (600 MHz, in DMSO-d_6) of NA; Figure S2: ^{13}C (300 MHz, in DMSO-d_6) spectrum of NA; Figure S3: DEPT spectrum (300 MHz, in DMSO-d_6) of NA; Figure S4: H-H COSY spectrum (600 MHz, in DMSO-d_6) of NA; Figure S5: Representation of docked ligand with BAX (PDB-ID: 2K7W); Figure S6: Ramachandran plot confirming that 96 % amino acids are in the allowed regions for the phi (φ) and psi (ψ) angles; Figure S7: Representation of docked ligand with BCL-2 (PDB-ID 1K3K); Figure S8: Ramachandran plot confirming that 96 % amino acids are in the allowed regions for the phi (φ) and psi (ψ) angles; Figure S9: Representation of docked ligand with NF-κB (PDB-ID 1NFK); Figure S10: Ramachandran plot confirming that 96 % amino acids are in the allowed regions for the phi (φ) and psi (ψ) angles; Figure S11: Representation of docked ligand with P53 (PDB-ID 1AIE); Figure S12: Ramachandran plot confirming that 96 % amino acids are in the allowed regions for the phi (φ) and psi (ψ) angles; Figure S13: Representation of docked ligand with BAX (PDB-ID: 2K7W); Figure S14: Representation of docked ligand with BCL-2 (PDB-ID 1K3K); Figure S15: Representation of docked ligand with NF-κB (PDB-ID 1NFK); Figure S16: Representation of docked ligand with P53 (PDB-ID 1AIE); Figure S17: Histological examination for the protecting proficiency of NA on testes in rat. Note: 40× Hematoxylin-eosin stain.

Author Contributions: M.M. performed all the experiments and wrote the manuscript. A.F. and M.I.A. checked the manuscript for errors. M.R.K. performed the docking. S.S.u.H. prepared the draft and S.B. managed the manuscript. I.-u.H. supervised the work. All authors have read and agreed to the published version of the manuscript.

Funding: This research work of current project was funded by Higher Education Commission Pakistan, through National Research Program for Universities (HEC/NRPU-QAU7528).

Institutional Review Board Statement: We confirm that any aspect of the work covered in this manuscript that involved experimental animals was conducted by strictly following the guidelines approved by the ethical committee of Quaid-i-Azam University, Islamabad, Pakistan (letter no. QAU-PHM-017/2016 for animal care and letter no. QAU-PHM-023/2016 for experimentation, dated 24 October 2016) for this study.

Informed Consent Statement: Not applicable.

Data Availability Statement: The data presented in this study are available in supplementary material.

Acknowledgments: The authors thank the University of Oradea, Oradea, Romania, for financial support in publishing this paper.

Conflicts of Interest: The authors declare no conflict of interest.

References

1. Siegel, R.L.; Miller, K.D.; Fuchs, H.E.; Jemal, A. Cancer statistics. *CA Cancer J. Clin.* **2022**, *72*, 7–33. [CrossRef]
2. Mapelli, P.; Ghezzo, S.; Samanes Gajate, A.M.; Preza, E.; Brembilla, G.; Cucchiara, V.; Ahmed, N.; Bezzi, C.; Presotto, L.; Bettinardi, V. Preliminary Results of an Ongoing Prospective Clinical Trial on the Use of 68Ga-PSMA and 68Ga-DOTA-RM2 PET/MRI in Staging of High-Risk Prostate Cancer Patients. *Diagnostics* **2021**, *11*, 2068. [CrossRef] [PubMed]
3. Hou, Z.; Huang, S.; Li, Z. Androgens in prostate cancer: A tale that never ends. *Cancer Lett.* **2021**, *516*, 1–12. [CrossRef] [PubMed]
4. Crawford, E.D. Understanding the epidemiology, natural history, and key pathways involved in prostate cancer. *Urology* **2009**, *73*, 4–10. [CrossRef] [PubMed]
5. Abdel-Moneim, A.M.; Al-Kahtani, M.A.; El-Kersh, M.A.; Al-Omair, M.A. Free Radical-Scavenging, Anti-Inflammatory/Anti-Fibrotic and Hepatoprotective Actions of Taurine and Silymarin against CCl4 Induced Rat Liver Damage. *PLoS ONE* **2015**, *10*, e0144509. [CrossRef] [PubMed]
6. Mínguez-Alarcón, L.; Hauser, R.; Gaskins, A.J. Effects of bisphenol A on male and couple reproductive health: A review. *Fertil. Steril.* **2016**, *106*, 864–870. [CrossRef]
7. Shams ul Hassan, S.; Ishaq, M.; Zhang, W.; Jin, H.-Z. An overview of the mechanisms of marine fungi-derived antiinflammatory and anti-tumor agents and their novel role in drug targeting. *Curr. Pharm. Des.* **2021**, *27*, 2605–2614. [CrossRef]
8. Sumanasuriya, S.; De Bono, J. Treatment of Advanced Prostate Cancer—A Review of Current Therapies and Future Promise. *Cold Spring Harb. Perspect. Med.* **2017**, *12*, a030635. [CrossRef]
9. Ali, S.; Khan, M.R.; Shah, S.A.; Batool, R.; Maryam, S.; Zahra, Z. Protective aptitude of Periploca hydaspidis Falc against CCl(4) induced hepatotoxicity in experimental rats. *Biomed. Pharmacother.* **2018**, *105*, 1117–1132. [CrossRef]
10. Memariani, Z.; Abbas, S.Q.; ul Hassan, S.S.; Ahmadi, A.; Chabra, A. Naringin and naringenin as anticancer agents and adjuvants in cancer combination therapy: Efficacy and molecular mechanisms of action, a comprehensive narrative review. *Pharmacol. Res.* **2021**, *171*, 105264. [CrossRef]
11. Glevitzky, I.; Dumitrel, G.A.; Glevitzky, M.; Pasca, B.; Otrisal, P.; Bungau, S.; Cioca, G.; Pantis, C.; Popa, M. Statistical analysis of the relationship between antioxidant activity and the structure of flavonoid compounds. *Rev. Chim.* **2019**, *70*, 3103–3107. [CrossRef]
12. Petronelli, A.; Pannitteri, G.; Testa, U. Triterpenoids as new promising anticancer drugs. *Anti-Cancer Drugs* **2009**, *20*, 880–892. [CrossRef] [PubMed]
13. Majid, M.; Nasir, B.; Zahra, S.S.; Khan, M.R.; Mirza, B.; ul Haq, I. *Ipomoea batatas* L. Lam. ameliorates acute and chronic inflammations by suppressing inflammatory mediators, a comprehensive exploration using in vitro and in vivo models. *BMC Complement. Altern. Med.* **2018**, *18*, 216. [CrossRef] [PubMed]
14. Dong, G.; Xu, N.; Wang, M.; Zhao, Y.; Jiang, F.; Bu, H.; Liu, J.; Yuan, B.; Li, R. Anthocyanin Extract from Purple Sweet Potato Exacerbate Mitophagy to Ameliorate Pyroptosis in Klebsiella pneumoniae Infection. *Int. J. Mol. Sci.* **2021**, *22*, 11422. [CrossRef]
15. Naomi, R.; Bahari, H.; Yazid, M.D.; Othman, F.; Zakaria, Z.A.; Hussain, M.K. Potential Effects of Sweet Potato (*Ipomoea batatas*) in Hyperglycemia and Dyslipidemia—A Systematic Review in Diabetic Retinopathy Context. *Int. J. Mol. Sci.* **2021**, *22*, 10816. [CrossRef]
16. Dinda, B.; Ghosh, B.; Arima, S.; Sato, N.; Harigaya, Y. Chemical constituents of Evolvulus nummularius. *Indian J. Chem.* **2007**, *46*, 492–498. [CrossRef]
17. Sato, H.; Macchiarulo, A.; Thomas, C.; Gioiello, A.; Une, M.; Hofmann, A.F.; Saladin, R.; Schoonjans, K.; Pellicciari, R.; Auwerx, J. Novel potent and selective bile acid derivatives as TGR5 agonists: Biological screening, structure-activity relationships, and molecular modeling studies. *J. Med. Chem.* **2008**, *51*, 1831–1841. [CrossRef]
18. Ramachandran, S.; Kaushik, I.S.; Srivastava, S.K. Pimavanserin: A Novel Autophagy Modulator for Pancreatic Cancer Treatment. *Cancers* **2021**, *13*, 5661. [CrossRef]
19. Narożna, M.; Krajka-Kuźniak, V.; Kleszcz, R.; Baer-Dubowska, W. Indomethacin and Diclofenac Hybrids with Oleanolic Acid Oximes Modulate Key Signaling Pathways in Pancreatic Cancer Cells. *Int. J. Mol. Sci.* **2022**, *23*, 1230. [CrossRef]
20. Ayyanar, M.; Subash-Babu, P. *Syzygium cumini* (L.) Skeels: A review of its phytochemical constituents and traditional uses. *Asian Pac. J. Trop. Biomed.* **2012**, *2*, 240–246. [CrossRef]
21. Younis, T.; Khan, M.I.; Khan, M.R.; Rasul, A.; Majid, M.; Adhami, V.M.; Mukhtar, H. Nummularic acid, a triterpenoid, from the medicinal plant Fraxinus xanthoxyloides, induces energy crisis to suppress growth of prostate cancer cells. *Mol. Carcinog.* **2018**, *57*, 1267–1277. [CrossRef] [PubMed]
22. Limami, Y.; Pinon, A.; Leger, D.V.; Pinault, E.; Delage, C. The P2Y2/Src/p38/COX-2 pathway is involved in the resistance to ursolic acid-induced apoptosis in colorectal and prostate cancer cells. *Biochimie* **2012**, *94*, 1754–1763. [CrossRef] [PubMed]
23. Ul Hassan, S.S.; Muhammad, I.; Abbas, S.Q.; Hassan, M.; Majid, M.; Jin, H.Z.; Bungau, S. Stress driven discovery of natural products from actinobacteria with anti-oxidant and cytotoxic activities including docking and admet properties. *Int. J. Mol. Sci.* **2021**, *22*, 11432. [CrossRef] [PubMed]

24. Cruz, J.V.; Serafim, R.B.; da Silva, G.M.; Giuliatti, S.; Rosa, J.M.C.; Araújo Neto, M.F.; Leite, F.H.A.; Taft, C.A.; da Silva, C.; Santos, C.B.R. Computational design of new protein kinase 2 inhibitors for the treatment of inflammatory diseases using QSAR, pharmacophore-structure-based virtual screening, and molecular dynamics. *J. Mol. Model.* **2018**, *24*, 225. [CrossRef] [PubMed]
25. Chen, Y.; Tian, Y.; Gao, Y.; Wu, F.; Luo, X.; Ju, X.; Liu, G. In silico Design of Novel HIV-1 NNRTIs Based on Combined Modeling Studies of Dihydrofuro[3,4-d]pyrimidines. *Front. Chem.* **2020**, *8*, 164. [CrossRef] [PubMed]
26. Spengler, E.K.; Kleiner, D.E.; Fontana, R.J. Vemurafenib-induced granulomatous hepatitis. *Hepatology* **2017**, *65*, 745–748. [CrossRef] [PubMed]
27. Cele, N.D.; Sangweni, N.F.; Mosa, R.A.; Penduka, D.; Lazarus, G.G.; Singh, M.; Zharare, G.E.; Opoku, A.R. Testicular Dysfunction Ameliorative Effect of the Methanolic Roots Extracts of Maytenus procumbens and Ozoroa paniculosa. *Evid.-Based Complement. Altern. Med.* **2017**, *2017*, 8204816. [CrossRef]
28. Olasantan, O.; Areola, J.O.; Ayannuga, O.A.; Babalola, O.O. Evaluation of the gonadoprotective effects of *Allanblackia floribunda* Oliver (Clusiacea) on testes and accessory organs of Wistar rats. *J. Med. Biol. Sci. Res.* **2015**, *1*, 134–144.
29. Majid, M.; Ijaz, F.; Baig, M.W.; Nasir, B.; Khan, M.R.; Haq, I.-U. Scientific Validation of Ethnomedicinal Use of Ipomoea batatas L. Lam. as Aphrodisiac and Gonadoprotective Agent against Bisphenol A Induced Testicular Toxicity in Male Sprague Dawley Rats. *BioMed Res. Int.* **2019**, *2019*, 8939854. [CrossRef]
30. Willard, L.; Ranjan, A.; Zhang, H.; Monzavi, H.; Boyko, R.F.; Sykes, B.D.; Wishart, D.S. VADAR: A web server for quantitative evaluation of protein structure quality. *Nucleic Acids Res.* **2003**, *31*, 3316–3319. [CrossRef]
31. Ahmed, A.; Saeed, A.; Ali, O.M.; El-Bahy, Z.M.; Channar, P.A.; Khurshid, A.; Tehzeeb, A.; Ashraf, Z.; Raza, H.; Ul-Hamid, A.; et al. Exploring Amantadine Derivatives as Urease Inhibitors: Molecular Docking and Structure–Activity Relationship (SAR) Studies. *Molecules* **2021**, *26*, 7150. [CrossRef] [PubMed]
32. Shan, J.; Xuan, Y.; Ruan, S.; Sun, M. Proliferation-inhibiting and apoptosis-inducing effects of ursolic acid and oleanolic acid on multi-drug resistance cancer cells in vitro. *Chin. J. Integr. Med.* **2011**, *17*, 607–611. [CrossRef] [PubMed]
33. Majid, M.; Khan, M.R.; Shah, N.A.; Ul Haq, I.; Farooq, M.A.; Ullah, S.; Sharif, A.; Zahra, Z.; Younis, T.; Sajid, M. Studies on phytochemical, antioxidant, anti-inflammatory and analgesic activities of Euphorbia dracunculoides. *BMC Complement. Altern. Med.* **2015**, *15*, 349. [CrossRef] [PubMed]
34. Maryam, S.; Khan, M.R.; Shah, S.A.; Zahra, Z.; Majid, M.; Sajid, M.; Ali, S. In vitro antioxidant efficacy and the therapeutic potential of Wendlandia heynei (Schult.) Santapau & Merchant against bisphenol A-induced hepatotoxicity in rats. *Toxicol. Res.* **2018**, *7*, 1173–1190. [CrossRef]
35. Khan, D.; Qindeel, M.; Ahmed, N.; Asad, M.I.; ullah Shah, K.; Rehman, A. Development of an intelligent, stimuli-responsive transdermal system for efficient delivery of Ibuprofen against rheumatoid arthritis. *Int. J. Pharm.* **2021**, *610*, 121242. [CrossRef]

molecules

MDPI

Review

Biotechnological Innovations from Ocean: Transpiring Role of Marine Drugs in Management of Chronic Disorders

Saurabh Bhatia [1,2], Rashita Makkar [3], Tapan Behl [3,*], Aayush Sehgal [3], Sukhbir Singh [3], Mahesh Rachamalla [4], Vasudevan Mani [5], Muhammad Shahid Iqbal [6] and Simona Gabriela Bungau [7,8,*]

1. Natural and Medical Sciences Research Center, University of Nizwa, Birkat Al Mauz 616, Nizwa P.O. Box 33, Oman; sbsaurabhbhatia@gmail.com
2. School of Health Science, University of Petroleum and Energy Studies, Dehradun 248007, India
3. Chitkara College of Pharmacy, Chitkara University, Patiala 140401, India; rashitamakker32@gmail.com (R.M.); aayushsehgal00@gmail.com (A.S.); sukhbir.singh@chitkara.edu.in (S.S.)
4. Department of Biology, University of Saskatchewan, 112 Science Place, Saskatoon, SK S7N 5E2, Canada; maheshgupta65@gmail.com
5. Department of Pharmacology and Toxicology, College of Pharmacy, Qassim University, Buraydah 51452, Saudi Arabia; vasumpharmacol@gmail.com
6. Department of Clinical Pharmacy, College of Pharmacy, Prince Sattam bin Abdulaziz University, Alkharj 11942, Saudi Arabia; m.javed@psau.edu.sa
7. Department of Pharmacy, Faculty of Medicine and Pharmacy, University of Oradea, 410073 Oradea, Romania
8. Doctoral School of Biomedical Sciences, University of Oradea, 410087 Oradea, Romania

* Correspondence: tapan.behl@chitkara.edu.in (T.B.); sbungau@uoradea.ro (S.G.B.)

Abstract: Marine drugs are abundant in number, comprise of a diverse range of structures with corresponding mechanisms of action, and hold promise for the discovery of new and better treatment approaches for the management of several chronic diseases. There are huge reserves of natural marine biological compounds, as 70 percent of the Earth is covered with oceans, indicating a diversity of chemical entities on the planet. The marine ecosystems are a rich source of bioactive products and have been explored for lead drug molecules that have proven to be novel therapeutic targets. Over the last 70 years, many structurally diverse drug products and their secondary metabolites have been isolated from marine sources. The drugs obtained from marine sources have displayed an exceptional potential in the management of a wide array of diseases, ranging from acute to chronic conditions. A beneficial role of marine drugs in human health has been recently proposed. The current review highlights various marine drugs and their compounds and role in the management of chronic diseases such as cancer, diabetes, neurodegenerative diseases, and cardiovascular disorders, which has led to the development of new drug treatment approaches.

Keywords: marine drugs; diabetes mellitus; cancer; cardiovascular disorders; neurodegeneration

Citation: Bhatia, S.; Makkar, R.; Behl, T.; Sehgal, A.; Singh, S.; Rachamalla, M.; Mani, V.; Iqbal, M.S.; Bungau, S.G. Biotechnological Innovations from Ocean: Transpiring Role of Marine Drugs in Management of Chronic Disorders. *Molecules* **2022**, *27*, 1539. https://doi.org/10.3390/molecules27051539

Academic Editor: Jesus Simal-Gandara

Received: 23 December 2021
Accepted: 21 February 2022
Published: 24 February 2022

1. Introduction

Drug molecules derived from marines sources are highly heterogenous in nature due to the abundant coverage of oceans, which thereby host the lives of a wide diversity of species [1,2]. In a pharmacological preclinical study, about 75 natural compounds that were isolated from marine organisms significantly showed biological and therapeutic activities [3,4]. The first marine drug from the cone snail peptide—namely, ziconotide, under the trade name Prialt—was approved in 2004 in the United States for the management of spinal cord injury, mediating chronic pain. Later, in 2007, another marine product named trabectedin, a sea squirt metabolite, was also approved by the European Union for the treatment of soft tissue sarcoma [5]. The drugs obtained from marine sources have displayed an exceptional potential in the management of various types of chronic diseases, including cancer, due to their potent anticancer activities [5–7]. These drugs have also gained the great interest of those developing new antimicrobial agents. Sponges belonging to the phylum *Porifera*, are the oldest and most prolific marine organisms on the planet. Demospongiae, a class of *Porifera*, account for 83% of the species with the highest number of bioactive compounds [8]. Another marine genus named *Lendenfeldia* is rich in sulfated steroids, and its metabolites possess anti-HIV, anti-inflammatory, antitumor, and antifouling activities. Drug discovery programs have significantly increased their exploration of lead molecules from marine natural products, and a higher number of bioactive products have been screened for their activity and are under development in clinical trials [9]. Terrestrial plants such as digitalis, morphine, and many other natural compounds have served as drug molecule since olden times. However, the modern pharmaceutical industry has expanded a keen interest in developing drug molecules from marine sources, as they are believed to provide more novel and potent drug compounds since they can survive under extreme conditions, such as the photodynamic and extreme temperatures, pressure, and oxidative stress of the ocean [10]. The basic scientific research in pharmacology and chemistry of marine-derived natural products mainly began in the early 1970s and has now finally begun to bear fruit due to advances in analytical and screening techniques that have fastened the drug discovery process [11]. The current review highlights the drugs obtained from marine natural products and their role in the treatment and management of chronic diseases such as cardiovascular and neurodegenerative diseases, diabetes mellitus, and cancer. These drugs offer a look into the future of promising products that can be obtained from the sea.

The development of drugs from marine sources is a highly tedious process, as it is difficult to procure and manufacture quantities of novel drug leads from marine sources [12]. For instance, the marine sponges are highly chemically versatile in nature and act as a resource of developmental compounds, such as hemiasterlin and discodermolide, which are extracted from primitive metals found in marine habitats. Sponges are extremely treasured since they are difficult to extract, and specimens are mostly collected by hand during deep and shallow water scuba diving, but also with the help of submarines equipped with robotic arms. These techniques are highly expensive and result in an uncertain yield, thereby posing a great challenge to those developing medicines and the pharmaceutical industry. Nonetheless, the interest of researchers in marine products has remained intact, which has led to the budding of innovative solutions to overcome the challenges [13,14]. The story of discovery of therapeutic marine compounds begins with the identification of a marine source and the further isolation of a therapeutic target by analytical techniques. Once a bioactive compound is isolated, it is further studied for its structure and is presented into preclinical and clinical studies for an estimation of its activity. The discovery of drugs from marine sources has also led to the development of genome mining techniques, which have a tendency to improve future discovery processes.

2. Identification and Isolation of Bioactive Compound from Marine Natural Extract

Natural products along with their structural analogues have contributed to pharmacotherapy throughout history, especially in the management of infectious and cancerous diseases [15,16]. However, the major challenge associated with natural products has been drug discovery, due to poor techniques of screening, isolation, characterization of the drug, and its optimization—all of which has now been overcome by advances in analytical tools such as gene mining and advanced techniques of microbial culturing, which has revitalized the interest of the pharmaceutical industry in identifying drug leads from marine sources, opening several new treatment opportunities [17,18]. The natural drugs are typically products with higher molecular mass [19,20], and they present several advantages which are discussed in this article, yet they have several drawbacks, which have led pharmaceutical companies to reduce their efforts in the discovery of natural product-derived molecular leads [21]. The screening of natural products typically consists of an extract library mainly derived from the natural sources that may not necessarily be compatible with traditional target-based assays, making it tedious to identify bioactive compounds of interest [22]. Several tools and techniques are applied to assess whether a new molecule has been discovered or whether it is merely a rediscovery of already known compounds—a process that can be very challenging [23]. In addition, a major hurdle faced by the pharmaceutical industry is obtaining intellectual property rights for an unmodified natural product that has relevant bioactivity, as the naturally occurring compounds cannot always be patented in their native forms; however, the simpler molecules with biological activity can be patented easily [24]. The complex structures of natural products are advantageous in generating structural analogues for exploring the structure's activity relationships and optimizing them for targeted mechanism of action [25]. The modern techniques—which include genome mining, genome engineering, and advances in analytical procedures and systems of cultivation—have led to an increased emphasis on drug development from marine sources, as they have helped overcome many of the major challenges that were being faced by the researchers—techniques that have proven to be promising [26,27].

The application of advanced analytical techniques begins with the screening of crude drug extracts, followed by further isolation and identification of a bioactive molecule that is fractionated to obtain the active moiety from the natural product [28]. The isolation of a bioactive compound is a laborious task and is highly challenging. The molecule isolated is run through the extract libraries and further exposed to high throughput screening so that the crude extract can be pre-fractionated into subfractions that are more suitable for the system that handles automated liquids [3]. The methods of fractionation can be altered to obtain preferential subfractions with active compound drugs that are alike in nature. This process can increase the number of hits compared with the compounds obtained from crude extracts, which enables more efficient and promising hits [6]. An advance in the instrumentation used in analytical methods, combined with advanced computational approaches, can lead to isolation of possible analogue structures of natural products [29]. The precise information about the metabolic composition of the crude marine extract can be obtained through metabolomics, which helps in prioritizing the isolation of a compound and its dereplication to annotate the structural analogues of the newly derived product [30] (Figure 1).

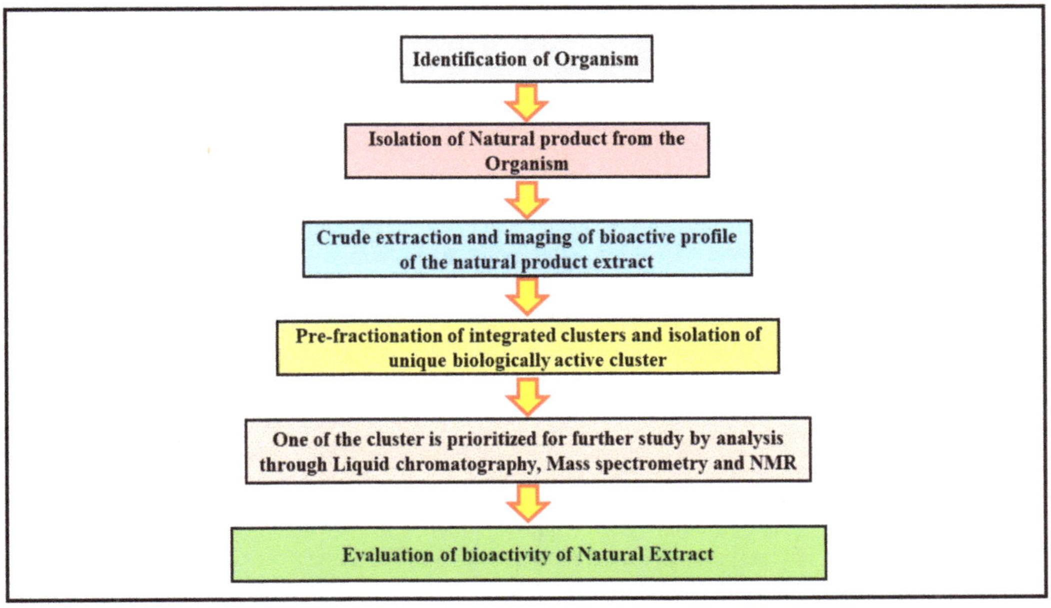

Figure 1. The figure illustrates the process of isolation of a bioactive molecule from a marine source. The identification of a specific organism is a first step in determining the bioactive lead moiety. Once the marine source with a desired therapeutic effect is discovered, the natural products within it are

obtained by extraction processes. The crude extract collected at the end is further studied using imaging techniques to understand its bioactive profile. The integrated clusters of the bioactive compound in the crude extract solution are further pre-fractionated to obtain unique clusters that are biologically active. The active clusters are analyzed through advanced analytical techniques such as NMR, mass spectroscopy, and liquid chromatography for their structure elucidation and bioactivity determination, and the desired cluster is carried forward for further studies. The bioactive lead molecule selected is evaluated for its therapeutic activity through pre-clinical and further clinical trial studies.

The current review highlights the role of marine drugs in the management of cancer, diabetes mellitus, cardiovascular disorders, and neurodegenerative diseases which are listed below.

3. Marine Drugs in the Management of Cancer

Cancer is a chronic disease and can be found in almost all multicellular organisms. The disease is strongly associated with aging because, with increasing age, mutations within the somatic cells accumulate and promote unregulated growth and invasion of dysfunctional cells, which leads to altered functions in the body and works against the organism's health. Apart from the uncontrolled cellular growth, this disease also displays genetic instability, evasion of growth suppressors, immortality of replication, resistance to cell death, angiogenesis initiation, energy metabolism reprogramming, prevention of immune destruction, tumor-derived inflammation, and metastasis [31–33]. Advanced research and technology combined with efforts to determine and understand the hallmarks of cancer have led to improvement in the clinical outcomes of cancer patients by developing newer and more novel diagnostic techniques and therapeutic medications. The higher costs of these therapies and the sometimes extension of patient survival by just few months pose a major challenge to ongoing treatment, thereby provoking a dire need to identify promising new therapies for the management of this life threating disease. There are two general classes of cancer therapies: cancer therapies derived from natural compounds and cancer therapies derived from synthetic compounds—each further subcategorized into small molecules or low molecular weight substances that elicit biological responses by entering cells readily and biologics or large molecular weight substances such as ribonucleic acid (RNA) or monoclonal antibodies that are penetrated across cells with the help of delivery systems [34]. The majority of the cancer treatment drugs are naturally derived substances. For instance, the most primarily used chemotherapeutic drugs for the management of prostrate, breast, and other cancers—namely, docetaxel and paclitaxel—are derived from the taxanes plant. Cabazitaxel, another naturally derived anticancer compound was derived by chemical diversification of taxanes [35]. Factually, most of the anticancer drugs derived from natural resources are derived from terrestrial ecosystems, and about 100,000 compounds have been isolated from plants alone. About 99% of the total living space on earth is deep ocean, and oceans are where80% of the entire species in the world live. In the recent years, the interest of researchers has progressively focused on the marine environment, and researchers have successfully isolated over 2000 compounds over the past three decades [36]. The richness in species of the ocean and its extraordinary diversity with a large temperature and pressure tolerance window, presence of variety of chemicals and metals, saline nature, low to bright light, and allelopathic defenses has attracted the pharmaceutical industries towards the ocean, despite the small number of compounds isolated from it to date. Marine sources are believed to have treasurable therapeutic potential based on the unique dwelling inhabitant and hence the ocean is being explored for its hidden potential. Its noteworthy that the products derived from the ocean are extremely potent and act through multiple molecular pathways and collectively have an ability to target different hallmarks of cancer [37]. Some of the anticancer compounds isolated from marine sources are listed below.

Over 1000 compounds have been isolated from marine sources and are being tested for their activity in preclinical studies; 23 marine derived compounds are under clinical trials between phases I to III, and 7have been approved for marketing. Four compounds out of the total number of marine-derived molecules—namely, trabectedin, cytarabine, eribulin mesylate, and brentuximab vedotin, a conjugated antibody—are being used clinically for their anticancer properties [38].

3.1. *Cytarabine (Cytosar)*

Cytarabine is the debutant lead molecule that has been isolated from the ocean for the management of cancer. Cytarabine was developed by the synthesis of analogs of natural arabino nucleosides and cytosine arabinose from the Caribbean sponge *Cryptotethya crypta*. The chemical structure of this anticancer compound has been found to be related to Spong uridine and Spong thymidine, which are natural products isolated from the marine sponge *Tectitethya cripta*. Cytarabine acts by rapidly converting into its respective triphosphates arabinonucleoside by phosphorylation in a sequential manner. It is an antimetabolite drug with a structure that is sufficiently similar to natural metabolites of the body and acts by interfering with their functioning and hence preserving normal cellular metabolism [39]. The triphosphocytarabine formed upon phosphorylation becomes a substrate for DNA polymerase and subsequently is amalgamated in place of cytosine within the DNA. The arabinose is implanted instead of deoxyribose upon binding of DNA with cytarabine triphosphate and promotes elongation of the DNA strain by preventing phosphodiester bonding between the two pentose sugars, thereby prohibiting synthesis of DNA and hampering abnormal cellular growth. This drug was first approved for its clinical applications in 1969 and has been used in the management of wide array of leukemias, such as non-Hodgkin's lymphoma, acute lymphocytic leukemia, chronic myelogenous leukemia, etc. Cytarabine has been claimed to be the foremost example of a commercially available marine-derived drug, even though the molecule itself is not a natural product but a structural analog [40]. The use of adenine arabinose analogs has also been noted in the development of vidarabine, an antiviral drug used in the management of varicella zoster and herpes simplex virus [41]. Furthermore, another antiviral drug, azidothymidine, has also reportedly been found to be extremely effective in the management of acquired immune deficiency syndrome (AIDS) by blocking the activity of the reverse transcriptase enzyme of the virus, which is highly crucial for replication of the human immune virus (HIV).

3.2. *Trabectedin*

Trabectedin was identified as one of the most abundant structurally related alkaloids that has been isolated from the Caribbean ascidian *Ecteinascidia turbinate*. The drug was first isolated with a very complex procedure in 1996. This molecule acts as an alkylator of DNA but differs from the usual alkylating agents. Trabectedin binds with the guanine residues of the double helix DNA and generates specific sequences, causing bending of strands in a direction opposite to the site of alkylation. These trabectedin abducts arrest the activity of RNA polymerase II and prevent the transcription of DNA, thereby preventing abnormal cellular growth. The most prominent effect of this drug is the inhibition of transcription of MDR1 genes, which are chiefly responsible for producing P glycoproteins and initiating the detoxification processes in the cells. The drug also successfully prevents the repair of DNA lesions due to the rest of RNA polymerase II, thereby producing a significant impact on the tumor microenvironment. Trabectedin also activates caspase 8 protein, which further induces apoptosis in macrophages and monocytes, thereby prohibiting the release of inflammatory mediators and the growth of angiogenic factors and preventing metastasis. Trabectedin acts via DNA alkylation by binding to the guanine residues, further generating exclusive sequences that cause bending of the double helix strands in a direction opposite to that of alkylation and differing greatly from other alkylating agents. The action of RNA polymerase II is arrested with simultaneous DNA transcription prohibition, hence preventing the abnormal growth of the cells. The drug prominently acts by inhibiting

MDR1 genes transcription, which chiefly indulging in the synthesis of P glycoproteins, thereby stimulating the process of cellular detoxification [42]. The repair of the lesions in DNA is also counteracted by this drug due to RNA polymerase II arrest, hence producing a significant effect in the deterioration of the environment of the tumor.

3.3. Eribulin Mesylate

Eribulin is a mesylate salt prototype of high molecular weight halichondrin B, obtained from a class of halichondrins, a progression of macrocyclic polyethersthathave anticancer activity. The drug is obtained from *Halichondria okadai* sponges found on the coast of the Miura Peninsula in Japan. The macrocyclic ring of halichondrin B is responsible for the anticancer activity, as determined by the structure–activity relationship. This drug is believed to be the most complex drug synthesized. This naturally derived potent antimitotic drug mainly acts by inhibiting the microtubule. Tubulin is the ultimate target of this compound, and it prevents polymerization by binding to it and further arresting the microtubular extension. This irreversibly blocks the mitosis of the cells, and its prolongation finally induces apoptosis-mediated cell death. The drug eribulin mesylate was approved in 2010 by the FDA for the management of advanced metastatic breast cancer, and it was approved as second-line treatment for liposarcoma therapy in 2016 [43].

3.4. Brentuximab Vedotin

In 1972, anticancer activity was discovered from the extract of *Dolabella auriculria*, a gastropod mollusk found in the Indian Ocean. After 15 years of continuous research, the peptides—namely, dolastatins—were identified as being the chief active compounds that solely possess potent antiproliferative activity against tumor cells [43]. These peptides led to microtubule blockage and polymerization, consequently preventing rapid tumor cell division and hence prohibiting the growth and proliferation of tumor cells in the body. Brentuximab vedotin has an antibody drug conjugate structure and is mainly developed with a dolastatin10-derived natural product molecule, monomethylauristatin E, isolated from the mollusk *Dolabella auricularia*. In 2012, Adcetris had been approved for the treatment and management of Hodgkin's lymphoma, and monomethylauristatin E had also been studied in several clinical trials based on its ability to form complexes with antibodies and proteins on the membrane. Another analogue—namely, glembatumumabvedotin—has also been recognized for its activity in the management of melanomas, especially those with metastatic breast cancer, due to its association with transmembrane glycoproteins [44]. Polatuzumabvedotin and pinatuzumabvedotin are also being evaluated for their activity in the management of lymphomas and leukemias, due to their direct antibody-directed action on CD 22 and CD 79b proteins (Table 1).

Table 1. The below table lists the marine drugs used in cancer treatment and their respective structures.

Name	Structure
Cytarabine	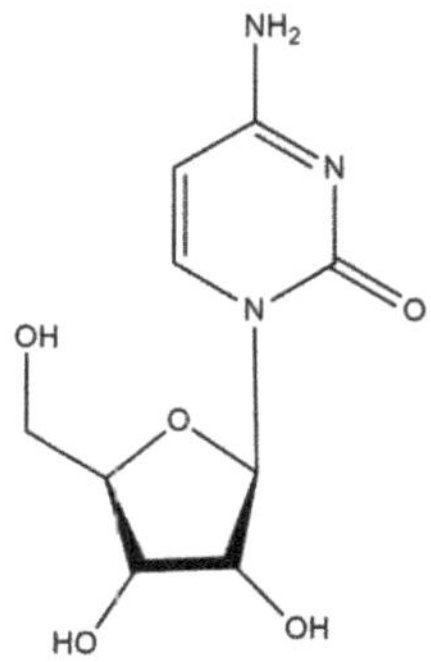

Table 1. *Cont.*

Name	Structure
Trabectedin	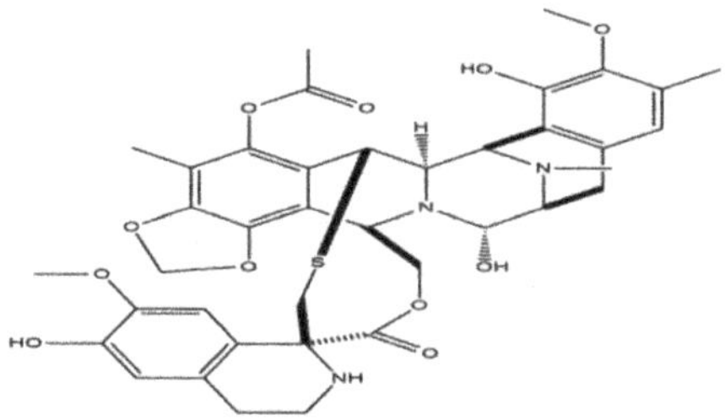
Eribulin Mesylate	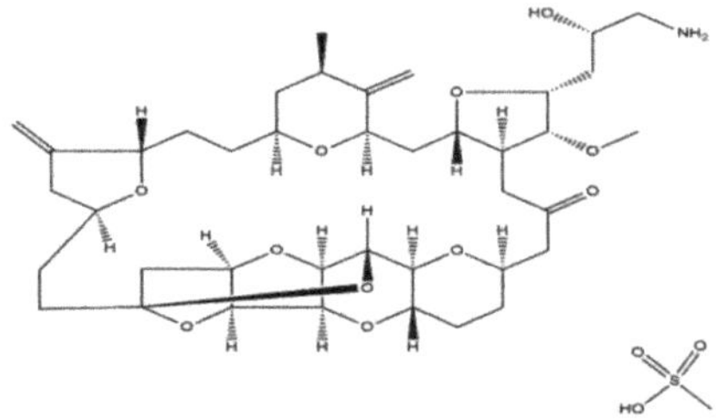
Brentuximab vedotin	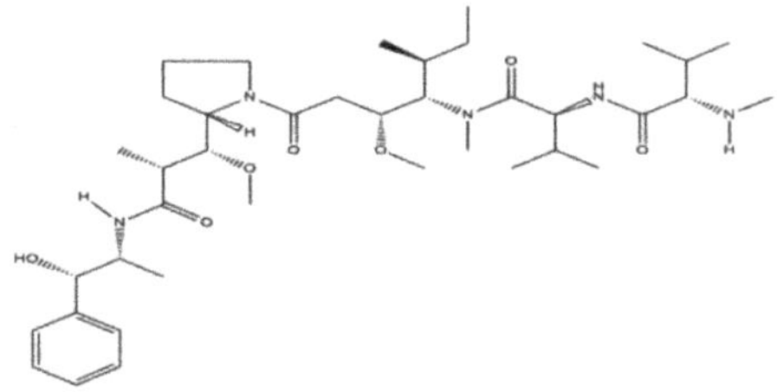

4. Marine Drugs in the Management of Diabetes Mellitus

Diabetes mellitus is a chronic metabolic disease that is mainly characterized by blood sugar level elevation and abnormal metabolism of sugar, and it contributes in a major way to mortality and morbidity rates across several developing and developed nations, including India. This disease includes a defect in the functioning of insulin, including its secretion and site of action. The destruction of other microstructures such as neurons, nephrons, and retina depicts the severity of the disease, highlighting its role in affecting nerves, kidney, and eyes. Several cardiovascular disorders and other conditions also occur upon the emergence of diabetes [45]. An increase in reactive oxygen species and oxidative stress is considered as playing a crucial role in the development of diabetes and its associated complications [46]. Food acts as a major source of energy in the form of sugar and helps in maintaining the physiological functioning of the millions of cells in the body. The body regulates sugar by moving it through the cell membrane via two mechanisms: a receptor, which acts as a door, and insulin, which targets the receptors. Type 1 diabetes is a hyperglycemic condition with a defect in insulin, while type 2 diabetes is mainly characterized by increased levels of sugar, mostly due to a defect in receptors [45]. The prevalence of type 2 diabetes is higher globally in comparison with type 1diabetes. It has been estimated that more than 20 million people worldwide have been diagnosed with type 1diabetes, with a predicted annual increase of 2% to 5% every year in several countries. However, type 2 diabetes accounts for approximately 90% to 95% of diabetic cases globally. In the year 2011, over 280 million people were estimated to be affected by type 2 diabetes and the number is assumed to rise to up to 500 million by 2030 [47]. It has become necessary to adopt preventive measures to reduce the burden of this disease on the health and economy of a country. Several synthetic drugs are available in the market to treat this disease, but the medications are not fully effective and are also costly. Moreover, continued use of these drugs can lead to undesirable adverse reactions in the patients. Patients suffering with type 1 diabetes are solely dependent on external insulin injections for their survival and for maintaining a normal life, but it is not comfortable to inject insulin daily. However, type 2 diabetes can be initially managed or controlled by modifications in lifestyle and diet, but type 2 diabetes often requires treatment with oral antidiabetic drugs in the disease's later stages, and at the end, treatment typically requires insulin injections, which is the most severe scenario. Antioxidants and immune therapy, islet therapy, inhibitors of alpha glucosidase, and other antidiabetic drugs are some of the available therapies to control and manage type 2 diabetes mellitus, but a wide range of side effects also come alongside, which has led to the continued development of novel preventive and regenerative therapies for preventing deficiencies in beta cells mass and for prolong the earlier stage of this disease. Marine sources are being explored due to their promising potential as therapeutic agents in the management of various medical disorders so that they can be employed as a novel or adjuvant therapies. Several marine drugs have been identified in recent years for their exceptional potential for treating or curing diseases. Algae and fish have acted as chief sources of several peptide molecules that possess lipid lowering, anticancer, and anticoagulation properties [48]. In addition, principal antioxidants such as phenolics, carotenoids, and omega 3 fatty acids are also derived from marine seaweeds, crustaceans and fish oil, and bacteria, respectively.

Over 500 marine and freshwater cyanobacteria have been studied for anti-glucosidase and anti-amylase activity, and 38 interesting candidates have also been determined to be fruitful in the management of diabetes [49]. A marine sponge-related bacterium known as *Coralliphaga* has been found to possess major activity in polysaccharide degradation and in the processing of glycolipids and glycoproteins, as it produces a number of glucosidase inhibitors, presenting it as a good target for development in the treatment of diabetes, as well as obesity. Strains of Streptomyces bacteria such as *Streptomyces corchorusii* subspecies rhodomarinus presented fascinating antidiabetic properties by inhibiting the activity of enzyme amylase, while other species of a similar strain led to the production of two novel compounds having N-acetyl-glucosaminidase inhibition properties—namely, Pyrostatins A and B [50]. In addition, to the antidiabetic activity of bacteria, cyanobacteria, and actinomycetes, marine fungi have also been screened to assess whether they might have antidiabetic action. A protein tyrosine phosphatase (PTP1B) inhibitor—namely, aquastatin B—has been obtained from the marine fungus *Cosmospora* species SF-5060, which was isolated from the sediment collected from inter-tides at Gejae Island in Korea [51]. Photosynthetic eukaryotic microalgae, which form a major part of freshwater and marine phytoplankton, have also been confirmed to have significant activity in the management of diabetes mellitus [52,53].

Recent biotechnological advances in aquatic technology have successfully identified promising antidiabetic agents in microalgal species, by virtue of their anti-glycation function. The green microalgal species named *Chlorella* and the diatom *Nitzschia laevis* have been found to have maximum inhibitory effects against the formation of total advanced glycated end products (AGEs)—specifically, N-carboxymethyllysine and pentosidine [54]. The presence of carotenoids such as neoxanthin, antheraxanthin, violaxanthin, and lutein account for the strong AGEs inhibitory activity of Chlorella, while linoleic acid, eicosapentaenoic acid, and arachidonic fatty acids have presented similar bioactivity in *Nitzschia laevis*. The different extracts of these microalgae—mainly *Chlorella zofingiensis*—were tested for their antiglycation activity, and it was determined that the extracts rich in astaxanthin possessed the highest antiglycative and antioxidant activity and hence can be used as a food supplement for the prevention of diabetes in patients [55]. Three strains of microalgae—namely,

Chlorella protothecoides, *Chlorella zofingiensis*, and the diatom *N. laevis*—have been evaluated to possess protective action against the exogenous and endogenous AGEs in an ARPE-19 cell-based model due to the presence of nutritional ingredients such as carotenoids and omega 3 fatty acids within them. These microalgae also decrease the levels of mRNA expression of vascular endothelial growth factor (VEGF) and matrix metalloproteinase 2 (MMP-2), which are key factors in the etiology of diabetes-induced retinopathy and therefore can be used as a food supplement in the prevention and management of diabetic retinopathy due to presence of carotenoids and omega 3 fatty acids, while also preventing cataract development and further macular degeneration [56].

Corals, sea grasses, fish, shark fusion proteins, sea anemones, salmon skin, and even fish and shellfish wastes have also been evaluated for their antidiabetic properties in the past 15 years. For centuries, macroalgae have been consumed by coastal people as a readily available food, even when its nutritional properties were veiled. In todays' time, they have been adopted as a healthy lifestyle approach in several countries and are being consumed entirely in the form of extracts or complete foods for their beneficial properties [57]. A number of green, red, and brown algae such as *Palmaria, Ecklonia cava, Alaria, Rhodomela confervoides*, and *Ascophyllum* have presented significant antidiabetic properties. Amylase inhibitory activity was observed in red algae *Palmaria* species phenolic extracts, in combination with antidiabetic properties presented by protein hydrolysates from *Palmaria palmata*. The methanolic extracts of the brown algae *Ecklonia cava, Pelvetica siliquosa* also presented diminished levels of glucose in the plasma of the diabetic rats. *Sinularia firma* and *Sinularia erecta*, two soft corals, have also been proven to exert blood glucose lowering activity in diabetic rats exposed to their methanolic extracts, which also prohibited a postprandial rise in hyperglycemia in normal rats [47,58]. Free fatty acids were significantly decreased upon consumption of collagen peptides from marine wild fishes, and they also regulated the nuclear receptors in patients with insulin dependence or type-2 diabetes.

Therefore, it can be concluded that marine products and byproducts, if explored judicially, can be the source of several promising and novel lead molecules in the treatment and management of diabetes (Table 2).

Table 2. The below table lists the marine drugs used in diabetes mellitus and their respective structures.

Name	Structure
Pyrostatins A and B	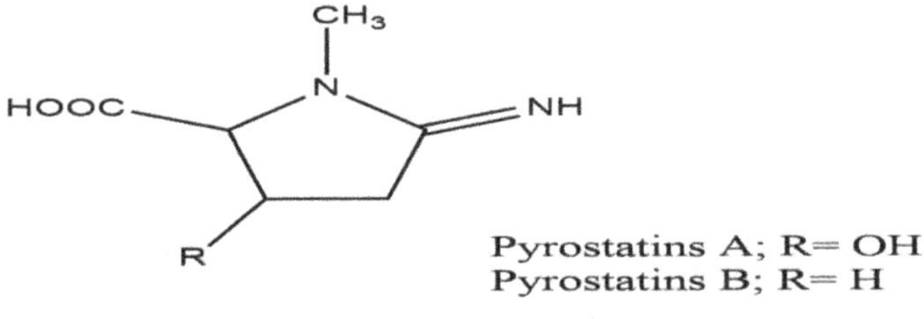
N-carboxymethyllysine	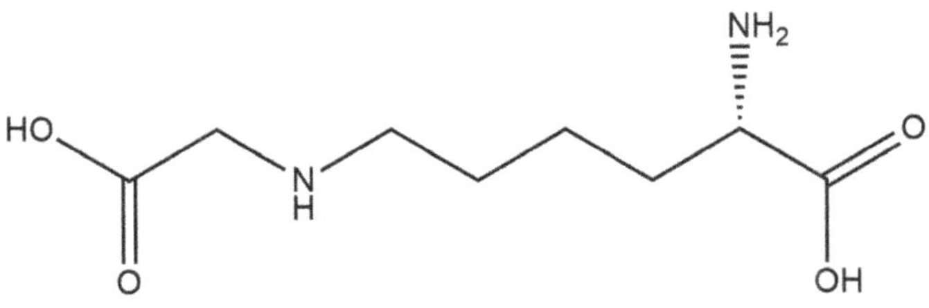

Table 2. *Cont.*

Name	Structure
Pentosidine	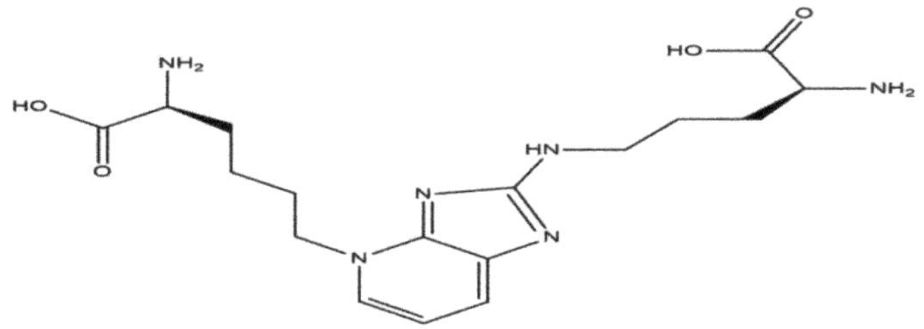
neoxanthin	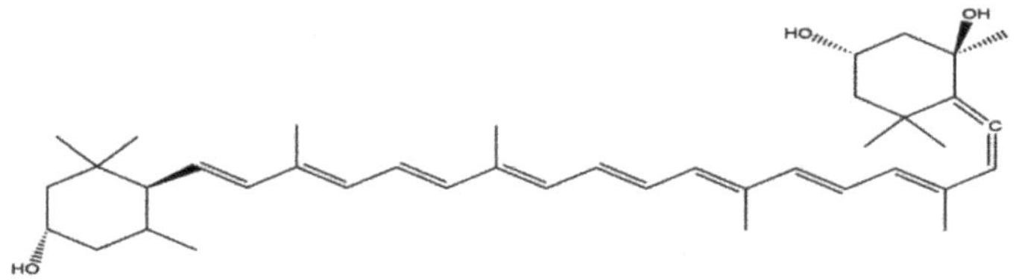
antheraxanthin	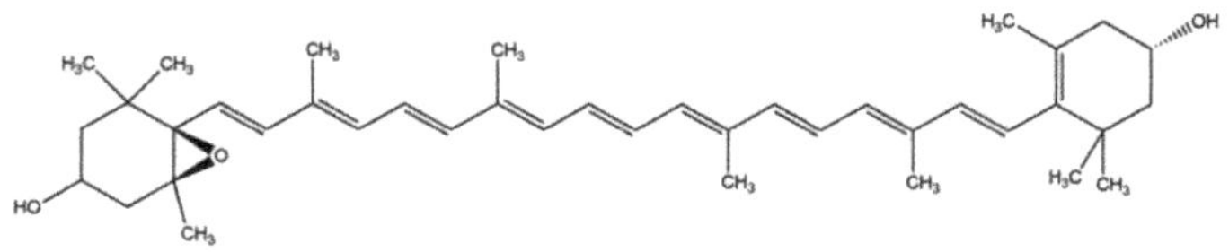
violaxanthin	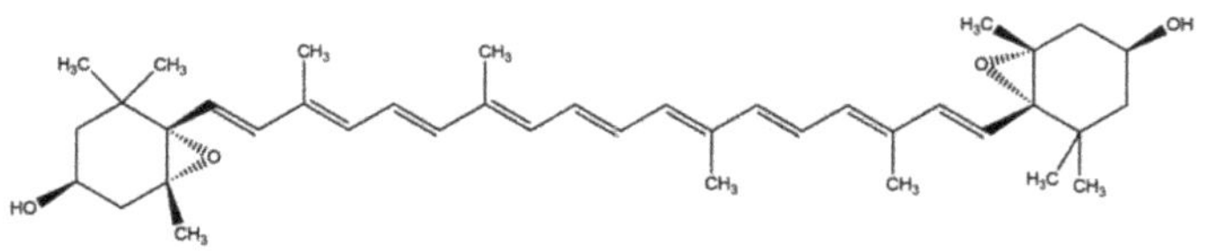
Lutein	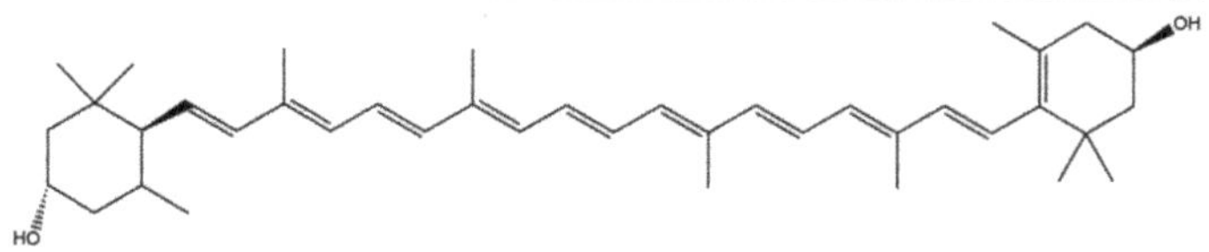

5. Marine Natural Products and Cardiovascular Diseases

Cardiovascular disorders (CVDs) are a group of diseases of the blood vessels and heart and include coronary artery disease, atherosclerosis, cerebrovascular disease, hypercholesterolemia, rheumatic heart disease, and several other conditions. They are primarily the leading cause of death worldwide, and 4 out of every 5 deaths related to CVDs are reported to be due to cardiac arrest or stroke. About 17.9 million people die from CVDs every year, which is double the number of deaths due to cancer, and many of these deaths are premature, occurring in patients below 70 years of age. Unhealthy diet, lack of exercise, smoking, alcohol, and other altered lifestyle conditions are the primary causes that initiate the risk of CVDs in the healthy population. The effects may be experienced by patients as increased blood pressure, hypercholesterolemia, obesity, increased glucose in blood, and weight gain. These initial or intermediate risks can be diagnosed in primary healthcare facilities, and if not taken care of or ignored, they can lead to serious conditions of stroke, cardiac arrest, heart failure, and other complications. Increased levels of cholesterol in the body lead to atherosclerosis, which causes a narrowing and clogging of the arteries due to cholesterol and fat deposition, which further leads to coronary artery diseases (CAD). These conditions can also result in hypoxia and myocardial ischemia. The CVDs mainly affect the younger generations and their families in a destructive context and are worthy of more attention to their efficient control and management [59]. The pharmacological interventions used in the management of these diseases include treatment with antiplatelet drugs, statins, or other lipid lowering drugs, anticoagulant drugs, beta blockers, renin angiotensin aldosterone (ACE) inhibitors, diuretics, etc. The use of lipid-lowering agents significantly improves the survival rate in patients affected by coronary artery diseases. The existing therapies that are being used for the prevention of CVDs possess a certain set of limitations and adverse effects and do not provide comprehensive treatment. Hence, owing to a greater population exposed to and affected with these disorders, there comes a dire need to explore and develop other health-friendly treatment approaches that can prevent the disease with minimal side effects. The natural products obtained from marine sources have been considered to be a valuable source for the discovery of new and novel lead molecules by virtue of the diverse chemical structures and pharmacological properties associated with them.

Some of the marine drugs that have been proven to be efficacious in the management of cardiovascular diseases, including coronary artery diseases, are discussed below.

5.1. Fucoxanthin

Fucoxanthin is an oxygen-containing carotenoid compound isolated from brown algae. This compound has been studied as a treatment to prevent the over-oxidation of lipids and restrain their accumulation [60]. Fucoxanthin also downregulates transcription factors such as peroxisome proliferator activated receptor (PPAR) and sterol regulatory element-binding protein 1c, which are play a role in adipogenesis. This compound decreases the expression of fatty acid synthase and increases the activity of adipose triglyceride lipase, further leading to the production and phosphorylation of hormones that are sensitive to lipolysis. Furthermore, fucoxanthin also represses the genetic expression of interleukin- 6 and acetyl coenzyme A carboxylase (ACC) and ultimately prohibits the accumulation of adipose tissue, improves resistance to insulin, reduces the thickness of cholesterol, and also affects the concentration of triglycerides in the body. The compound successfully upgraded the functioning of glucose transporter 4 (GLUT 4) and enabled it to consume more energy by oxidizing fatty acids and producing heat. A remarkable fall in the levels of blood glucose was also achieved by its property of enhancing the manifestation of glucokinase (GCK) mRNA and curbing the expression of mRNA phosphoenolpyruvate carboxy kinase. All of these factors help in reducing cardiovascular risks by decreasing lipid concentrations, and fucoxanthin can be used in the management of cardiovascular diseases such as hyperlipidemia, atherosclerosis, coronary artery disease, hypertension, etc. [61]. The illustration of proteins PPAR-α, carnitine, and p-ACC was promoted by fucoxanthin, which led to decreased expression of fatty acid synthase (FAS) on the liver, thereby decreasing the levels of lipids in the bloodstream. Advanced ongoing investigations revealed the beneficial effects of fucoxanthin in preventing lipid increase, and hence controlling and managing cardiovascular diseases.

5.2. Saponins

Sea cucumber saponins (SCS) are a glycoside group that comprises of spirostane and triterpene aglycone compounds. They are the secondary metabolites that are obtained from the sea and have been demonstrated to possess anti-atherosclerotic activity. They mainly act by regulating the metabolism of lipids and glucose in the body. The main functional constituent of sea cucumber saponins is adiponectin, which is believed to restore lipid and glucose metabolism by the promotion of sirtuin 1 (SIRT1), which further restricts the activity of sterol regulatory element-binding protein (SREBP)-1c and FAS. All of these factors ultimately promote hepatic fatty acid oxidation. The mRNA levels of GCK can be stimulated by SIRT 1 in the liver, which results in the catalysis of phosphorylation of glucose and is the first move towards glycolysis [62]. To conclude, the SCSs ameliorate the metabolism of lipids and glucose by altering the activity of SIRT 1, FAS, GCK, PPAR-α upon secretion of adiponectin and hence suppressing lipid production and synthesis and enhancing oxidation of fatty acids, thereby reducing the risk of cardiovascular diseases.

5.3. Astaxanthin

Astaxanthin, also classified as xanthophyll, is a naturally occurring microalgae and marine compound renowned for its bioactivity. The unique structure of the astaxanthin molecule accounts for its characteristically strong antioxidant activity and enables it to quench oxygen in singlet states and release free radicals. The molecule depicts its beneficial role in several biological activities such as amelioration of oxidative stress, decrease in inflammation, and altered lipid and glucose concentrations. Astaxanthin induces the oxidation of high-density lipoproteins (HDL) and adiponectin levels and suppresses the oxidation of low-density lipoproteins (LDL), i.e., promotes good cholesterol. The molecule possesses atherosclerosis-protective effects, as it can lighten the proportion of plaque formation and cholesterol in the aortic sinus region and aorta in mice, respectively. The levels of thiobarbituric acid (TBARS) were also decreased owing to the antioxidant activity of the compound, and hence it acts as a strong oxidation resistance substance [63]. The macrophage activation was suppressed, while granulocytosis of neutrophils was promoted by astaxanthin to induce phagocytosis and sterilization. Hence, this compound is inferred as being able to sustain less damage in body and to control the levels of lipid upon its elevation.

5.4. Xyloketal B

Xyloketal Bis a new style of an extraordinary structure containing marine compound isolated from the *Xylaria* species. This compound prevents endotheliocytes from suffering oxidative injury by inducing the oxidation of low-density lipoprotein due to suppression of NADPH oxidase-derived reactive oxygen species synthesis, and it facilitates the further release of nitric oxide. The endothelial dysfunction due to oxidative stress is also controlled. The compound has been determined to be effective in suppressing the levels of oxidative stress in vascular tissues, to improve the integrity of endothelial cells that were previously injured due to conditions of stress, and to promote vaso relaxation. Xyloketal B also reduced the levels of lipids by promoting their oxidation and by suppressing their accumulation [64].

The consumption and demand of marine products has increased gradually with development in the recent years. Marine products are believed to contain more bioactive substances, as they have a tendency to survive in the complex environment of the ocean. Extensive research in products obtained from marine resources has confirmed the role of marine drugs in the management of cardiovascular diseases, which has also encouraged researchers to hunt for and discover more safe and novel lead molecules (Table 3).

Table 3. The below table lists the marine drugs used in cardiovascular disorders with their respective structures.

Name	Structure
Fucoxanthin	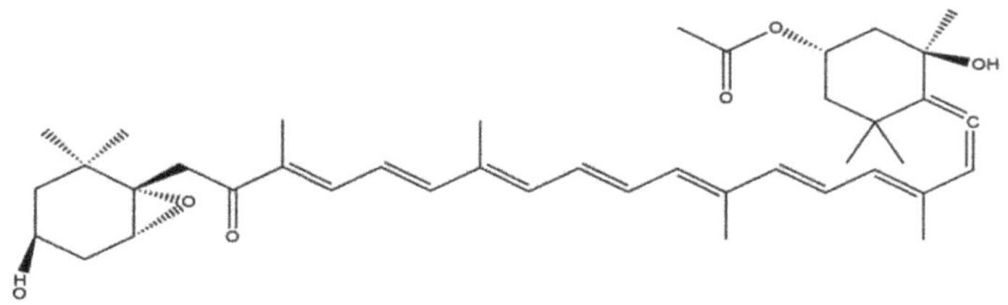

Table 3. *Cont.*

Name	Structure
adiponectin	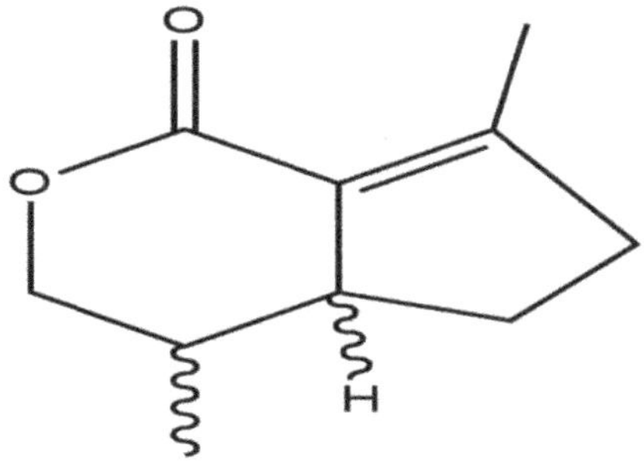
Astaxanthin	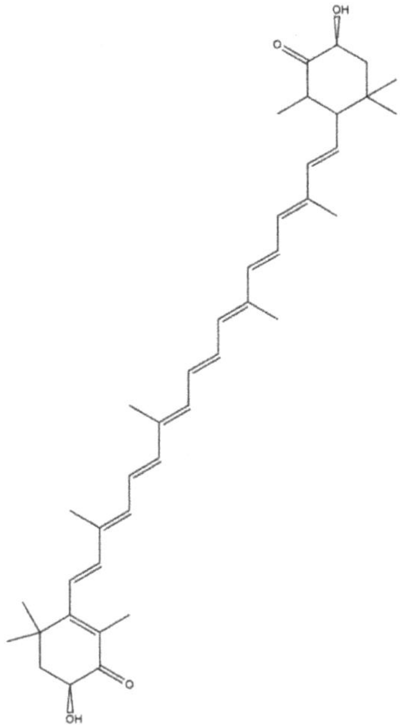
Xyloketal B	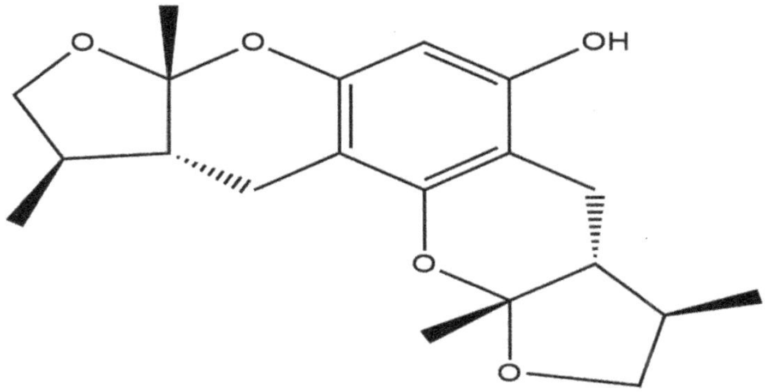

6. Marine Drugs in Neurodegeneration

Neurodegenerative diseases are the greatest medical challenge faced by researchers in the 21st century. They are a group of heterogeneous late onset disorders that are caused by progressive dysfunction of neuronal cells and their death, leading to cognitive impairment and altered movement. With increasing age, the incidence of neurodegenerative diseases also increases, and they are more common in the elderly age group of patients. The main factor that is associated with the process of neurodegeneration is aging [65,66]. These diseases are mainly characterized by loss in the functioning of neurons and the formation of misfolded proteins, accompanied by their aggregation, extracellular intracellular deposits, and neuronal cell death. Excessive release of reactive oxygen species (ROS) and neuroinflammation contribute chiefly to the pathophysiology of different neurodegenerative diseases and depict a direct consequence of perturbation in the homeostasis of the central nervous system (CNS) [67]. Effective treatments are being explored, and efforts are being made to manage age-related degeneration and promote stability in patients. The current therapies employed in the management of neurodegenerative disorders improve quality of life to only a certain extent. Hence, the field calls for research and development that explores and discovers new and safer lead molecules with novel mechanisms of action by targeting certain physiological pathways and improving the condition of the patients. As the interest rises in natural products, the attention also drifts towards the sea, as it has the greatest diversity of plants and animals and is the least explored. Hence, below are described some of the promising compounds obtained from the sea and a description of their consumption as supplemental therapies for a role in the management and prevention of neurodegenerative diseases [68,69] (Figure 2).

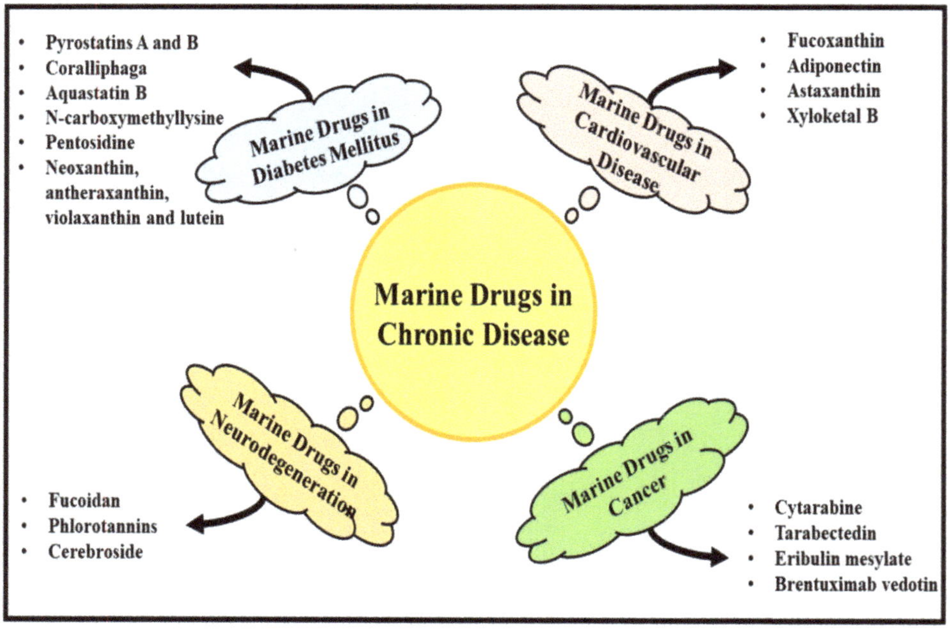

Figure 2. This figure provides a descriptive overview of the marine drugs used in diabetes mellitus, cancer, neurodegenerative disease, and cardiovascular disorders.

The venom of sea snail *Conus Magus* led to the derivation of a synthetic form of the peptideziconotide, which is a selective N-type calcium channel blocker and is used in conditions of chronic pain owing to its antinociceptive effect without the development of tolerance, as is witnessed in other opioids and morphine. This fact represents a significant advantage of this marine-derived product for use in the long-term management of conditions of pain. Many other drugs are also being identified from the ocean and extracts of marine invertebrates. Several compounds isolated from marine sources have been studied and found to possess modulatory action on the CNS, and can thus be involved in the management of neurodegenerative conditions. For instance, potent cholinesterase inhibitors were isolated from sponges, mollusks, algae, bryozoans, cnidarians echinoderms, and tunicates. Xestospongia, xestosaprols, Indonesian marine sponges, tasiamide B cyanobacteria, carteriospongia-derived carteriosulfonic acids, petrosamine sponges, *Reniera sarai* sponges, octocorals, and lophotoxin are some of the marine-derived bioactive products that are assumed to be hypothetical candidates for development as novel drugs for themanagement of CNS disorders. Marine source-derived cholinesterase inhibitors can be used in the management of mild and moderate atopic dermatitis; however, their use can only reduce symptoms in a patient and cannot arrest the progression of the disease [70,71]. Another marine compound named conotoxins is an antagonist of CNS receptors and exhibits a significant role in various physiological processes such as memory, learning, and attention. Some of the marine drugs used prominently in the management of neurodegenerative disease are listed below.

6.1. *Fucoidan*

Fucoidan is a polysaccharide isolated from brown algae—namely, *Saccharina japonica*—and comprises sugar, fucose, and sulfate content [72]. This compound exhibits a protective effect in the management of Parkinson's disease due to its interaction with 1-methyl-4-phenyl-1,2,3,6- tetrahydropyridine (MPTP), and it thereby effectively improves motor impairment. The compound also successfully was able to counteract the depletion of dopaminergic neurons in the striatal region. Upon treatment with this compound, the morphology of neuronal cells was preserved and mitochondrial activity was found to be increased. The pharmacological mechanism of this protective action is unclear; however, it is demonstrated to be due to its antioxidant activity, as it prevents the generation of reactive oxygen species and other mechanisms such as anti-inflammation [73].

6.2. *Seaweeds*

Seaweed is another marine compound that is believed to have exceptional bioactive properties due to its antioxidant nature. The different extracts of seaweeds were tested, and their neuroprotective effect was elucidated as being owed to their anti-apoptotic mechanism. The extracts substantially prevented dopaminergic neurotoxicity, and an increase in cell viability was noted, leading to its use in the management of Parkinson's disease. Phlorotannins, exclusively found in brown seaweeds, are responsible for the antioxidant activity of this compound. *Codium Tomentosum* has emerged as a promising neuroprotective seaweed that has anti-genotoxic and anti-oxidative properties. The compound readily scavenges both nitrogen and oxygen free radical species and contains different organic acids, such as malic acid, aconitic acid, oxalic acid, malonic and fumaric acids, phenolic compounds, and secondary metabolites [74,75].

6.3. *Cerebrosides*

Cerebrosides are a category of neutral glycosphingolipids that are mainly present in marine species. These compounds are extensively found in the brain and convert into ceramides, which are further transitioned into sphingomyelins, sulfatides, and other associated glycosphingolipids. These compounds are highly crucial in maintaining functioning and homeostasis in the brain. The unique structure of this compound is composed of three units—a polar monosaccharide head, long chain of sphingoid base, and amide-linked fatty acids—that contribute to the biological activities of this compound that has attracted the attention of the pharmaceutical industry. Sea cucumber cerebrosides are believed to improve cognitive functioning and synaptic plasticity and to attenuate hyperphosphorylation of tau proteins by regulating the neurotrophic pathway. Administering sea cucumber cerebrosides as a food supplement may present neuroprotective effects and may even improve cognitive functioning. The compound has also been reported to have activity useful in the management of Alzheimer's and Parkinson's diseases. Certain marine compounds can be further synthesized through this molecule to produce more specific biological effects [76–78] (Tables 4 and 5), Figure 2.

Table 4. The below table lists the marine drugs used in neurodegenerative diseases with their respective structures.

Name	Structure
Fucoidan	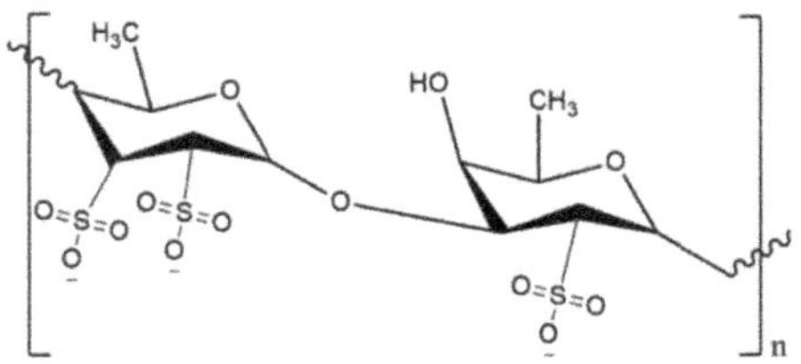
Phlorotannins	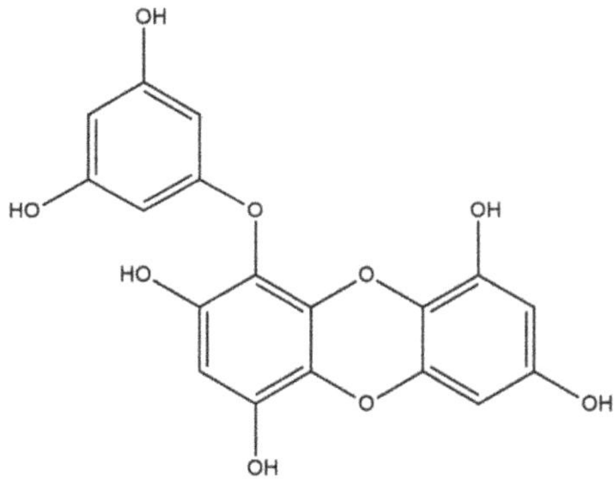
Cerebrosides	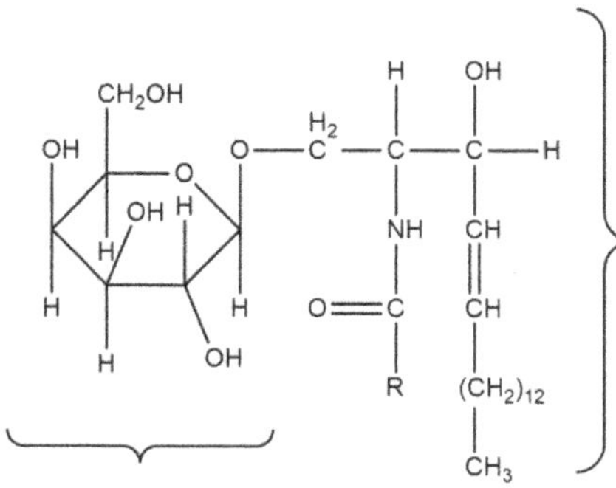 Carbohydrate residue; Ceramide Residue

Table 5. The below table lists compounds obtained from marine sources and their uses.

Name of the Compound	Source	Scientific Name	Uses	References
Ziconotide	Cone snails	*Conidae*	Management of spinal cord injury-mediated chronic pain	[4]

Table 5. *Cont.*

Name of the Compound	Source	Scientific Name	Uses	References
Hemiasterlin, discodermolide	Marine sponges	*Spongia officinalis*	Anticancer, anti-inflammatory, antibiotic	[11]
Cytarabine	Caribbean sponge	*Cryptotethya crypta*	Anticancer	[32–34]
Trabectedin	Caribbean ascidian	*Ecteinascidia turbinate*	Anticancer	[35]
Eribulin Mesylate	Sponge	*Halichondria okadai*	Anticancer	[36]
Brentuximab vedotin	Gastropod mollusk	*Dolabella auriculria*	Anticancer	[36,37]
	marine sponge bacterium	*Coralliphaga*	Antidiabetic	[43]
Aquastatin B	Marine fungi	*Cosmospora* species SF-5060	Antidiabetic	[44]
	Chlorella and diatom	*Nitzschia laevis*	Antidiabetic	[47]
Astaxanthin		*Chlorella zofingiensis*	Antidiabetic	[48]
	Green, red and brown algae	*Palmaria, Ecklonia cava, Alaria, Rhodomela confervoides* and *Ascophyllum*	Antidiabetic	[40,51]
Fucoxanthin	Brown algae	*Phaeophyceae*	Antihyperlipidemic	[53]
Spirostaneand triterpene aglycone compounds	Sea cucumber saponins	*Holothurialessoni*	Anti-atherosclerotic	[55]
Xyloketal B		*Xylaria* species	Antioxidant, antihyperlipidemic	[57]
Fucoidan	Brown algae	*Saccharina japonica*	Parkinson disease, anti-inflammation	[63]
	Seaweed	*Codium Tomentosum*	Neuroprotective, anti-apoptotic	[64,65]
Aplidin	Marine tunicate	*Aplidium albicans*	Anticancer	[76]
Tetrodotoxin	Puffer fish	*Tetraodontidae*	Analgesic	[79,80]
Plinabulin	Marine fungus	belonging to species of *Aspergillus*	Under investigation in clinical trials phase III as antitumor	[81–84]
Salinosporamide A or Marizomib	Marine bacteria	*Salinispora arenicola* and *Salinispora tropica*	Proteasome inhibitor	[85–87]

7. Future Prospects of Marine Drugs and Drug Candidates Obtained from the Ocean

For decades, a strong effort has been undertaken by the pharmaceutical industry and academia to derive therapeutically active marine drugs [78]. Eleven drugs with their origin in marine natural sources have successfully navigated their way through development and into the market, among which at least five can be used in the management of cancer, such as cytarabine (Cytosar-U), eribulin mesylate (Halaven), and trabectedin (ET-473, Yondelis) and including antibody drug conjugates such as polatuzumabvedotin (Polivy) and brentuximab vedotin (Adcetris). Other drugs such as Epanova, Lovaza, and Vascepa have also been used in treatment of hypertriglyceridemia, while vidarabine (Vira-A) and iota-carageenan (Carragelose) have been used as antiviral drugs and zinconotide (Prialt) has been used for ameliorating chronic pain. More than 20 drug candidates have been obtained from marine natural products as of now and which are under investigation in different phases of clinical trials, including several drug leads that are being studied extensively for their activity in pre-clinical studies [79–82]. The link between a direct therapeutic target (also called the binding partner) and a mechanism of bioactive action associated with the disease serves as an incentive for conducting preclinical and clinical studies. The use of novel techniques in the identification of targets and the capture of associated mechanisms of action act as a game changer and can lead to the implementation of early stages of discovery efforts for new drug leads [83,84]. Marine natural products are highly diverse and thus present a greater opportunity for discovering a wide number of first-in-class therapeutic agents that comprise proteins and other unusual biological targets, such as lipid membranes, sterols, etc. [85]. These targets further lead to the production of the effective antibody drug conjugates that are commonly used in the management of several chronic diseases [86]. The marine environment offers a rich arsenal of bioweapons that possess diverse structure and functional properties and that are under investigation in different stages of clinical trials. Aplidin is a first-in-class cyclic depsipeptideplitidepsin isolated from marine tunicate named *Aplidium albicans,* and it interacts strongly with eukaryotic elongation factor 1A2 (eEF1A2) in the cells of a tumor [87]. This marine drug is being studied in combination with dexamethasone for its efficacy in the management and treatment of relapsed and refractory multiple myeloma and has successfully reached phase III clinical trials [88].

Tectin or tetrodotoxin is a derivative of guanidine and comprises a skeleton of highly oxygenated carbon obtained from puffer fish, which produce one of the most popular marine toxins [89]. It is mainly isolated from the liver, skin, and ovaries of puffer fish belonging to family *Tetraodontidae.* The toxin accumulates in the body through the food chain in marine oceans and is derived from specific endosymbiotic bacteria, mainly pseudomonas and vibrio species. Tetrodotoxin acts by competitively and selectively hindering the activity of voltage-gated sodium channels (VGSCs) in the cell membrane of the nerve cells, which prevents propagation of action potential due to depolarization and causes loss of sensation [90]. The toxin is being used to an extensive degree due to its chemical nature in characterizing voltage-gated sodium channels, and it is also applied as an analgesic to cure pain. Tetrodotoxin successfully completed two phase III safety and efficacy studies in 2012 for its use in the management of moderate-to-severe inadequately controlled pain related to cancer, the results of which are still awaited [91].

Another marine drug that is commonly used for its therapeutic use is plinabulin, a synthetic analogue of a naturally occurring aromatic alkaloid—namely, halimide—mainly obtained from a marine fungus belonging to a species of *Aspergillus* [92,93]. It is mainly collected from the waters of the Philippine islands. Plinabulin targets tubulin monomer near the colchicine binding site and inhibits polymerization of tubulin, which further causes tumor vascular network destabilization and reduces blood flow to the tumor region, thereby preventing tumor progression [94]. The marine drug is being investigated in phase III clinical trials for its apoptotic effects in tumor cells and vascular disruption activity in the management and treatment of non-small cell lung cancer [84,95].

The marine bacteria named *Salinispora arenicola* and *Salinispora tropica* led to isolation of a potent proteasome inhibitor, Salinosporamide A or Marizomib. Proteasomes received a huge amount of attention after the success of the first proteasome inhibitor, the drug bortezomib, which was approved for the management of cancer and has been developed as a class of proteins in anticancer therapy [96]. In the year 2012, an analog of epoxomicin—namely, carfilzomib (Kyprolis)—was approved as an anticancer drug [97,98]. Carfilzomib was isolated from strains of actinomyces and acted as a potent proteasome inhibitor, suggesting its eminent role as a marine-derived therapeutic agent. Within 3 years of its discovery, Salinosporamide A successfully entered phase I clinical trials for investigation of its activity in the management of multiple myeloma, and it was further transitioned into phase 2 trials that have also been completed (Figure 3).

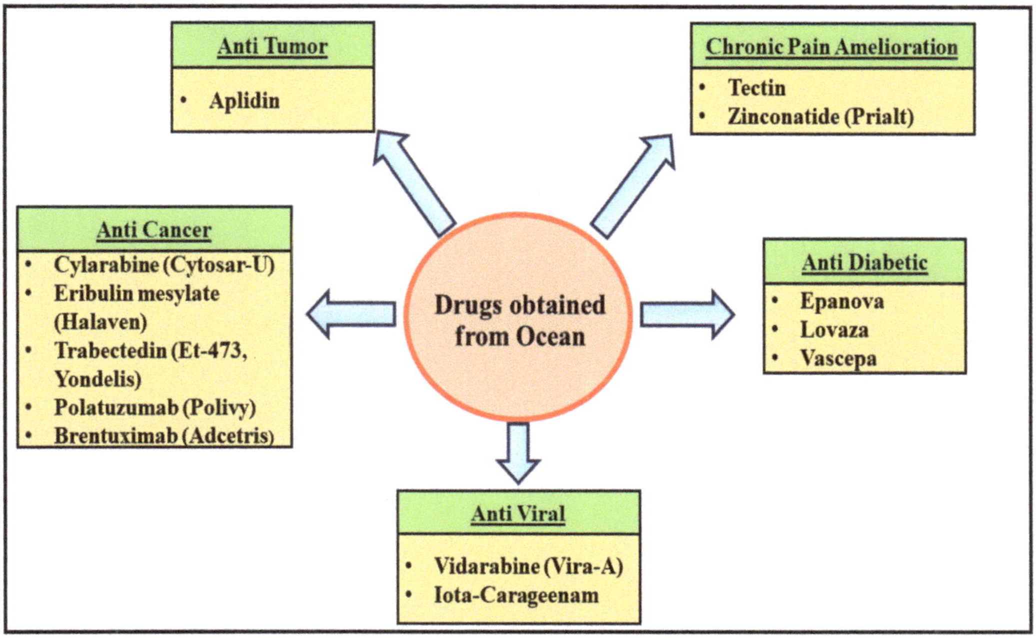

Figure 3. Marine drugs obtained from the ocean and their therapeutic uses in the management of conditions, starting from pain to life threatening diseases such as cancer.

8. Conclusions

The structure–activity relationship (SAR) of the isolated marine natural products and their derivatives revealed a scaffold that increased the activity of the chemical compounds obtained for the therapeutic management of several disorders. The marine products obtained can also be optimized further in laboratories to synthesize more precise and potent lead molecules that can be effectively employed for their disease targeting activity and also to fight against pathogens. Advances in techniques have led to an increased understanding and broadening of the biological activities of these drugs in the management of chronic diseases, including cancers. The interest of researchers in marine products is flourishing, which has led to the budding of innovative solutions for overcoming the challenges of their extraction and isolation. Marine drugs exert significant therapeutic effects such as antioxidant, anti-inflammatory, anti-apoptotic, antimicrobial, antidiabetic, and antihyperlipidemic properties. The consequences of administering these compounds are extremely less severe when considered in the context of the intensity of a disease and other alternative therapies.

Author Contributions: Conceptualization, S.B., R.M., and T.B.; methodology, T.B. and R.M.; investigation, S.B., R.M., and T.B.; resources, T.B. and S.G.B.; data curation, M.R., A.S., and S.S.; writing—original draft preparation, S.B., R.M., and T.B.; writing—review and editing, T.B. and V.M.; visualization, A.S. and M.S.I.; supervision, T.B. and S.G.B. All authors have read and agreed to the published version of the manuscript.

Funding: Funding for publication of this paper was provided by the University of Oradea, Oradea, Romania, by an Internal Project.

Institutional Review Board Statement: Not applicable.

Informed Consent Statement: Not applicable.

Data Availability Statement: Not applicable.

Acknowledgments: The authors would like to thank the Chitkara College of Pharmacy, Chitkara University, Punjab, India, for providing the basic facilities for the completion of the present article.

Conflicts of Interest: The authors declare no conflict of interest.

References

1. Jiménez, C. Marine Natural Productsin Medicinal Chemistry. *ACS Med. Chem. Lett.* **2018**, *9*, 959–961. [CrossRef]
2. Newman, D.J.; Cragg, G.M. Natural Productsas Sources of New Drugs from 1981 to 2014. *J. Nat. Prod.* **2016**, *79*, 629–661. [CrossRef]
3. Patridge, E.; Gareiss, P.; Kinch, M.S.; Hoyer, D. An Analysis of FDA-Approved Drugs: Natural Products and Their Derivatives. *Drug Discov. Today* **2016**, *21*, 204–207. [CrossRef] [PubMed]
4. Ghosh, S.; Sarkar, T.; Pati, S.; Kari, Z.A.; Edinur, H.A.; Chakraborty, R. Novel Bioactive Compounds From Marine Sourcesas a Tool for Functional Food Development. *Front. Mar. Sci.* **2022**, *76*, 832957. [CrossRef]
5. Blunt, J.W.; Copp, B.R.; Hu, W.P.; Munro, M.H.G.; Northcote, P.T.; Prinsep, M.R. Marine Natural Products. *Nat. Prod. Rep.* **2009**, *26*, 170–244. [CrossRef] [PubMed]
6. Newman, D.J.; Cragg, G.M. Drugs and Drug Candidates from Marine Sources: An Assessment of the Current "StateofPlay". *Planta Med.* **2016**, *82*, 775–789. [CrossRef]
7. Glevitzky, I.; Dumitrel, G.A.; Glevitzky, M.; Pasca, B.; Otrisal, P.; Bungau, S.; Cioca, G.; Pantis, C.; Popa, M. Statistical Analysis of the Relationship between Antioxidant Activity and the Structure of Flavonoid Compounds. *Rev. De Chim.* **2019**, *70*, 3103–3107. [CrossRef]
8. Altmann, K.H. Drugs from the Oceans: Marine Natural Products as Leads for Drug Discovery. *Chimia* **2017**, *71*, 646–652. [CrossRef]
9. Romano, G.; Costantini, M.; Sansone, C.; Lauritano, C.; Ruocco, N.; Ianora, A. Marine Microorganisms as a Promising and Sustainable Source of Bioactive Molecules. *Mar. Environ. Res.* **2017**, *128*, 58–69. [CrossRef]
10. Anjum, K.; Abbas, S.Q.; Shah, S.A.A.; Akhter, N.; Batool, S.; Hassan, S.S.U. Marine Sponges as a Drug Treasure. *Biomol. Ther.* **2016**, *24*, 347. [CrossRef] [PubMed]
11. Moumbock, A.F.A.; Li, J.; Mishra, P.; Gao, M.; Günther, S. Current Computational Methods for Predicting Protein Interactions of Natural Products. *Comput. Struct. Biotechnol. J.* **2019**, *17*, 1367. [CrossRef] [PubMed]
12. Moumbock, A.F.A.; Simoben, C.V.; Wessjohann, L.; Sippl, W.; Günther, S.; Ntie-Kang, F. Computational studies and biosynthesis of natural products with promising anticancer properties. In *Natural Products and Cancer Drug Discovery*; InTech Open: London, UK, 2017. [CrossRef]
13. Tirumala, M.G.; Anchi, P.; Raja, S.; Rachamalla, M.; Godugu, C. Novel Methods and Approaches for Safety Evaluation of Nanoparticle Formulations: A Focus towards in Vitro Models and Adverse Outcome Pathways (AOP). *Front. Pharmacol.* **2021**, *12*, 2157. [CrossRef] [PubMed]
14. Walsh, C.T.; Fischbach, M.A. Natural Products Version 2.0: Connecting Genes to Molecules. *J. Am. Chem. Soc.* **2010**, *132*, 2469–2493. [CrossRef] [PubMed]
15. Grover, M.; Behl, T.; Sehgal, A.; Singh, S.; Sharma, N.; Virmani, T.; Rachamalla, M.; Farasani, A.; Chigurupati, S.; Alsubayiel, A.; et al. In Vitro Phytochemical Screening, Cytotoxicity Studies of Curcuma Longa Extracts with Isolation and Characterisation of Their Isolated Compounds. *Molecules* **2021**, *26*, 7509. [CrossRef] [PubMed]
16. Puppala, E.R.; Jain, S.; Saha, P.; Rachamalla, M.; Syamprasad, N.P.; Yalamarthi, S.S.; Abubakar, M.; Chaudhary, A.; Chamundeswari, D.; Murty, U.S.N.; et al. Perillyl Alcohol Attenuates Rheumatoid Arthritis via Regulating TLR4/NF-K Band Keap1/Nrf2 Signaling Pathways: A Comprehensive Study On in-Vitro and in-Vivo Experimental Models. *Phytomedicine* **2022**, *97*, 153926. [CrossRef]
17. Molinski, T.F.; Dalisay, D.S.; Lievens, S.L.; Saludes, J.P. Drug Development from Marine Natural Products. *Nat. Rev. Drug Discov.* **2009**, *8*, 69–85. [CrossRef]
18. Dinarvand, M.; Spain, M. Identification of Bioactive Compounds from Marine Natural Products and Exploration of Structure—Activity Relationships (Sar). *Antibiotics* **2021**, *10*, 337. [CrossRef]
19. Pereira, F.; Aires-de-Sousa, J. Computational Methodologies in the Exploration of Marine Natural Product Leads. *Mar. Drugs* **2018**, *16*, 236. [CrossRef]
20. Di Masi, J.A.; Grabowski, H.G.; Hansen, R.W. Innovation in the Pharmaceutical Industry: New Estimates of R&D Costs. *J. Health Econ.* **2016**, *47*, 20–33. [CrossRef]
21. Nantasenamat, C.; Prachayasittikul, V. Maximizing Computational Tools for Successful Drug Discovery. *Expert Opin. Drug Discov.* **2015**, *10*, 321–329. [CrossRef]
22. Gasteiger, J. Chemoinformatics: Achievements and Challenges, a Personal View. *Molecules* **2016**, *21*, 151. [CrossRef]
23. Mueller, R.; Dawson, E.S.; Meiler, J.; Rodriguez, A.L.; Chauder, B.A.; Bates, B.S.; Felts, A.S.; Lamb, J.P.; Menon, U.N.; Jadhav, S.B.; et al. Discovery of (2-(2-Benzoxazoyl Amino)-4-Aryl-5-Cyanopyrimidine MGlu 5NAMs: From Artificial Neural Network Virtual Screen to in Vivo Tool Compound. *Chem. Med. Chem* **2012**, *7*, 406. [CrossRef] [PubMed]
24. Leelananda, S.P.; Lindert, S. Computational Methods in Drug Discovery. *BeilsteinJ. Org. Chem.* **2016**, *12*, 2694–2718. [CrossRef] [PubMed]

25. Katsila, T.; Spyroulias, G.A.; Patrinos, G.P.; Matsoukas, M.T. Computational Approaches in Target Identification and Drug Discovery. *Comput. Struct. Biotechnol. J.* **2016**, *14*, 177–184. [CrossRef] [PubMed]
26. Rodriguez, A.L.; Grier, M.D.; Jones, C.K.; Herman, E.J.; Kane, A.S.; Smith, R.L.; Williams, R.; Zhou, Y.; Marlo, J.E.; Days, E.L.; et al. Discovery of Novel Allosteric Modulators of Metabotropic Glutamate Receptor Subtype5 Reveals Chemical and Functional Diversity and in Vivo Activity in Rat Behavioral Models of Anxiolytic and Antipsychotic Activity. *Mol. Pharmacol.* **2010**, *78*, 1105–1123. [CrossRef]
27. Gaudêncio, S.P.; Pereira, F. Dereplication: Racing to Speed up the Natural Products Discovery Process. *Nat. Prod. Rep.* **2015**, *32*, 779–810. [CrossRef]
28. Pérez-Victoria, I.; Martín, J.; Reyes, F. Combined LC/UV/MS and NMR Strategies for the Dereplication of Marine Natural Products. *Planta Med.* **2016**, *82*, 310. [CrossRef]
29. Vijayakrishnan, R. Structure-Based Drug Design and Modern Medicine. *J. Postgrad. Med.* **2009**, *55*, 310. [CrossRef]
30. Talele, T.; Khedkar, S.; Rigby, A. Successful Applications of Computer Aided Drug Discovery: Moving Drugs from Concept to the Clinic. *Curr. Top. Med. Chem.* **2010**, *10*, 127–141. [CrossRef]
31. Jones, P.A.; Baylin, S.B. The Epigenomics of Cancer. *Cell* **2007**, *128*, 683–692. [CrossRef]
32. Chauthe, S.K.; Mahajan, S.; Rachamalla, M.; Tikoo, K.; Singh, I.P. Synthesis and Evaluation of Linear Furanocoumarinsas Potential Anti-Breast and Anti-Prostate Cancer Agents. *Med. Chem. Res.* **2015**, *24*, 2476–2484. [CrossRef]
33. Varun, K.; Mahesh, R.; Prajwal, N.L.; Khatik, G.T.; Sangamwar, A.; Kulbhushan, T.A.; Nair, V. Design and Synthesis of Optically Pure3-Aryl-6-Methyl-2-Thioxotetrahydropyrimidin-4(1H)-Ones as Anti-ProstateCancerAgents. *RSC Adv.* **2014**, *4*, 37868–37877. [CrossRef]
34. Lujambio, A.; Lowe, S.W. The Microcosmos of Cancer. *Nature* **2012**, *482*, 347–355. [CrossRef] [PubMed]
35. Bouchet, B.P.; Galmarini, C.M. Cabazitaxel, a New Taxane with Favorable Properties. *Drugs Today* **2010**, *46*, 735–742. [CrossRef]
36. Rocha, J.; Peixe, L.; Gomes, N.C.M.; Calado, R. Cnidarians as a Source of New Marine Bioactive Compounds An Overview of the Last Decade and Future Steps for Bioprospecting. *Mar. Drugs* **2011**, *9*, 1860–1886. [CrossRef] [PubMed]
37. Jensen, P.R.; Fenical, W. Marine microorganisms and drug discovery: Current status and future potential. In *Drugs from the Sea;* Karger: Basel, Switzerland, 2000; pp. 6–29.
38. Dayanidhi, D.L.; Thomas, B.C.; Osterberg, J.S.; Vuong, M.; Vargas, G.; Kwartler, S.K.; Schmaltz, E.; Dunphy-Daly, M.M.; Schultz, T.F.; Rittschof, D.; et al. Exploring the Diversity of the Marine Environment for New Anti-Cancer Compounds. *Front. Mar. Sci.* **2021**, *7*, 1184. [CrossRef]
39. Saeed, A.F.U.H.; Su, J.; Ouyang, S. Marine-Derived Drugs: Recent Advances in Cancer Therapy and Immune Signaling. *Biomed. Pharmacother.* **2021**, *134*, 111091. [CrossRef]
40. Patel, R.S.; Rachamalla, M.; Chary, N.R.; Shera, F.Y.; Tikoo, K.; Jena, G. Cytarabine Induced Cerebellar Neuronal Damage in Juvenile Rat: Correlating Neurobehavioral Performance with Cellular and Genetic Alterations. *Toxicology* **2012**, *293*, 41–52. [CrossRef]
41. Dyshlovoy, S.A.; Honecker, F. Marine Compounds and Cancer: The First Two Decades of XXI Century. *Mar. Drugs* **2020**, *18*, 20. [CrossRef]
42. Dyshlovoy, S.A.; Honecker, F. Marine Compounds and Cancer: Updates 2020. *Mar. Drugs* **2020**, *18*, 643. [CrossRef]
43. Jimenez, P.C.; Wilke, D.V.; Costa-Lotufo, L.V. Marine Drugs for Cancer: Surfacing Biotechnological Innovations from the Oceans. *Clinics* **2018**, *73*, 42. [CrossRef] [PubMed]
44. Van de Donk, N.W.C.J.; Dhimolea, E. Brentuximab Vedotin. *mAbs* **2012**, *4*, 458–465. [CrossRef]
45. American Diabetes Association. Diagnosis and Classification of Diabetes Mellitus. *Diabetes Care* **2014**, *37*, 62–67. [CrossRef]
46. Newsholme, P.; Cruzat, V.F.; Keane, K.N.; Carlessi, R.; de Bittencourt, P.I.H. Molecular Mechanisms of ROS Production and Oxidative Stress in Diabetes. *Biochem. J.* **2016**, *473*, 4527–4550. [CrossRef]
47. Barde, S.R.; Sakhare, R.S.; Kanthale, S.B.; Chandak, P.G.; Jamkhande, P.G. Marine Bioactive Agents: A Short Review on New Marine Antidiabetic Compounds. *Asian Pac. J. Trop. Dis.* **2015**, *5*, 209–213. [CrossRef]
48. Skyler, J.S. Diabetes Mellitus: Pathogenesis and Treatment Strategies. *J. Med. Chem.* **2004**, *47*, 4113–4117. [CrossRef] [PubMed]
49. Lordan, S.; Ross, R.P.; Stanton, C. Marine Bioactives as Functional Food Ingredients: Potential to Reduce the Incidence of Chronic Diseases. *Mar. Drugs* **2011**, *9*, 1056–1100. [CrossRef] [PubMed]
50. Imada, C. Enzyme Inhibitors and Other Bioactive Compounds from Marine Actinomycetes. *Antonie Leeuwenhoek* **2005**, *87*, 59–63. [CrossRef]
51. Debbab, A.; Aly, A.H.; Lin, W.H.; Proksch, P. Bioactive Compounds from Marine Bacteria and Fungi: Mini review. *Microb. Biotechnol.* **2010**, *3*, 544–563. [CrossRef]
52. Carotenuto, Y.; Esposito, F.; Pisano, F.; Lauritano, C.; Perna, M.; Miralto, A.; Ianora, A. Multi-Generation Cultivation of the Copepod *Calanus Helgolandicusina* Re-Circulating System. *J. Exp. Mar. Biol. Ecol.* **2012**, *418*, 46–58. [CrossRef]
53. Ianora, A.; Miralto, A.; Poulet, S.A.; Carotenuto, Y.; Buttino, I.; Romano, G.; Casotti, R.; Pohnert, G.; Wichard, T.; Colucci-D'Amato, L.; et al. Aldehyde Suppression of Copepod Recruitmentin Blooms of a Ubiquitous Planktonic Diatom. *Nature* **2004**, *429*, 403–407. [CrossRef] [PubMed]
54. Sun, Z.; Peng, X.; Liu, J.; Fan, K.W.; Wang, M.; Chen, F. Inhibitory Effects of Microalgal Extracts on the Formation of Advanced Glycation Endproducts (AGEs). *Food Chem.* **2010**, *120*, 261–267. [CrossRef]

55. Sun, Z.; Liu, J.; Zeng, X.; Huangfu, J.; Jiang, Y.; Wang, M.; Chen, F. Astaxanthin Is Responsible for Antiglycoxidative Properties of Microalga *Chlorella zofingiensis*. *Food Chem.* **2011**, *126*, 1629–1635. [CrossRef]
56. Sun, Z.; Liu, J.; Zeng, X.; Huangfu, J.; Jiang, Y.; Wang, M.; Chen, F. Protective Actions of Microalgae against Endogenous and Exogenous Advanced Glycation Endproducts (AGEs) in Human Retinal Pigment Epithelial Cells. *Food Funct.* **2011**, *2*, 251–258. [CrossRef] [PubMed]
57. Sharifuddin, Y.; Chin, Y.X.; Lim, P.E.; Phang, S.M. Potential Bioactive Compounds from Seaweed for Diabetes Management. *Mar. Drugs* **2015**, *13*, 5447–5491. [CrossRef] [PubMed]
58. Shi, D.; Guo, S.; Jiang, B.; Guo, C.; Wang, T.; Zhang, L.; Li, J. HPN, a Synthetic Analogue of Bromophenol from Red Alga *Rhodomela Confervoides*: Synthesis and Anti-Diabetic Effects in C57BL/KsJ-Db/Db Mice. *Mar. Drugs* **2013**, *11*, 350–362. [CrossRef]
59. Gaziano, T.; Reddy, K.S.; Paccaud, F.; Horton, S.; Chaturvedi, V. Cardiovascular Disease. *Dis. Control. Priorities Dev. Ctries.* **2006**, *2*, 1–44.
60. Zhao, J.; Cao, Q.; Xing, M.; Xiao, H.; Cheng, Z.; Song, S.; Ji, A. Advances in the Study of Marine Products with Lipid-Lowering Properties. *Mar. Drugs* **2020**, *18*, 390. [CrossRef]
61. Liang, B.; Cai, X.-Y.; Gu, N. Marine Natural Products and Coronary Artery Disease. *Front. Cardiovasc. Med.* **2021**, *8*, 739932. [CrossRef]
62. Cao, Q.; Zhao, J.; Xing, M.; Xiao, H.; Zhang, Q.; Liang, H.; Ji, A.; Song, S. Current Research Landscape of Marine-Derived Anti-Atherosclerotic Substances. *Mar. Drugs* **2020**, *18*, 440. [CrossRef]
63. Kishimoto, Y.; Yoshida, H.; Kondo, K. Potential Anti-Atherosclerotic Properties of Astaxanthin. *Mar. Drugs* **2016**, *14*, 35. [CrossRef] [PubMed]
64. Chen, W.L.; Qian, Y.; Meng, W.F.; Pang, J.Y.; Lin, Y.C.; Guan, Y.Y.; Chen, S.P.; Liu, J.; Pei, Z.; Wang, G.L. A Novel Marine Compound Xyloketal B Protects against Oxidized LDL-Induced Cell Injury in Vitro. *Biochem. Pharmacol.* **2009**, *78*, 941–950. [CrossRef]
65. Thakur, M.; Rachamalla, M.; Niyogi, S.; Datusalia, A.K.; Flora, S.J.S. Molecular Mechanism of Arsenic-Induced Neurotoxicity Including Neuronal Dysfunctions. *Int. J. Mol. Sci.* **2021**, *22*, 10077. [CrossRef] [PubMed]
66. Perry, G. Promise from the Sea. *Mar. Drugs* **2016**, *14*, 178. [CrossRef] [PubMed]
67. Bhardwaj, S.; Kesari, K.K.; Rachamalla, M.; Mani, S.; Ashraf, G.M.; Jha, S.K.; Kumar, P.; Ambasta, R.K.; Dureja, H.; Devkota, H.P.; et al. CRISPR/Cas9 Gene Editing: New Hope for Alzheimer's Disease Therapeutics. *J. Adv. Res.* **2021**. [CrossRef]
68. Rehni, A.K.; Singh, N.; Rachamalla, M.; Tikoo, K. Modulation of Histone Deacetylase AttenuatesNaloxone-Precipitated Opioid Withdrawal Syndrome. *Naunyn Schmiedebergs Arch. Pharmacol.* **2012**, *385*, 605–619. [CrossRef]
69. Alonso, D.; Castro, A.; Martinez, A. Marine Compounds for the Therapeutic Treatment of Neurological Disorders. *Expert Opin. Ther. Pat.* **2005**, *15*, 1377–1386. [CrossRef]
70. Grosso, C.; Valentão, P.; Ferreres, F.; Andrade, P.B. Review: Bioactive Marine Drugs and Marine Biomaterials for Brain Diseases. *Mar. Drugs* **2014**, *12*, 1377–1386.
71. Catanesi, M.; Caioni, G.; Castelli, V.; Benedetti, E.; d'Angelo, M.; Cimini, A. Benefits under the Sea: The Role of Marine Compounds in Neurodegenerative Disorders. *Mar. Drugs* **2021**, *19*, 24. [CrossRef]
72. Olasehinde, T.A.; Olaniran, A.O.; Okoh, A.I. Sulfated Polysaccharides of Some Seaweeds Exhibit Neuroprotection via Mitigation of Oxidative Stress, Cholinergic Dysfunction and Inhibition of Zn-Induced Neuronal Damage in HT-22 Cells. *BMC Complement. Med. Ther.* **2020**, *20*, 251. [CrossRef] [PubMed]
73. Luo, D.; Zhang, Q.; Wang, H.; Cui, Y.; Sun, Z.; Yang, J.; Zheng, Y.; Jia, J.; Yu, F.; Wang, X.; et al. Fucoidan Protects against Dopaminergic Neuron Death in Vivo and in Vitro. *Eur. J. Pharmacol.* **2009**, *617*, 33–40. [CrossRef]
74. Hannan, M.A.; Dash, R.; Haque, M.N.; Mohibbullah, M.; Sohag, A.A.M.; Rahman, M.A.; Uddin, M.J.; Alam, M.; Moon, I.S. Neuroprotective Potentials of Marine Algae and Their Bioactive Metabolites: Pharmacological Insights and Therapeutic Advances. *Mar. Drugs* **2020**, *18*, 347. [CrossRef] [PubMed]
75. Celikler, S.; Vatan, O.; Yildiz, G.; Bilaloglu, R. Evaluation of Anti-Oxidative, Genotoxic and Antigenotoxic Potency of *Codium Tomentosum* Stackhouse Ethanolic Extract in Human Lymphocytes in Vitro. *Food Chem. Toxicol.* **2009**, *47*, 796–801. [CrossRef]
76. Possede Chaves, E.; Sipione, S. Sphingolipids and Gangliosides of the Nervous System in Membrane Function and Dysfunction. *FEBS Lett.* **2010**, *584*, 1748–1759. [CrossRef]
77. Yamaguchi, R.; Kanie, Y.; Kanie, O.; Shimizu, Y. A Unique Structural Distribution Pattern Discovered for the Cerebrosides from Starfish Asterias Amurensis. *Carbohydr. Res.* **2019**, *473*, 115–122. [CrossRef] [PubMed]
78. Lago, J.; Rodriguez, L.P.; Blanco, L.; Vieites, J.M.; Cabado, A.G. Tetrodotoxin, an Extremely Potent Marine Neurotoxin: Distribution, Toxicity, Origin and Therapeutical Uses. *Mar. Drugs* **2015**, *13*, 6384–6406. [CrossRef]
79. Bane, V.; Lehane, M.; Dikshit, M.; O'Riordan, A.; Furey, A. Tetrodotoxin: Chemistry, Toxicity, Source, Distribution and Detection. *Toxins* **2014**, *6*, 693–755. [CrossRef] [PubMed]
80. Nieto, F.R.; Cobos, E.J.; Tejada, M.Á.; Sánchez-Fernández, C.; González-Cano, R.; Cendán, C.M. Tetrodotoxin (TTX) as a Therapeutic Agent for Pain. *Mar. Drugs* **2012**, *10*, 281–305. [CrossRef]
81. Kanoh, K.; Kohno, S.; Asari, T.; Harada, T.; Katada, J.; Muramatsu, M.; Kawashima, H.; Sekiya, H.; Uno, I. (−)-Phenylahistin: A New Mammalian Cell Cycle Inhibitor Produced by *Aspergillus ustus*. *Bioorganic Med. Chem. Lett.* **1997**, *7*, 2847–2852. [CrossRef]
82. Gomes, N.G.M.; Lefranc, F.; Kijjoa, A.; Kiss, R. Can Some Marine-Derived Fungal Metabolites Become Actual Anticancer Agents? *Mar. Drugs* **2015**, *13*, 3950–3991. [CrossRef]

83. Behl, T.; Rachamalla, M.; Najda, A.; Sehgal, A.; Singh, S.; Sharma, N.; Bhatia, S.; Al-Harrasi, A.; Chigurupati, S.; Vargas-De-La-Cruz, C.; et al. Applications of Adductomics in Chemically Induced Adverse Outcomes and Major Emphasis on DNA Adductomics: A Pathbreaking Tool in Biomedical Research. *Int. J. Mol. Sci.* **2021**, *22*, 10141. [CrossRef] [PubMed]
84. Nicholson, B.; Lloyd, G.K.; Miller, B.R.; Palladino, M.A.; Kiso, Y.; Hayashi, Y.; Neuteboom, S.T.C. NPI-2358 Is a Tubulin-Depolymerizing Agent: In-Vitro Evidence for Activity as a Tumor Vascular-Disrupting Agent. *Anti-Cancer Drugs* **2006**, *17*, 25–31. [CrossRef] [PubMed]
85. Singh, A.V.; Bandi, M.; Raje, N.; Richardson, P.; Palladino, M.A.; Chauhan, D.; Anderson, K.C. A Novel Vascular Disrupting Agent Plinabulin Triggers JNK-Mediated Apoptosis and Inhibits Angiogenesisin Multiple Myeloma Cells. *Blood* **2011**, *117*, 5692–5700. [CrossRef] [PubMed]
86. Feling, R.H.; Buchanan, G.O.; Mincer, T.J.; Kauffman, C.A.; Jensen, P.R.; Fenical, W. Salinosporamide A: A Highly Cytotoxic Proteasome Inhibitor from a Novel Microbial Source, a Marine Bacterium of the New Genus Salinospora. *Angew. Chem. Int. Ed.* **2003**, *42*, 355. [CrossRef]
87. Gulder, T.A.M.; Moore, B.S. Salinosporamide Natural Products: Potent 20S Proteasome Inhibitors as Promising Cancer Chemotherapeutics. *Angew. Chem. Int. Ed.* **2010**, *49*, 9346–9367. [CrossRef]
88. Meng, L.; Mohan, R.; Kwok, B.H.B.; Elofsson, M.; Sin, N.; Crews, C.M. Epoxomicin, a Potent and Selective Proteasome Inhibitor, Exhibits in Vivo Antiinflammatory Activity. *Proc. Natl. Acad. Sci. USA* **1999**, *96*, 10403–10408. [CrossRef]
89. Chauhan, D.; Catley, L.; Li, G.; Podar, K.; Hideshima, T.; Velankar, M.; Mitsiades, C.; Mitsiades, N.; Yasui, H.; Letai, A.; et al. A Novel Orally Active Proteasome Inhibitor Induces Apoptosis in Multiple Myeloma Cells with Mechanisms Distinct from Bortezomib. *Cancer Cell* **2005**, *8*, 407–419. [CrossRef]
90. Richardson, P.G.; Spencer, A.; Cannell, P.; Harrison, S.J.; Catley, L.; Underhill, C.; Zimmerman, T.M.; Hofmeister, C.C.; Jakubowiak, A.J.; Laubach, J.P.; et al. Phase 1 Clinical Evaluation of Twice-Weekly Marizomib (NPI-0052), a Novel Proteasome Inhibitor, in Patients with Relapsed/Refractory Multiple Myeloma (MM). *Blood* **2011**, *118*, 302. [CrossRef]
91. Andersen, R.J. Sponging off Nature for New Drug Leads. *Biochem. Pharmacol.* **2017**, *13*, 3–14. [CrossRef]
92. Liu, Y.; Salvador, L.A.; Byeon, S.; Ying, Y.; Kwan, J.C.; Law, B.K.; Hong, J.; Luesch, H. Anticolon Cancer Activity of Largazole, a Marine-Derived Tunable Histone Deacetylase Inhibitor. *J. Pharmacol. Exp. Ther.* **2010**, *335*, 351–361. [CrossRef]
93. Ying, Y.; Taori, K.; Kim, H.; Hong, J.; Luesch, H. Total Synthesis and Molecular Target of Largazole, a Histone Deacetylase Inhibitor. *J. Am. Chem. Soc.* **2008**, *130*, 8455–8459. [CrossRef] [PubMed]
94. Hong, J.; Luesch, H. Largazole: From Discovery to Broad-Spectrum Therapy. *Nat. Prod. Rep.* **2012**, *29*, 449–456. [CrossRef] [PubMed]
95. Engene, N.; Tronholm, A.; Salvador-Reyes, L.A.; Luesch, H.; Paul, V.J. *Caldora Penicillata* Gen. Nov., Comb. Nov. (Cyanobacteria), a Pantropical Marine Species with Biomedical Relevance. *J. Phycol.* **2015**, *51*, 670–681. [CrossRef] [PubMed]
96. Salvador-Reyes, L.A.; Engene, N.; Paul, V.J.; Luesch, H. Targeted Natural Products Discovery from Marine Cyanobacteria Using Combined Phylogenetic and Mass Spectrometric Evaluation. *J. Nat. Prod.* **2015**, *78*, 486–492. [CrossRef]
97. Chen, Q.Y.; Chaturvedi, P.R.; Luesch, H. Process Development and Scale-up Total Synthesis of Largazole, a Potent Classi Histone Deacetylase Inhibitor. *Org. Process Res. Dev.* **2018**, *22*, 190–199. [CrossRef]
98. Dybdal-Hargreaves, N.F.; Risinger, A.L.; Mooberry, S.L. Eribulin Mesylate: Mechanism of Action of a Unique Microtubule Targeting Agent. *Clin. Cancer Res.* **2015**, *21*, 2445–2452. [CrossRef] [PubMed]